Conjugated Linoleic Acids and Conjugated Vegetable Oils

RSC Catalysis Series

Editor-in-Chief:
Professor James J Spivey, *Louisiana State University, Baton Rouge, USA*

Series Editors:
Professor Chris Hardacre, *Queen's University Belfast, Northern Ireland*
Professor Zinfer Ismagilov, *Boreskov Institute of Catalysis, Novosibirsk, Russia*
Professor Umit Ozkan, *Ohio State University, USA*

Titles in the Series:
 1: Carbons and Carbon Supported Catalysts in Hydroprocessing
 2: Chiral Sulfur Ligands: Asymmetric Catalysis
 3: Recent Developments in Asymmetric Organocatalysis
 4: Catalysis in the Refining of Fischer–Tropsch Syncrude
 5: Organocatalytic Enantioselective Conjugate Addition Reactions:
    A Powerful Tool for the Stereocontrolled Synthesis of Complex Molecules
 6: *N*-Heterocyclic Carbenes: From Laboratory Curiosities to Efficient
    Synthetic Tools
 7: *P*-Stereogenic Ligands in Enantioselective Catalysis
 8: Chemistry of the Morita–Baylis–Hillman Reaction
 9: Proton-Coupled Electron Transfer: A Carrefour of Chemical Reactivity
    Traditions
10: Asymmetric Domino Reactions
11: C-H and C-X Bond Functionalization: Transition Metal Mediation
12: Metal Organic Frameworks as Heterogeneous Catalysts
13: Environmental Catalysis Over Gold-Based Materials
14: Computational Catalysis
15: Catalysis in Ionic Liquids: From Catalyst Synthesis to Application
16: Economic Synthesis of Heterocycles: Zinc, Iron, Copper, Cobalt,
    Manganese and Nickel Catalysts
17: Metal Nanoparticles for Catalysis: Advances and Applications
18: Heterogeneous Gold Catalysts and Catalysis
19: Conjugated Linoleic Acids and Conjugated Vegetable Oils

How to obtain future titles on publication:
A standing order plan is available for this series. A standing order will bring
delivery of each new volume immediately on publication.

For further information please contact:
Book Sales Department, Royal Society of Chemistry, Thomas Graham House,
Science Park, Milton Road, Cambridge, CB4 0WF, UK
Telephone: +44 (0)1223 420066, Fax: +44 (0)1223 420247
Email: booksales@rsc.org

Visit our website at www.rsc.org/books

Conjugated Linoleic Acids and Conjugated Vegetable Oils

Edited by

Bert Sels
Centre for Surface Chemistry, Leuven, Belgium
Email: bert.sels@biw.kuleuven.be

An Philippaerts
Centre for Surface Chemistry, Leuven, Belgium
Email: an.philippaerts@biw.kuleuven.be

THE QUEEN'S AWARDS
FOR ENTERPRISE:
INTERNATIONAL TRADE
2013

RSC Catalysis Series No. 19

Print ISBN: 978-1-84973-900-9
PDF eISBN: 978-1-78262-021-1
ISSN: 1757-6725

A catalogue record for this book is available from the British Library

Published by The Royal Society of Chemistry,
Thomas Graham House, Science Park, Milton Road,
Cambridge CB4 0WF, UK

Registered Charity Number 207890

For further information see our web site at www.rsc.org

Printed and bound in Great Britain by CPI Group (UK) Ltd, Croydon, CR0 4YY

Foreword: *Exegi Monumentum Aere Perennius*[†]

At the occasion of the appearance of the book *Conjugated Linoleic Acids and Conjugated Vegetable Oils*, edited by experts in heterogeneous catalysis, a few words clarifying and justifying the initiative might be of interest to the reader.

At the beginning of the second millennium, at the Center of Surface Science and Catalysis, KU Leuven, a project was started on shape selective transformations of vegetable oils with (heterogeneous) zeolite-based catalysts. Stimulated initially by Albert J. Dijkstra,[1] the concept of essentially trans-free partial hydrogenation of fatty acid esters and edible vegetable oils with shape selective zeolites was developed. The editors of the present book on Conjugated Linoleic Acids (CLAs) were the driving force behind this research. Major scientific hurdles had to be overcome in order to achieve the prestigious goal. It took almost a decade developing an appropriate catalyst, namely a ZSM-5 zeolite with the Pt metal load fully incorporated in the zeolite crystalline lattice.[2] So, the intuition of A.J.D. was proven to be correct. Thank you Albert! This regioselective hardening of soy bean oil yielded nutritive almost trans-free shortenings with unprecedented physical properties.[3]

Meanwhile, Bert Sels accepted an academic chair at KU Leuven on 'Catalytic Conversion of Biomass and Bio-platform Molecules', while An Philippaerts made a major effort in accumulating postdoctoral expertise in the same area. Equipped with that catalyst design knowledge, it took only a minor leap to achieve hydrogen-free production of conjugated linoleic acids

[†]*Horace: I have erected a monument more lasting than bronze.*

RSC Catalysis Series No. 19
Conjugated Linoleic Acids and Conjugated Vegetable Oils
Edited by Bert Sels and An Philippaerts
© The Royal Society of Chemistry 2014
Published by the Royal Society of Chemistry, www.rsc.org

and esters using Ru totally encapsulated in the zeolite intracrystalline voids.[4] This was just another recent 'discovery' in the area of shape selective catalytic transformations of vegetable oils. Although it was realized at the initial discussion phase of this book project that the American Oil Chemists Society (AOCS) has been publishing overviews dealing with different aspects of CLAs on a regular basis, the last update being made in 2006,[5] it was decided that there was need for a textbook covering the topic from many different angles. Therefore, with the contribution of the experts at the basis of science and technology development in the area, the editors succeeded in less than 12 months in producing a textbook on all aspects of CLA science and technology. In my opinion, it will appeal not only to industrialists and researchers in this interdisciplinary field but also will attract the attention of the general public.

Pierre A. Jacobs
Prof. Em. Mand. KU Leuven
Leuven, Belgium

References

1. A.J. Dijkstra, *Eur. J. Lipid Sci. Technol.*, 2006, **108**, 249.
2. A. Philippaerts, S. Paulussen, A. Breesch, S. Turner, O.I. Lebedev, G. Van Tendeloo, B. Sels and P.A. Jacobs, *Angew. Chem. Int. Ed.*, 2011, **50**, 3947.
3. A Philippaerts, A. Breesch, G. Annika, P.-J. De Cremer, J. Kayaert, G. Hofkens, P. De Mooter, Jacobs and B. Sels, *J. Am. Oil Chem. Soc.*, 2011, **88**, 2023.
4. A. Philippaerts, S. Goossens, W. Vermandel, M. Tromp, S. Turner, J. Geboers, G. Van Tendeloo, P. A. Jacobs and B. F. Sels, *ChemSusChem*, 2011, **4**, 757.
5. *Advances in Cojugated Linoleic Acid Research*, ed. M. P. Yurawecz, J. K. G. Kramer, O. Gudmundson, M. W. Pariza and S. Banni, AOCS Press, Chamapign, Ill., 2006, vol. 3.

Preface

Recent interest from academia, nutritionists, the paint and polymer industry, and the feed and food industry in conjugated linoleic acids (CLAs) and conjugated vegetable oils has grown spectacularly in the last few years. CLA isomers, either in their natural or synthetic forms, have not only been associated with different health effects but they are also interesting renewable compounds in the production of industrial products such as paints, inks and polymers.

Likewise, as a result of the interest in CLA, its literature has recently increased enormously and may come across as chaotic, especially due to the fact that often conflicting results have been published. Therefore, the objective of this book is to provide a comprehensive and up-to-date overview of all the various aspects of CLA, which will be easily understood by a wide variety of readers with some chemical background, both from industry as well as from academia.

This book is organized into six chapters, each chapter covering a key aspect of CLA that has been studied increasingly in recent years.

In Chapter 1, Shingfield and Wallace review the synthesis pathway of CLA in ruminants and humans. CLAs are naturally present in meats and milk products of ruminants, where they are synthesized by rumen bacteria during the biohydrogenation process. CLA can also be synthesized endogenously in tissues and the mammary gland in ruminants, and recent studies have shown that CLA can be synthesized endogenously in humans. As CLA has been associated with numerous positive health benefits, and CLA obtained from ruminants are the primary source of CLA in human nutrition, a lot of research has been conducted to increase the CLA content in milk and meats from ruminants. Chapter 1 provides a comprehensive evaluation of the most recent evidence on the biochemical, microbial, nutritional and physiological factors influencing the amount and distribution of CLA isomers formed.

RSC Catalysis Series No. 19
Conjugated Linoleic Acids and Conjugated Vegetable Oils
Edited by Bert Sels and An Philippaerts
© The Royal Society of Chemistry 2014
Published by the Royal Society of Chemistry, www.rsc.org

As CLA has been associated with health-promoting properties, there is a lot of interest in enriching feed and food ingredients with CLA. In Chapter 2, the use of CLA in animal feed is discussed by Everaert, Koppenol and Buyse. This chapter provides a comprehensive review of literature concerning various outcomes of dietary CLA supplementation in livestock, *i.e.* ruminants (diary and meat-type), pigs, poultry (laying and meat-type), and some fish species. The effects covered comprise zootechnical and reproductive performance, CLA enrichment and fatty acid profile in tissues and animal products, and immune status. Chapter 3 by Park and Wu assesses the diverse health benefits of CLA in humans. In this chapter the current knowledge of the influence of CLA on body fat regulation, cancer and cardiovascular diseases prevention, modulation of immune and inflammatory responses, and benefits on bone health is reviewed. Also the potential health concerns of CLA supplementation are considered. Finally, biological activities of CLA metabolites and other conjugated fatty acids are discussed.

The bioactivity of CLA was only first reported in the 1980s. Since then, there has been growing interest in CLA for feed and food applications. However, CLA has been used since the early 1930s in chemical applications, mainly as drying oil used, for instance, in paints, glues and inks. For these applications CLA is produced on a commercial scale via either homogeneous base-catalysed isomerisation of vegetable oils enriched in linoleic acid or via dehydration of castor oil. Nowadays, conjugated vegetable oils are largely used as building blocks for the synthesis of various bio-based thermosetting materials for applications in the automobile and construction industries. These topics, relating to the industrial production of CLA and conjugated vegetable oils and their chemical use, are covered in Chapter 4 by Quirino.

Due to the increasing interest in CLA and conjugated vegetable oils, many studies arose describing new synthesis procedures of CLA and conjugated vegetable oils, aiming at higher yields of total CLA isomers or even of very specific CLA isomers. The recent advances in the production of CLA and conjugated vegetable oils are presented in Chapters 5 and 6. Chapter 5 focuses on the microbial and enzymatic production of conjugated fatty acids. Ogawa, Takeuchi and Kishino clearly describe the use of a bioprocess for the production of CLA with high isomer-selectivity, which has a high potential for medicinal and nutraceutical purposes. Besides the microbial and enzymatic production of CLA, a great deal of research has also been focused on the production of CLA and conjugated vegetable oils via metal catalysis. The use of metal catalysts, both homogeneous and heterogeneous, is tackled in Chapter 6 by Belkacemi, Chorfa and Hamoudi. It is clearly illustrated that it is nowadays possible to produce both CLA and its derivates as conjugated vegetable oils via homogeneous and heterogeneous metal catalysts with appealing CLA yields and productivities.

The analysis of different conjugated fatty acids isomers, usually in a complex matrix containing plenty of other non-conjugated fatty acids, is very challenging. In Chapter 7, Kramer, Fardin-Kia, Aldai, Mossoba and

Delmonte give a clear overview of the various analytical techniques available, focusing especially on gas and high pressure liquid chromatography. The benefits and limitations of different GC and HPLC columns are evaluated. Nevertheless, up to now it is not possible to analyse all the possible geometric and positional isomers of fatty acids in one single analysis. Therefore, the authors propose various combinations of methods in order to analyse as many conjugated and non-conjugated fatty acids as possible.

Finally, we would like to acknowledge our colleagues throughout the world who, through their research on conjugated fatty acids, contributed to this book. We wish in particular to thank all the contributing authors, for their enthusiasm and efforts in providing a comprehensive and up-to-date overview of the CLA topic in their expertise field and making this book an indispensable reference work for everybody who is interested in conjugated fatty acids and oils and their applications. Lastly, we would like to thank the editorial staff of the Royal Chemical Society, for their collaboration and the final editing.

An Philippaerts
Bert Sels
Leuven

Contents

RSC Catalysis Series No. 19
Conjugated Linoleic Acids and Conjugated Vegetable Oils
Edited by Bert Sels and An Philippaerts
© The Royal Society of Chemistry 2014
Published by the Royal Society of Chemistry, www.rsc.org

CHAPTER 1

Synthesis of Conjugated Linoleic Acid in Ruminants and Humans

K. J. SHINGFIELD*[a,b] AND R. J. WALLACE[c]

[a] MTT Agrifood Research Finland, Animal Production Research, FI-31600, Jokioinen, Finland; [b] Institute of Biological, Environmental and Rural Sciences, Aberystwyth University, Aberystwyth, Ceredigion, SY23 3EB, UK; [c] Rowett Institute of Nutrition and Health, University of Aberdeen, Bucksburn, Aberdeen, AB21 9SB, UK
*Email: kes14@aber.ac.uk

1.1 Introduction

There is increasing evidence that nutrition plays an important role in the development of human chronic diseases including cancer, cardiovascular disease, insulin resistance and obesity. Developing foods and diets that promote human health is central to public health initiatives for preventing and lowering the economic and social impact of chronic disease.[1] Direct and indirect costs of cardiovascular disease (CVD) have been estimated at $445 billion in the United States[2] and €200 billion within the European Union.[3] Global costs of CVD in 2010 totalled US$863 billion.[4] These costs are projected to increase several fold by 2030, reaching unsustainable levels due to people living longer and the rapid increase in obesity in developed and developing countries.[2,5]

Following the identification of the anti-mutagenic properties of conjugated linoleic acid (CLA) isomers in cooked beef,[6-8] numerous studies have

RSC Catalysis Series No. 19
Conjugated Linoleic Acids and Conjugated Vegetable Oils
Edited by Bert Sels and An Philippaerts
© The Royal Society of Chemistry 2014
Published by the Royal Society of Chemistry, www.rsc.org

investigated the biological activity of CLA isomers in cell culture and animal models (http://fri.wisc.edu/cla.php). Much of the research has focused on the effects of *cis*-9, *trans*-11 18:2 (trivial name rumenic acid) or *trans*-10, *cis*-12 18:2 due to the cost and availability in a range of mammalian and avian species. In addition to the inhibition of mutagenesis, specific isomers of CLA have been demonstrated to modulate energy metabolism, immunity, inflammation, insulin resistance and bone metabolism in several animal models.[9–21] However, evidence that the physiological effects described *in vitro* or in other mammalian species are also replicated in humans remains inconclusive.[22–28]

The optimal intake of one or more isomers of CLA in humans remains to be established. Direct or exponential extrapolation of data from studies in the rat model of carcinogenesis implicate intakes of *cis*-9, *trans*-11 18:2 between 95 and 3500 mg per day being required for significant decreases in cancer risk in human populations.[29,30] Estimates of *cis*-9, *trans*-11 18:2 consumption in human populations vary between 15 and 1500 mg per day[29,31–39] depending on the methodology used to estimate dietary intakes, with marked differences between countries, gender and socio-economic groups. Isomers of CLA are present in a wide range of foods including milk, beef and lamb, and to a much lesser extent in pork, poultry, fish and eggs, with trace amounts in some vegetable sources.[32,35–43] Milk and dairy products are the major source of CLA in the human diet contributing to between 66 and 80% of total intake.[31,35–39] Typically concentrations of CLA in pork, chicken and fish are lower than 0.1 g per 100 g lipid,[44] whereas the consumption of lamb, beef and other ruminant meat products account for 15–32% of average daily CLA intakes in developed countries.[34–36,38,44]

The CLA status of humans can be increased using oral supplements or fortification of foods with synthetic sources, which typically contain equal amounts of *cis*-9, *trans*-11 18:2 and *trans*-10, *cis*-12 18:2, or from a higher consumption of ruminant-derived foods. In contrast to synthetic sources, meat and milk from ruminants contain numerous positional and geometric isomers of CLA with conjugated double bonds at positions 6,8 through to 13,15, with *cis*-9, *trans*-11 18:2 being the major isomer, and *trans*-7, *cis*-9 18:2 or *trans*-11, *cis*-13 18:2 as the second most abundant.[44–47]

Producing ruminant-derived foods containing higher amounts of CLA offers the opportunity to increase the consumption of CLA, principally *cis*-9, *trans*-11 18:2, without requiring major changes in the habitual diet or eating habits. For this reason, a considerable amount of research has been dedicated to understanding the nutritional, physiological and genetic factors influencing CLA concentrations in meat and milk. The present chapter provides a comprehensive review of the most recent evidence on the biochemical, microbial, nutritional and physiological factors influencing the biosynthesis of CLA isomers in ruminants and humans.

1.2   Lipid Metabolism in the Rumen

1.2.1   Substrate Supply

Ruminant diets vary in composition depending on species, physiological state, and the cost and availability of feed ingredients. Diets often contain forage species (grasses, legumes or forage maize) of variable maturity and nutritional value, and differ in composition from those containing forages as the sole feed to combinations of forage, cereals and protein supplements. By-products of the food industry or lipid supplements may also be included. A general characteristic of ruminant diets is the relatively high fibre content (>300 g cell wall constituents per kg dry matter (DM)) and low amounts of lipid (<50 g per kg DM).[45] Lipid in cereal grains, plant oils, marine lipids and by-products are predominantly in the form of triacylglycerol (TAG). Most of the lipid in grasses and legume forages is present as phospholipids (PL) and glycolipids (GL) located within thylakoid membranes of chloroplasts.[48–50] In forages, GL are the major lipid class, with galactolipids (mono- and di-galactosyl diacylglyerol) being the most prevalent. These differ from TAG in that one or more carbohydrate molecules are linked to one position of the glycerol backbone. Forages and oilseeds also contain several PL species, (phosphatidylcholine, phosphatidylglycerol and phosphatidyl-ethanolamine) as structural components of cell membranes. Most PL contain a diacylglycerol covalently bonded to a phosphate group, which is often esterified to a simple organic molecule such as choline. In general, both fatty acid moieties bound to glycerol in glycolipids or PL are un-saturated. Non-esterified fatty acids (NEFA) are minor components of most ruminant feeds, but are the major lipid class in ensiled grasses and forage legumes and in certain proprietary fat supplements. Changes arising during fermentation in silo are characterized by a substantial decrease in the relative abundance of polar membrane lipid and an increase in NEFA, TAG, diacylglycerol (DAG) and monoacylglycerol (MAG) fractions attributable to the activity of plant and microbial lipases.[51–54]

The amount and composition of constituent fatty acids differs substantially between ingredients in ruminant diets (Table 1.1). For grasses and legume forages fatty acid content is generally lower than 50 g per kg DM with *cis*-9, *cis*-12, *cis*-15 18:3 (α-linolenic acid; 18:3 n-3) as the major fatty acid (>50 g per 100 g fatty acids).[55–59] However, conservation of grass by drying results in substantial decreases in lipid content,[58,60,61] principally due to the disappearance of *cis*-9, *cis*-12 18:2 (linoleic acid; 18:2 n-6) and 18:3 n-3 by oxidation and leaf shatter.[49] Oxidation arises from the activity of plant lipoxygenases, which catalyse the incorporation of molecular oxygen in non-esterified 18:2 n-6 and 18:3 n-3, forming 9- and 13-hydroperoxy poly-unsaturated fatty acids, respectively, that are highly reactive and rapidly metabolized into a series of oxylipins including volatile leaf aldehydes and alcohols, hydroxy- and epoxy-fatty acids and jasmonates.[62,63] In contrast, 18:2 n-6 is the predominant fatty acid in forage maize, whole crop silages

Table 1.1 Typical lipid content and fatty acid composition of forages, oilseeds, plant and marine lipid supplements used in ruminant diets.

Ingredient	Oil (g/kg dry matter)	Fatty acid composition (g/100 g fatty acids)							Reference[a]
		16:0	18:0	cis-9 18:1	18:2n-6	18:3n-3	20:5n-3	22:6n-3	
Grass									
3 week growth	25.0	16.1	1.4	1.94	10.9	67.3	–	–	Dewhurst et al.[55]
6 week growth	15.2	19.4	2.0	2.40	11.9	60.7	–	–	Dewhurst et al.[55]
Early cut	33.2	15.1	2.3	4.4	18.2	49.9	–	–	Vanhatalo et al.[56]
Late cut	28.2	14.5	2.2	6.5	28.8	37.6	–	–	Vanhatalo et al.[56]
Wilted	20.5	19.0	0.3	1.09	3.7	18.0	–	–	Shingfield et al.[58]
Red clover									
Early cut	33.3	21.5	7.1	4.2	17.3	35.6	–	–	Vanhatalo et al.[56]
Late cut	30.2	19.5	3.8	3.5	21.4	38.8	–	–	Vanhatalo et al.[56]
Silage									
Grass	19.8	20.1	2.1	2.5	14.2	50.4	–	–	Shingfield et al.[57]
Maize	25.3	17.4	2.2	20.3	44.8	6.6	–	–	Shingfield et al.[57]
Red clover	30.4	20.6	6.5	3.4	17.4	40.0	–	–	Vanhatalo et al.[56]
Whole crop wheat	21.0	17.3	1.0	12.2	40.9	23.2	–	–	Noci et al.[64]
Hay									
Grass	8.1	35.0	0.6	2.6	5.4	15.6	–	–	Shingfield et al.[58]
Lucerne	10.3	22.9	4.1	4.3	17.7	22.3	–	–	AbuGhazaleh et al.[59]

									Reference
Rape									
Oil	963	6.0	2.3	48.1	27.4	10.3	–	–	Givens et al.[66]
Whole seeds	408	4.8	2.0	56.8	19.3	8.3	–	–	Givens et al.[66]
Sunflower									
Oil	962	6.1	3.6	26.5	60.4	0.1	–	–	Shingfield et al.[67]
Whole seeds	400	5.1	4.3	21.6	66.8	0.2	–	–	Woods and Fearon[65]
Linseed									
Oil	953	4.2	2.7	16.5	15.8	57.8	–	–	Shingfield et al.[68]
Whole seeds	360	6.1	3.4	18.8	16.3	54.4	–	–	Woods and Fearon[65]
Camelina									
Oil	95.4	5.6	2.4	11.6	15.7	37.0	–	–	Halmemies-Beauchet-Filleau et al.[70]
Whole seeds	37.8	8.3		17.8	27.6	46.3	–	–	Hurtaud and Peyraud[69]
Fish oil	950	15.0	2.6	11.0	1.2	0.9	16.5	10.5	Shingfield et al.[68]
Marine algae, *Schizochytrium* sp.	581	26.3	0.9	1.1	0.3	0.2	<0.1	37.8	Boeckaert et al.[71]

[a]Numbers refer to citations listed in the reference section.

and cereal grains.[45,57,64] Ruminant diets may also contain up to around 50 g per kg DM of additional lipid in the form of oils or oilseeds. Oils from rapeseeds, high oleic sunflowerseeds, olives and peanuts are a rich source of *cis*-9 18:1 (oleic acid),[65,66] cottonseeds, safflowerseeds, soyabeans and sunflowerseeds are abundant in 18:2 n-6,[65,67] whereas linseeds[65,68] and camelina[69,70] contain relatively high proportions of 18:3 n-3 (Table 1.1). In contrast, coconut oil is rich in 12:0 (lauric acid), whereas 16:0 (palmitic acid) is the major fatty acid in palm oil.[65] Ruminant diets may also be supplemented with fish oil containing *cis*-5, *cis*-8, *cis*-11, *cis*-14, *cis*-17 20:5 (eicosapentaenoic acid; 20:5 n-3) and *cis*-4, *cis*-7, *cis*-10, *cis*-13, *cis*-16, *cis*-19 22:6 (docosa-hexaenoic acid; 22:6 n-3)[68] or marine algae enriched in 22:6 n-3.[71]

1.2.2 Lipolysis

Following ingestion and mastication, the ester linkages of TAG, PL and GL are rapidly hydrolysed in the rumen.[72–74] Hydrolysis of dietary TAG occurs predominantly as a result of microbial lipases.[74] Forage plant tissues are also rich in endogenous galacto- and phospholipases, which remain active once ingested into the rumen for several hours, suggesting that senescence of the plant material itself in the rumen may contribute to ruminal lipolysis in grazing animals.[75] Dawson *et al.*[74] challenged this idea and concluded that microbial lipases were more important than plant enzymes. These conclusions were drawn using autoclaved grass as a substrate, which the authors noted was not ideal because of the many effects that autoclaving has in addition to enzyme denaturation. This debate has been revisited by Lee *et al.*,[76] who reported increased NEFA and decreased polar lipid abundance after 6 h incubation of fresh ryegrass leaves in buffer, confirming that plant-catalysed lipolysis occurred. Further, Van Ranst *et al.*[77] reported lipolysis of up to 60% after 8 h incubation of fresh red clover leaves. Both studies suggested the observed lipolysis to be due to active plant lipases that could contribute to overall ruminal lipolysis. However, until plant lipase activity is compared directly with that of ruminal micro-organisms, there will remain an uncertainty about their relative importance. Nonetheless, it may be useful to breed forage plants low in lipase activity in order to increase the amount of polyunsaturated fatty acids escaping the rumen. Plant lipids may also be compartmentalized in the ingested plant material, effectively protecting them from both endogenous and microbial hydrolysis.[78]

Among the various types of ruminal micro-organisms, the bacteria are considered to be most active in lipolysis.[48] The most active bacterial species isolated selectively using TAG as a substrate was *Anaerovibrio lipolytica*,[79] with which most research has been carried out. More recently, Jarvis *et al.*[80] isolated two bacteria from red deer that hydrolysed TAG and grew on glycerol. The bacteria, which were highly active against tallow, tripalmitin and olive oil, were most closely related to the genus *Propionibacterium* and clostridial cluster XIVa. Cirne *et al.*[81] isolated *Clostridium lundense* sp. nov. from the bovine rumen. *Clostridium lundense* exhibited lipolytic activity against olive oil, but neither its activity nor its numbers was reported, so it is difficult to assess its likely importance. Thus, a wider range of bacteria than is usually considered may be involved in the lipolysis of TAG in the rumen.

A. lipolyticus lacks the ability to hydrolyse galactolipids and PL and, therefore, other lipolytic species would be expected to predominate in grazing ruminants. The *Butyrivibrio* spp. appeared to contain all the phospholipase A, phospholipase C, lysophospholipase and phosphodiesterase activities typical of the mixed rumen contents.[48] Their lipase activity against TAG varies between different *Butyrivibrio* and *Pseudobutyrivibrio* strains, but not in a manner that corresponds to their position in the phylogenetic tree.[82] Little is known about how other lipases vary across different strains/species, nor whether other recently recognized species may possess such activities in the rumen of grazing ruminants.

There have been few recent studies that investigate protozoal lipolysis. Wright[83] suggested *Epidinium* spp. to be responsible for 30–40% of the lipolytic activity in the rumen. *Epidinium ecaudatum* was reported to liberate galactose from galactolipids, suggesting galactosidase activity, although lipase activity *per se* was not demonstrated.[84] Another protozoal species, *Entodinium caudatum*, was shown to have phospholipase activity,[85] but it is possible that this activity was more relevant to the intracellular metabolism of the protozoa than to the digestion of dietary lipids. The earlier studies to determine the contribution of protozoa to the lipolytic activity in the rumen were conducted using fractionated rumen fluid, with the possibility that lipolytic activity in protozoal fractions was more due to the activity of bacteria that the protozoa had ingested than that of the protozoa themselves. Once again, given that protozoa comprise up to half the microbial biomass present in the rumen, their lipolytic properties warrants further investigation.

Much is known about bacterial lipases in general. They comprise eight well-documented families and their modes of action are reasonably well characterized.[86] The lipase activity of *A. lipolyticus* was investigated in some detail by methods available at the time, some four decades ago.[87,88] Perhaps surprisingly, no cloning and sequencing studies seem to have been done when the technology became widely available, and the most detailed recent study[89] was derived from genomic analysis. Three enzymes from *A. lipolyticus* were identified from genomic analysis of *A. lipolyticus*. The alipA, alipB and alipC encoded 492-, 438- and 248-amino acid peptides, respectively. Phylogenetic analysis indicated that alipA and alipB clustered with the GDSL/ SGNH family II, and alipC clustered with lipolytic enzymes from family V. Subsequent expression and purification of the enzymes showed that they had esterase activities with substrate specificities favouring the hydrolysis of caprylate-, laurate- and myristate-containing substrates. Genomes are available for several species in the *Butyrivibrio* group, but to date no similar analysis appears to have been carried out for their lipase activities.

Metagenomic methods will be invaluable in order to understand the full complement of lipolytic enzymes that are present in the rumen. Expression libraries may be useful. Liu *et al.*[90] screened a metagenomic library from the rumen of grazing Holstein cows for lipase activity, using trioleoylglycerol as substrate. Out of 15 360 bacterial artificial chromosome (BAC) clones, only two were found to have high lipase activity, which seems surprisingly small. The likely origin of the genes was investigated, based on the other open reading frames (ORFs) present in the BAC clones. It was impossible to decipher the host for the first gene, *Rlip*1, but the second, *Rlip*2, gave most similar homologues from *Thermosinus carboxydivorans*, which has not previously been associated with the ruminal ecosystem. Nonetheless, by BLASTing their deposited lipase sequences it was discovered that the genes most likely encoded carboxyl esterases. Thus the prevalence of these lipases in the overall community is far from certain. Clearly many more lipase sequences must be analysed from the metagenome, and assignment to lipases assured, in order to understand the true nature of lipolytic enzymes active in the rumen.

Lipolysis is considered rate limiting for the biohydrogenation of dietary unsaturated fatty acids in the rumen.[48] During ruminal digestion of grasses, mono- and digalactosyl diacylglycerides are released following the rupture of chloroplasts.[73] Early studies involving incubations of ^{14}C-labelled substrate with strained rumen fluid indicated that lipolysis conforms to first-order kinetics, occurring at a high rate with a short lag time.[91] These observations led Hawke and Silcock[91] to conclude that hydrolysis of dietary acyl lipids in the rumen would be sufficiently rapid not to impede biohydrogenation. However, such conclusions were not supported by observations that the profile of products formed during intra-ruminal infusion of esterified or non-esterified 18:2 n-6 in sheep differed.[92,93]

Much of what is known about the factors influencing ruminal lipolysis is based on incubations of various substrates with rumen fluid. Laboratory-scale experiments have obvious drawbacks with respect to mimicking conditions *in vivo*. Accepting these limitations and the assumptions therein, a number of factors have been determined to influence the rate and extent of lipolysis in a simulated rumen environment (Figure 1.1). Increases in dietary nitrogen content were found to increase the rate at which triolein was hydrolysed,[94] whereas the rate of lipolysis was decreased by replacing fibre with starch[95] or when more mature forages were incubated.[96] As could be expected, lipolysis increases with incubation time,[97,98] but is often decreased at low rumen pH.[99,100] The extent of lipolysis is also decreased for lipids with a higher melting point containing relatively high proportions of saturated fatty acids.[45] Increasing the amount of substrate incubated also lowers progressively the rate and extent of lipolysis *in vitro*.[97] After 24 h incubations with buffered rumen fluid, between 73.7% and 89.5% of TAG in fish oil, linseed oil or sunflower oil was found to be hydrolysed.[98] Similarly, lipolysis of lipid in cocksfoot and red clover after 24 h incubations were reported to vary between 65.0% and 82.0%.[101,102]

Measurements of lipolysis *in vivo* are scarce. Studies in cattle,[54,61] goats[103] and sheep[104] all affirm that a small fraction of esterified dietary lipid escapes the rumen. In lactating cows fed diets based on grass silage or red clover silage, NEFA, PL, TAG, DAG and MAG fractions were found to account for 80%, 12%, 4.4%, 2.4% and 0.8% of total fatty acids in omasal digesta.[54]

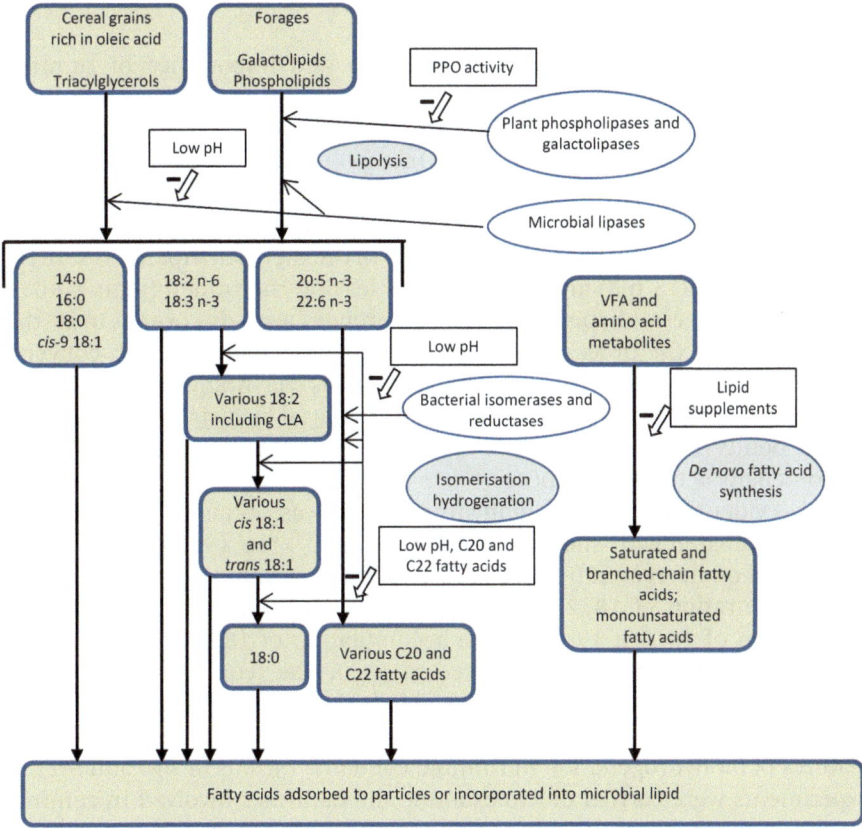

Figure 1.1 Schematic of the main transformation of dietary lipids in the rumen and factors influencing the relative rates of lipolysis, isomerization and biohydrogenation (adapted from Doreau *et al.*)[353] Following ingestion and mastication, dietary lipids are extensively hydrolysed in the rumen liberating non-esterified unsaturated fatty acids that are sequentially reduced to saturated end-products. Metabolism of unsaturated fatty acids proceeds via *cis* to *trans* isomerization, reduction or hydration of double bonds. Biohydrogenation to saturated end products is incomplete and numerous intermediates accumulate and escape the rumen.

The majority of 18:3 n-3 entered the omasum in the form of polar lipid (42–46%) originating from plant PL and GL, with smaller amounts as NEFA (22–26%), TAG (17–20%), DAG (8–14%) or MAG (2–3%). Between 85% and 93% of esterified lipid was reported to be hydrolysed in the rumen of lactating cows fed diets based on fresh grass or grass silage (forage : concentrate ratio 60 : 40), the extent of which was decreased to 80–86% and 70% when grass silage was replaced with grass hay[61] or red clover silage,[54] respectively. While incomplete, ruminal lipolysis is extensive, resulting in NEFA accounting for more than 80% of the total amount of fatty acids leaving the rumen.[48,54,61,103,104]

1.2.3 Biohydrogenation

From a historical perspective, the existence and importance of ruminal biohydrogenation have been known for more than 50 years. It has only been in recent years following the discovery of the biological activities of CLA isomers that research on ruminal lipid metabolism has focused more on the biochemical mechanisms involved in the formation of the less abundant fatty acid intermediates. Advances in lipid analysis allowed the marked differences in the fatty acid profile of lipids in the diet (principally *cis*-9 18:1, 18:2 n-6 and 18:3 n-3) and that of lipids leaving the rumen (mainly 16:0 and 18:0) to be established.,[105] These differences were discovered to be the consequence of the process we now call biohydrogenation, which converts unsaturated fatty acids to saturated fatty acids via firstly a *cis-trans* isomerization to *trans* fatty acid intermediates followed by hydrogenation of the double bonds (Figure 1.1).[48,106,107] Little if any oxidation or elongation of the carbon chain is thought to occur.[48]

Early evidence of ruminal biohydrogenation was obtained when linseed oil was incubated with sheep ruminal contents.[108] The 18:3 n-3 content of the linseed oil decreased from 30% to 5% with an accompanying increase in the concentration of 18:2. Shorland *et al.*[105] also noted that the ruminal metabolism of 18:3 n-3 caused the accumulation of 18:2 and 18:1 intermediates, which were then converted to 18:0. The requirement for a free carboxyl group for hydrogenation established that lipolysis must precede biohydrogenation.[91] After these initial studies demonstrated the overall features of biohydrogenation in ruminal contents, various *in vivo* and *in vitro* experiments were carried out to examine the pathways involved in ruminal digesta. Among these, the work of Dawson and his co-workers was particularly informative. Wilde and Dawson[109] constructed a metabolic scheme for the metabolism of 18:3 n-3 to 18:0 based on incubations of sheep ruminal contents with U-^{14}C labelled 18:3 n-3. The initial step was the isomerization of the *cis*-12 bond to either the Δ11 or Δ13 position. Thereafter, one of the bonds was hydrogenated to leave an 18:2, followed by hydrogenation of another bond producing an 18:1 intermediate. Hydrogenation of the 18:1 intermediate yielded 18:0 as the final product.

The other early contribution to delineating biohydrogenation pathways was the seminal work of Tove and his group with the bacterial species, *Butyrivibrio fibrisolvens*, that had been identified by Polan *et al.*[110] to possess high biohydrogenating activity. When incubated with *B. fibrisolvens*, 18:2 n-6 was initially isomerized to a conjugated 18:2, thought to be mainly *cis*-9, *trans*-11 18:2, but other dienoic isomers were also detected.[111] The *cis*-9, *trans*-11 18:2 intermediate was then hydrogenated to *trans*-11 18:1 (vaccenic acid) as the final product. However, in mixed ruminal bacteria, the 18:1 intermediate was further hydrogenated to 18:0.[112] Kepler and Tove[113] also incubated *B. fibrisolvens* with 18:2 n-6 and 18:3 n-3. They confirmed that 18:2 n-6 was first isomerized to *cis*-9, *trans*-11 18:2 followed by further hydrogenation to a mixture of *trans* 18:1 isomers. When 18:3 n-3 was used as a substrate, it was first isomerized to *cis*-9, *trans*-11, *cis*-15 18:3 (conjugated linolenic acid). This product was transient and further hydrogenated to a non-conjugated 18:2 intermediate as the final product. Slightly later, bacteria were isolated that converted monoenoic acids to 18:0 and hydrogenated 18:2 n-6 to 18:0, with traces of the intermediates that had been seen with *B. fibrisolvens* and mixed ruminal microbes.[114,115] The first step of 18:3 n-3 biohydrogenation is analogous to that of 18:2 n-6, namely the formation of a conjugated bond structure, *cis*-9, *trans*-11, *cis*-15 18:3.

Biohydrogenation of *cis*-6, *cis*-9, *cis*-12 18:3 (γ-linolenic acid, 18:3 n-6) during incubations with pure strains of *Butyrivibrio* and a bacterium identified as *Fusocillus babrahamensis* has also been investigated.[116] The group B bacterium (*Fusocillus*) hydrogenated 18:3 n-6 to 18:0, whereas incubations with the group A bacterium (*Butyrivibrio*) yielded *cis*-6, *trans*-11 18:2 as an end product. By analogy with the metabolic pathway described for 18:3 n-3, biohydrogenation of 18:3 n-6 is thought to proceed via isomerization of the *cis*-12 double bond to yield *cis*-6, *cis*-9, *trans*-11 18:3, that is sequentially reduced to *cis*-6, *trans*-11 18:2, *trans*-11 18:1 and 18:0.[116] Most feeds fed to ruminants do not contain 18:3 n-6. However, certain oilseeds including evening primrose and borage are relatively abundant in 18:3 n-6, but the effects of these lipids on ruminal lipolysis and biohydrogenation *in vivo* have not been investigated.

Several recent studies have investigated the metabolism of *cis*-6, *cis*-9, *cis*-12, *cis*-15 18:4 (stearidonic acid; 18:4 n-3) during incubations with rumen fluid.[117,118] Numerous intermediates were found to be formed. Based on the profile of fatty acids detected, the metabolism of 18:4 n-3 was proposed to proceed via the formation of 5,7,11,15 18:4 or 5,8,10,15 18:4.[118] Conjugated 18:4 products were reduced to yield 5,11,15 18:3 or 5,10,15 18:3 and hydrogenated yet further to 11,15 18:2 and 5,10 18:2, respectively.[118] *Trans*-6 to -12 18:1 were found to accumulate with *trans*-11 18:1 being the major intermediate formed.[117]

Much less is known about the fate of longer chain polyenoic fatty acids in the rumen. Detailed analysis of ruminal[119,120] and omasal digesta[121,122] have indicated that numerous 20- and 22-carbon biohydrogenation intermediates containing one or more *trans* double bonds are formed during the biohydrogenation of fatty acids in fish oil or marine algae. However, the biochemical pathways responsible are not known. The most recent investigations suggest that the first committed steps of 20:5 n-3 and 22:6 n-3 biohydrogenation in the rumen involve the reduction and/or isomerization of double bonds closest to the carboxyl group.[121,122] No conjugated ≥20-carbon fatty acids have been detected.

1.2.4 Metabolic Pathways

Ruminal biohydrogenation of *cis*-9 18:1, 18:2 n-6 and 18:3 n-3 *in vivo* varies from 58–87%, 70–95% and 85–100%, respectively.[107,123,124] The extent of ruminal fatty acid biohydrogenation is influenced by the composition and

amount of lipid in the diet, retention time in the rumen and the characteristics of the microbial population. Biohydrogenation is extensive, with
18:0 being the major fatty acid leaving the rumen on most diets. However,
the reduction of unsaturated 18-carbon fatty acids to 18:0 in the rumen is
incomplete and numerous 18:1, 18:2 and 18:3 intermediates accumulate.
The final reduction of *trans* 18:1 is thought to be the rate limiting step in the
complete biohydrogenation of 18-carbon unsaturated fatty acids.[125]
Decreases in pH from 6.5 to below 6.0 lowers the extent of *cis*-9
18:1,[126,127] 18:2 n-6[127–131] and 18:3 n-3[127,129,132] biohydrogenation and
inhibits the final reduction of *trans* 18:1 to 18:0[127,132] *in vitro*. The effects of
pH below normal physiological ranges in the rumen may be related to
disruption of bacterial cell membranes resulting in the inactivation of
membrane bound isomerases and reductases.[113]

Characterizing the biohydrogenation process *in vivo* represents a major
challenge. Relatively few studies have reported sufficiently detailed
measurements of post-ruminal fatty acid outflows to allow for inferences on
metabolic pathways. The situation is further complicated due to oils and
oilseeds being used to alter substrate supply rather than pure fatty acids. In
this regard incubations of fatty acids with rumen fluid or pure strains of
ruminal bacteria have proven invaluable in establishing the profile of
possible intermediates formed during the biohydrogenation of unsaturated
fatty acids (Table 1.2). However, intermediates formed *in vitro* may not
necessarily parallel biohydrogenation pathways *in vivo*. Discrepancies may
arise from the incubation of excessively high amounts of substrate, methods
used to introduce fatty acids into batch cultures, source of ruminal fluid and
potential loss of microbial activity over an extended incubation period.[107]

Early studies demonstrated that mixed or pure cultures of ruminal
bacteria were capable of hydrogenating *cis*-9 18:1 to 18:0.[48] More recent
experiments have indicated that *cis*-9 18:1 is also hydrated to 10-OH 18:0
(10-hydroxystearic acid), an intermediate that is further oxidized to 10-O
18:0 (10-ketostearic acid) during incubations with ruminal contents.[133,134]
Hydration of the *cis*-9 double bond involves the incorporation of one
hydrogen atom from water.[134] A select number of bacteria isolated from the
rumen, including species identified as *Fusocillus babrahamensis*,[135] *Seleno-
monas ruminantium*,[136] *Enterococcus faecalis*,[136] *Streptococcus bovis*[137] and
Propionibacterium acnes[134] are capable of hydrating *cis*-9 18:1. *P. acnes* is
known to catalyse the oxidation of 10-OH 18:0 to 10-O 18:0.[134] Incubations
of 1-[13]C *cis*-9 18:1 with rumen contents have indicated that *cis*-to *trans*
isomerization of the Δ9 double bond also takes place, resulting in the
formation of *trans*-6, -7, -9, -10, -11, -12, -13, -14, -15, and -16 18:1 intermediates.[138] Further investigations of the metabolism of 1-[13]C *cis*-9 18:1 by
mixed rumen microbes grown in continuous culture at pH 5.5 or pH 6.5 and
at different dilution rates indicated that decreases in pH and dilution rate
prevented the formation of *trans* 18:1 intermediates with a double bond
beyond the Δ10 position.[126] There is also evidence to suggest that *trans* to *cis*
isomerization may occur, following reports that incubations of *trans*-9 18:1

Table 1.2 Summary of intermediates formed during incubations of 18-carbon unsaturated fatty acids with rumen contents or pure cultures of rumen bacteria.

Substrate	Inoculum/ Bacterium	Intermediates and end-products	Reference[a]
cis-9 18:1	B. proteoclasticus	18:0	McKain et al.[134]
cis-9 18:1	E. faecalis	10-OH 18:0	Hudson et al.[136]
cis-9 18:1	P. acnes	10-OH 18:0, 10-0 18:0	McKain et al.[134]
cis-9 18:1	S. ruminantium	10-OH 18:0	Hudson et al.[136]
cis-9 18:1	Bovine rumen contents	trans-6, -7, -9, -10, -11, -12, -13, -14, -15, -16 18:1, 18:0	Mosley et al.[138]
cis-9 18:1	Bovine rumen contents	trans-6, -7, -9, -10, -11, -12, -13, -14, -15, -16 18:1, 18:0	AbuGhazaleh et al.[126]
cis-9 18:1	Bovine rumen fluid	10-OH 18:0, 10-0 18:0, 18:0	Jenkins et al.[133]
trans-9 18:1	Bovine rumen fluid	cis-9, -11 18:1, trans-6, -7, -11 18:1, 18:0	Proell et al.[139]
trans-10 18:1	B. proteoclasticus	18:0	McKain et al.[134]
trans-10 18:1	P. acnes	10-OH 18:0, 10-0 18:0	McKain et al.[134]
trans-11 18:1	B. proteoclasticus	18:0	McKain et al.[134]
cis-9, cis-12 18:2	B. fibrisolvens	trans-11 18:1	Maia et al.[207]
cis-9, cis-12 18:2	B. fibrisolvens	cis-9, cis-11 18:2, cis-9, trans-11 18:2, trans-9, cis-11 18:2, trans-9, trans-11 18:2	Wallace et al.[142]
cis-9, cis-12 18:2	B. hungatei	trans-11 18:1	Maia et al.[207]
cis-9, cis-12 18:2	B. proteoclasticus	cis-9, trans-11 18:2, trans-11 18:1	Maia et al.[207]
cis-9, cis-12 18:2	B. proteoclasticus	cis-9, cis-11 18:2, cis-9, trans-11 18:2, trans-9, cis-11 18:2, trans-9, trans-11 18:2	Wallace et al.[142]
cis-9, cis-12 18:2	C. aminophilum	cis-9 18:1	Maia et al.[207]
cis-9, cis-12 18:2	E. faecalis	10-OH 18:1, 13-OH 18:1	Hudson et al.[144]
cis-9, cis-12 18:2	F. succinogenes	16:0	Maia et al.[207]
cis-9, cis-12 18:2	M. multiacidus	cis-9 18:1	Maia et al.[207]
cis-9, cis-12 18:2	P. acnes	cis-10, trans-12 18:2, trans-10, cis-12 18:2, trans-10, trans-12 18:2	Wallace et al.[142]
cis-9, cis-12 18:2	S. bovis	13-OH 18:1	Hudson et al.[144]
cis-9, cis-12 18:2	Bovine rumen fluid	cis-6, cis-12 18:2, cis-7, cis-12 18:2, cis-8, cis-12 18:2, cis-9, cis-11 18:2, cis-10, cis-12 18:2, cis-9, trans-11 18:2, cis-9, trans-12 18:2, trans-8, cis-12 18:2, trans-8, cis-12 18:2, trans-9, cis-12 18:2, trans-10, cis-12 18:2, trans-9, trans-11 18:2, trans-10, trans-12 18:2, trans-9, trans-12 18:2, trans-6-8, -9, -10, -11, -12, -13-14 18:1, cis-9, -11, -12 18:1, 18:0	Honkanen et al.[141]

Table 1.2 (*Continued*)

Substrate	Inoculum/Bacterium	Intermediates and end-products	Reference[a]
cis-9, *cis*-12 18:2	Ovine rumen fluid	*cis*-9, *cis*-11 18:2, *cis*-9, *trans*-11 18:2, *trans*-10, *cis*-12 18:2, *trans*-11 18:1	Wąsowska *et al.*[146]
cis-9, *cis*-12 18:2	Ovine rumen fluid	*cis*-10, *cis*-12 18:2, *cis*-9, *trans*-11 18:2, *cis*-9, *trans*-12 18:2, *trans*-10, *cis*-12 18:2, *trans*-11 18:1, *trans*-9, *trans*-11 18:2, *trans* 4, 5, 6-8, -9, -10, -12 18:1, *cis*-10, -12, -13 18:1, 18:0	Jouany *et al.*[140]
cis-9, *cis*-12 18:2	Ovine rumen fluid	*cis*-9, *cis*-11 18:2, *cis*-12 18:2, *cis*-9, *trans*-11 18:2, *cis*-9, *trans*-11 18:2, *trans*-10, *cis*-12 18:2, *trans*-9, *trans*-11 18:2	Wallace *et al.*[142]
cis-9, *trans*-11 18:2	*B. fibrisolvens*	*trans*-11 18:1	McKain *et al.*[134]
trans-10, *cis*-12 18:2	*B. fibrisolvens*	*trans*-10, -12 18:1, *cis*-12 18:1	McKain *et al.*[134]
trans-9, *trans*-11 18:2	*B. fibrisolvens*	*trans*-11 18:1	McKain *et al.*[134]
trans-9, *trans*-11 18:2	*B. proteoclasticus*	*trans*-9, -11 18:1, *cis*-11 18:1	McKain *et al.*[134]
cis-9, *cis*-12, *cis*-15 18:3	Ovine rumen fluid	*cis*-9, *trans*-11, *cis*-15 18:3, *trans*-9, *trans*-11, *cis*-15 18:2, *cis*-15 18:2,	Wąsowska *et al.*[146]
cis-9, *cis*-12, *cis*-15 18:3	Ovine rumen fluid	*cis*-9, *cis*-11 18:2, *cis*-9, *cis*-15 18:2, *cis*-9, *trans*-13 18:2, *cis*-11, *trans*-13 18:2, *trans*-9, *cis*-12 18:2, *trans*-11, *cis*-15 18:2, *trans*-9, *trans*-11, *trans*-13 18:2, *trans* 6-8, -9, -11, -12, -13-14, -15, -16 18:1, *cis*-13, -15 18:1, 18:0	Jouany *et al.*[140]
cis-9, *cis*-12, *cis*-15 18:3	Bovine rumen contents[b]	*cis*-9, *cis*-11 18:2, *cis*-10, *cis*-12 18:2, *cis*-9, *trans*-11 18:2, *trans*-10, *cis*-12 18:2, *trans*-8, *trans*-10 18:2, *trans*-9, *trans*-11 18:2, *trans*-11, *trans*-13 18:2, *trans*-6-8, -9, -10, -11, -12 18:1, *cis*-9, -11 18:1, 18:0	Lee and Thomas[148]
cis-6, *cis*-9, *cis*-12, *cis*-15 18:4			Maia *et al.*[117]
cis-6, *cis*-9, *cis*-12, *cis*-15 18:4		5,7,11,15 18:4, 5,8,10,15 18:4, 5,10,15 18:3, 5,11,14 18:3, 5,11,15 18:3, *trans*-5, *trans*-10 18:2, *trans*-5, *trans*-11 18:2	Alves *et al.*[113]

[a]Numbers refer to citations listed in the reference section.

[b]Following 48 h incubations of U-^{13}C *cis*-9, *cis*-12, *cis*-15 18:3 significant ^{13}C enrichment was detected in numerous intermediates including conjugated 18:3 ($n=2$), non-methylene interrupted 18:3 ($n=12$) and non-methylene-interrupted 18:2 ($n=5$) isomers.

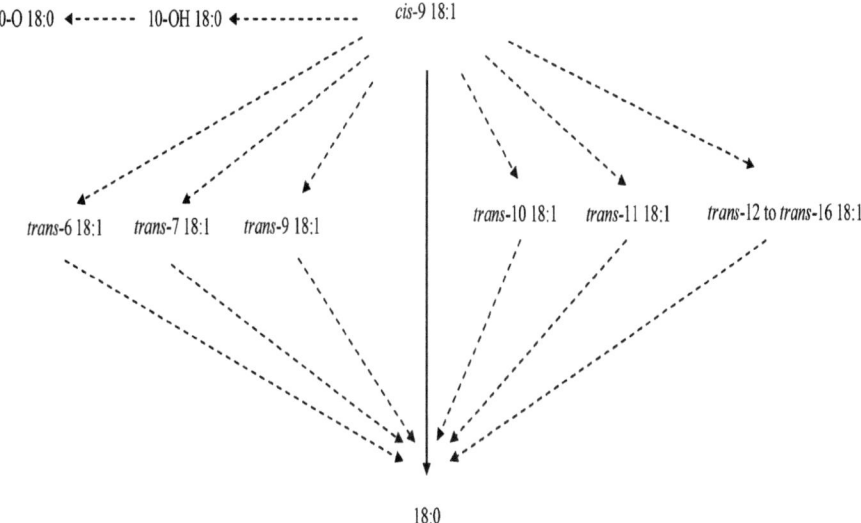

Figure 1.2 Pathways of *cis*-9 18:1 metabolism in the rumen. Formation of inter-mediates and end products illustrated is based on studies involving the incubation of labelled and unlabelled *cis*-9 18:1 with mixed and pure cultures of rumen bacteria. Arrows with solid lines highlight the major biohydrogenation pathway while arrows with dashed lines describe the formation of minor fatty acid metabolites.

with mixed ruminal bacteria resulted in the production of *cis*-9 18:1 and *cis*-11 18:1.[139] Incubations with pure or mixed cultures of ruminal bacteria have allowed several biochemical pathways of *cis*-9 18:1 metabolism to be characterized (Figure 1.2).

For many years isomerization of the *cis*-12 double bond leading to the formation of *cis*-9, *trans*-11 CLA has been considered the first committed step of 18:2 n-6 biohydrogenation in the rumen, followed by reduction to *trans*-11 18:1 and 18:0 (Figure 1.3). More recent studies have shown that geometric isomers of both 9,11 and 10,12 18:2 are formed during incubations of 18:2 n-6 with rumen fluid[127,132,140,141] and pure cultures of several ruminal bacteria.[142] The amounts of 9,11 and 10,12 CLA isomers formed during the isomerization of 18:2 n-6 also appear to be pH-dependent, but reports are conflicting. Some investigations demonstrated that decreases in pH below 6.0 lower the accumulation of *cis*-9, *trans*-11 18:2 and *trans*-10, *cis*-12 18:2,[127,129] whereas increased formation was reported in others.[130,131] Conjugated 18:2 products formed during the initial isomerization of 18:2 n-6 are transient and reduced to yield both *trans* (-4, -5, 6-8, -9, -10, -11, -12, -13 and -14) and *cis* (9, 10, 11, -12 and -13) 18:1 intermediates[134,140,141](Table 1.2). Decreases in pH from 6.4 to 5.6 have been shown to promote *trans*-10 18:1 formation, changes accompanied by a decrease in the accumulation of *trans*-11 18:1 accumulation.[127] Increases in the amounts of 18:2 n-6 incubated have also been shown to increase the ratio of *trans*-10 18:1/*trans*-11 18:1.[141,143]

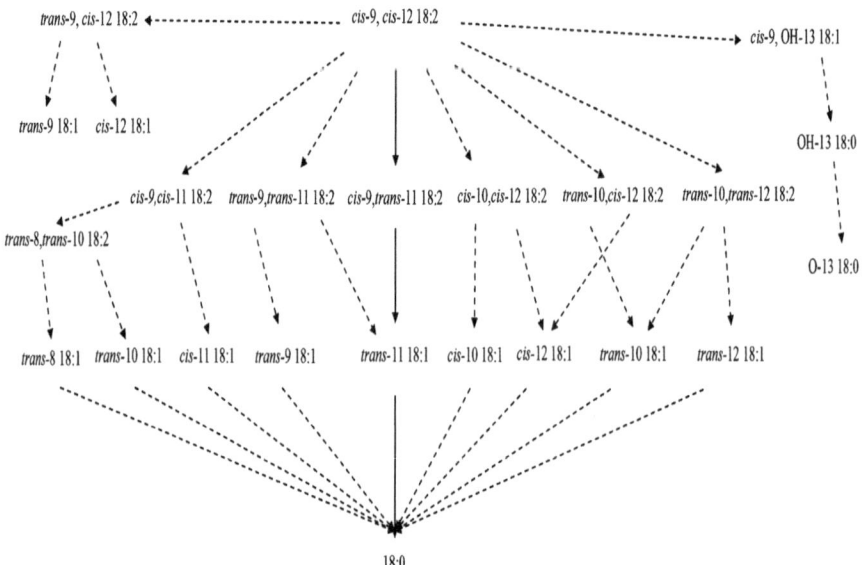

Figure 1.3   Pathways of *cis*-9, *cis*-12 18 : 2 metabolism in the rumen. Formation of intermediates and end products illustrated is based on studies involving the incubation of labelled and unlabelled *cis*-9, *cis*-12 18 : 2 with mixed and pure cultures of rumen bacteria. Arrows with solid lines highlight the major biohydrogenation pathway while arrows with dashed lines describe the formation of minor fatty acid metabolites.

Incubations with strains of *Enterococcus faecalis* isolated from the rumen have demonstrated that 18 : 2 n-6 may also be hydrated to yield 10-OH, *cis*-12 18 : 1 and *cis*-9, 13-OH 18 : 1.[144] Recent investigations provided the first evidence that biohydrogenation of 18 : 2 n-6 may also proceed by a mechanism involving the migration of the *cis*-9 double bond rather than hydrogen abstraction or isomerization of the *cis*-12 double bond resulting in the formation of *cis*-6, *cis*-12 18 : 2, *cis*-7, *cis*-12 18 : 2 and *cis*-8, *cis*-12 18 : 2.[141]

Owing to the higher number of double bonds the metabolic pathways responsible for the biohydrogenation of 18 : 3 n-3 in the rumen are far more complex than for *cis*-9 18 : 1 or 18 : 2 n-6. Most investigations indicate the metabolism of 18 : 3 n-3 in the rumen involves the formation of a single 18 : 3 intermediate *cis*-9, *trans*-11, *cis*-15 18 : 3, that is sequentially reduced to *trans*-11, *cis*-15 18 : 2, followed by *trans*-11 18 : 1 with 18 : 0 as an end product[48,109,113,145] (Figure 1.4).

Further studies have demonstrated that a diverse range of intermediates are formed during incubations of 18 : 3 n-3 with ruminal bacteria. In addition to *cis*-9, *trans*-11, *cis*-15 18 : 3, *trans*-9, *trans*-11, *cis*-15 18 : 3 is also produced during the initial isomerization of 18 : 3 n-3.[146] Biohydrogenation of 18 : 3 n-3 has been reported to result in the accumulation of numerous conjugated 18 : 2 (*cis*-9, *cis*-11 18 : 2, *cis*-11, *trans*-13 18 : 2 and *trans*-11, *trans*-13 18 : 2), non-methylene-interrupted 18 : 2 (*cis*-9, *cis*-15 18 : 2, *cis*-9, *trans*-13

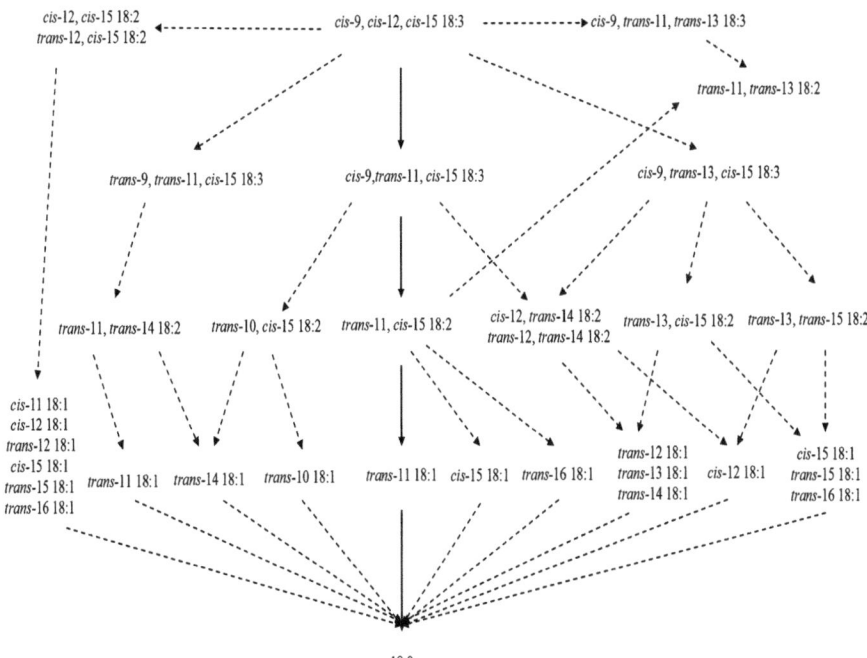

Figure 1.4 Pathways of *cis*-9, *cis*-12, *cis*-15 18:3 metabolism in the rumen. Formation of intermediates and end products illustrated is based on studies involving the incubation of labelled and unlabelled *cis*-9, *cis*-12, *cis*-15 18:3 with mixed and pure cultures of rumen bacteria. Arrows with solid lines highlight the major biohydrogenation pathway, while arrows with dashed lines describe the formation of minor fatty acid metabolites.

18:2, *trans*-9, *cis*-12 18:2, *trans*-11, *cis*-15 18:2, *trans*-9, *trans*-12 18:2), *trans* 18:1 (6-8,-9, -11, -12, -13-14, -15 and -16) and *cis* 18:1 (-13 and -15) intermediates.[140] Two strains of *B. fibrisolvens* have also been identified as capable of converting *trans*-11,*cis*-15 18:2 to *trans*-11, *cis*-13 18:2.[147] The complexity of the reactions responsible for the biohydrogenation of 18:3 n-3 was reinforced following reports that 14 positional and geometric isomers of 18:3 isomers were formed during incubations of U-[13]C 18:3 n-3 with ruminal contents.[148] After 48 h of incubation, ^{13}C enrichment was detected in *trans*-8, *trans*-10 18:2, *trans*-11, *trans*-13 18:2 and geometric isomers of 9, 11 and 10, 12 18:2. This is the first report that *cis*-9, *trans*-11 18:2 may also be formed from 18:3 n-3. In addition to studies *in vitro*, feeding diets enriched in 18:3 n-3 have been shown to increase *cis*-9, *trans*-13, *cis*-15 18:3, *cis*-9, *trans*-11, *trans*-13 18:3, *trans*-9, *trans*-11 18:2, *trans*-13, *trans*-15 18:2, 11,13 18:2 and 12,14 18:2 concentrations in bovine milk.[149–151] One or more of these fatty acids may originate from the rumen. While the major pathway describing the biohydrogenation of 18:3 n-3 in the rumen is well established, those responsible for the formation of minor intermediates remain incomplete and require further investigation.

1.3   Synthesis of CLA in the Rumen

1.3.1   Mechanisms of CLA Biosynthesis

We are indebted to the early work of Tove and his colleagues for establishing many of the properties of the first step in the metabolism of 18:2 n-6 and 18:3 n-3.[113,152,153] Linoleate isomerase (EC 5.2.1.5) is described as an enzyme that catalyses the conversion of 18:2 n-6 to other positional and geometric 18:2 isomers, including conjugated intermediates. The properties of this enzyme activity vary according to the host bacterial species and the product that is formed. Kepler and Tove[113] carried out a partial purification of linoleate isomerase activity from *B. fibrisolvens* and developed a convenient assay method based on the UV absorbance of the conjugated double bond system at 233 nm. The particulate cell-free preparation formed mainly *cis*-9, *trans*-11-18:2 from 18:2 n-6; the equilibrium was strongly in favour of *cis*-9,*trans*-11 18:2 formation. Perhaps significantly when it comes to understanding the mechanism, the authors failed to isolate a homogeneous preparation that was active. Furthermore, the reaction was short-lived in reaching an endpoint, and the reaction could be restored by the addition of more enzyme, suggesting that a cofactor was required. Isomerization of 18:3 n-3 occurred in a similar manner, but at a faster rate, to *cis*-9, *trans*-11, *cis*-15 18:3. Further refinement of the substrate specificity and mechanism of action followed,[152,153] but little further work was done for the following three decades.

In contrast, cloning, crystallization and structural analysis of the linoleate isomerase catalysing the formation of *trans*-10, *cis*-12 18:by non-ruminal *P. acnes* revealed at the atomic level, with a single protein, a mode of action that involves hydride abstraction by enzyme-bound FAD.[154,155] Isotope studies with a ruminal *P. acnes* were consistent with this mode of action.[142]

In non-ruminal lactic acid bacteria, which produce both isomers of CLA, it has been proposed that *cis*-9, *trans*-11 18:2 formation from 18:2 n-6 by lactic acid bacteria involves a hydration-dehydration mechanism via a 10-hydroxy,*cis*-12 18:1 intermediate.[156] Such a mechanism has been eliminated as a possible route of *cis*-9, *trans*-11 18:2 synthesis from 18:2 n-6 by *B. fibrisolvens*, since 10-hydroxy, *cis*-12 18:1 was not converted to *cis*-9, *trans*-11 18:2.[142] The *Lactobacillus* linoleate isomerase seems, like the *B. fibrisolvens* enzyme, to be a multi-component enzyme,[157] one component derived from the membrane fraction and another from the soluble cell content.

Two publications have claimed to have identified a gene encoding a linoleate isomerase that forms *cis*-9, *trans*-11 18:2. Park *et al.*[158] published a sequence purporting to be a linoleate isomerase from *B. fibrisolvens* A-38. The gene sequence of linoleate isomerase in non-ruminal lactic acid bacteria has been published within a patent.[159] Our own *in silico* analysis suggests that both claims may be mistaken. Furthermore, BLASTing the sequences against the recently available genomes of human intestinal species that produce *cis*-9, *trans*-11 18:2, including *B. fibrisolvens, Roseburia inulinivorans* and *R. hominis* (the last two are *cis*-9, *trans*-11 18 : 2 producers from the human intestine), failed to identify credible isomerase genes. In other preliminary studies, we have attempted to use published data from studies of the CLA reductase of *B. fibrisolvens* to identify neighbouring genes as possible candidates for linoleate isomerase, based on the assumption that CLA reductase and linoleate isomerase might be coordinately expressed. Fukuda *et al.*[160] isolated a mutant of *B. fibrisolvens* that formed exceptionally high concentrations of *cis*-9, *trans*-11 18 : 2 from 18 : 2 n-6. This resulted from the loss of the CLA reductase activity.[161] The mutant gene was identified, sequenced and expressed in *E. coli*, reportedly producing active linoleate isomerase. However, when we and others (G.T. Attwood, personal communication) BLASTed the deposited sequence against protein and nucleotide databases, a high similarity to oxaloacetate decarboxylase was revealed. Indeed, when we investigated the genomes of the previously described human bacteria, greater than 80% identity was obtained with a gene annotated as an oxaloacetate decarboxylase alpha subunit OadA. Clustering with this gene, rather than finding an ORF of unknown function, we found genes encoding transport and other genes relating to oxaloacetate metabolism. Oxaloacetate decarboxylase is very well characterized. Once again, therefore, the published sequence seems to be in error.

1.3.2   Microbiology of CLA Biosynthesis

There is little doubt that bacteria are the main organisms responsible for biohydrogenation in the rumen, but equally there is no debate that ruminal fungi and particularly protozoa have roles in determining the amounts of CLA and *trans* 18 : 1 isomers leaving the rumen.[106]

1.3.2.1   Role of Bacteria

Polan *et al.*[162] identified *B. fibrisolvens* as a bacterium that rapidly hydro-genates 18 : 2 n-6, forming *cis*-9, *trans*-11 18 : 2 and *trans*-11 18 : 1 as inter-mediates. Biohydrogenation of *B. fibrisolvens* did not result in the formation of 18 : 0. Subsequent studies[115,163] identified other bacteria that were capable of biohydrogenation, but the results did not provide much information about relative activities. The method used radio-labelled fatty acids as sub-strates and was therefore highly sensitive, but the concentrations of labelled acids used as substrates were very low (2 μg ml⁻¹), making comparisons of specific activity and relative contribution of different species difficult. Even minor activity could have led to a positive result. Among the bacteria capable of forming 18 : 0, two were identified as '*Fusocillus*' spp.[115] *Fusocillus* is not a genus that has endured in bacterial taxonomy, nor did any of the cultures survive into the DNA-sequencing era. Van de Vossenberg and Joblin[164] isolated a bacterium from a cow at pasture that could also form 18 : 0 from 18 : 2 n-6. It was phenotypically similar to '*Fusocillus*' and 16S rRNA analysis indicated that it was phylogenetically close to *Butyrivibrio hungatei*. Subsequently, a species named *Clostridium proteoclasticum* was identified as a 18:0 producer with morphological and metabolic properties that were indistinguishable from those reported for *Fusocillus*.[165] Moon *et al.*[166] reclassified *C. proteoclasticum* as *Butyrivibrio proteoclasticus* from its 16S rRNA gene sequence. The bacteria involved in the different steps of the biohydrogenation process were classified for many years as group A and B.[48] Group A bacteria hydrogenated 18:2 n-6 and 18:3 n-3 to *trans*-11 18:1, whereas group B bacteria converted both fatty acids to 18:0. It is now more appropriate to describe the bacteria based on their correct taxonomy (Figure 1.5).

Phenotypically, the *B. hungatei* and *B. proteoclasticus* groups are much more sensitive to the toxic effects of unsaturated fatty acids than the rest of the *Butyrivibrio/Pseudobutyrivibrio* cluster, such that their isolation from media containing unsaturated fatty acids is made much more difficult.[82,165] They can also be distinguished phenotypically and genetically based on the mechanism by which they form 4:0 (butyrate), their second most abundant fermentation product after 2:0 (acetate). *B. hungatei* and *B. proteoclasticus* had a butyrate kinase activity >600 U per mg protein, while the others had much lower activity[82] (Figure 1.5). The butyrate kinase gene was present in *B. hungatei* and *B. proteoclasticus* but not in the other group. It has been suggested that the different sensitivities to the toxic effects of CLA isomers and *trans*-11 18:1 may be linked to the enzyme mechanism by which butyrate is produced. Intracellular acyl-CoA concentrations, principally the precursors of 4:0, in *B. fibrisolvens* are depleted when the bacterium is exposed to 18:2 n-6.[167]

Metabolism of 18:2 n-6 by *Butyrivibrio* results in the formation of *cis*-9, *trans*-11 18:2, with smaller amounts of *trans*-9, *trans*-11 18:2, and *trans*-11 18:1, but no *trans*-10,*cis*-12 18:2 or *trans*-10 18:1 is formed.[134] The bacteria responsible for *trans*-10 18:1 must therefore be different to the *Butyrivibrio* spp. The formation of *trans*-10, *cis*-12 18:2 also occurs by a different enzymic mechanism to that of *cis*-9, *trans*-11 18:2.[142] Enrichment cultures with starch were reported to isomerize 18:2 n-6 to *trans*-10, *cis*-12 18:2,[168] but these also contained abundant large cocci identified as *Megasphaera elsdenii*. Authors concluded that *M. elsdenii* as capable of *trans*-10, *cis*-12 CLA production in the rumen. Subsequent studies indicated that *P. acnes* could be responsible for the formation of *trans*-10, *cis*-12 18:2.[165] Wallace *et al.*[165] found no lineate isomerase activity in any pure culture of *M. elsdenii*. Furthermore, when digesta samples from cows producing high amounts of *trans*-10, *cis*-12 18:2 were analysed for *M. elsdenii* by qPCR of 16S rRNA genes, numbers were <10^3 g^{-1}, while much larger numbers of *P. acnes* were detectable (R.J. Wallace and S. Muetzel, unpublished observation). *P. acnes* does not, however, convert *trans*-10, *cis*-12 18:2 to *trans*-10 18:1.[134] *Butyrivibrio* spp. are capable of reducing *trans*-10, *cis*-12 18:2 and *trans*-10, *trans*-12 18:2.[134] However, it remains difficult to understand how *Butyrivibrio* would catalyse the reduction of *trans*-10 18:1 but not the isomerization of 18:2 n-6 to *trans*-10, *cis*-12-18:2. Shifts in biohydrogenation pathways

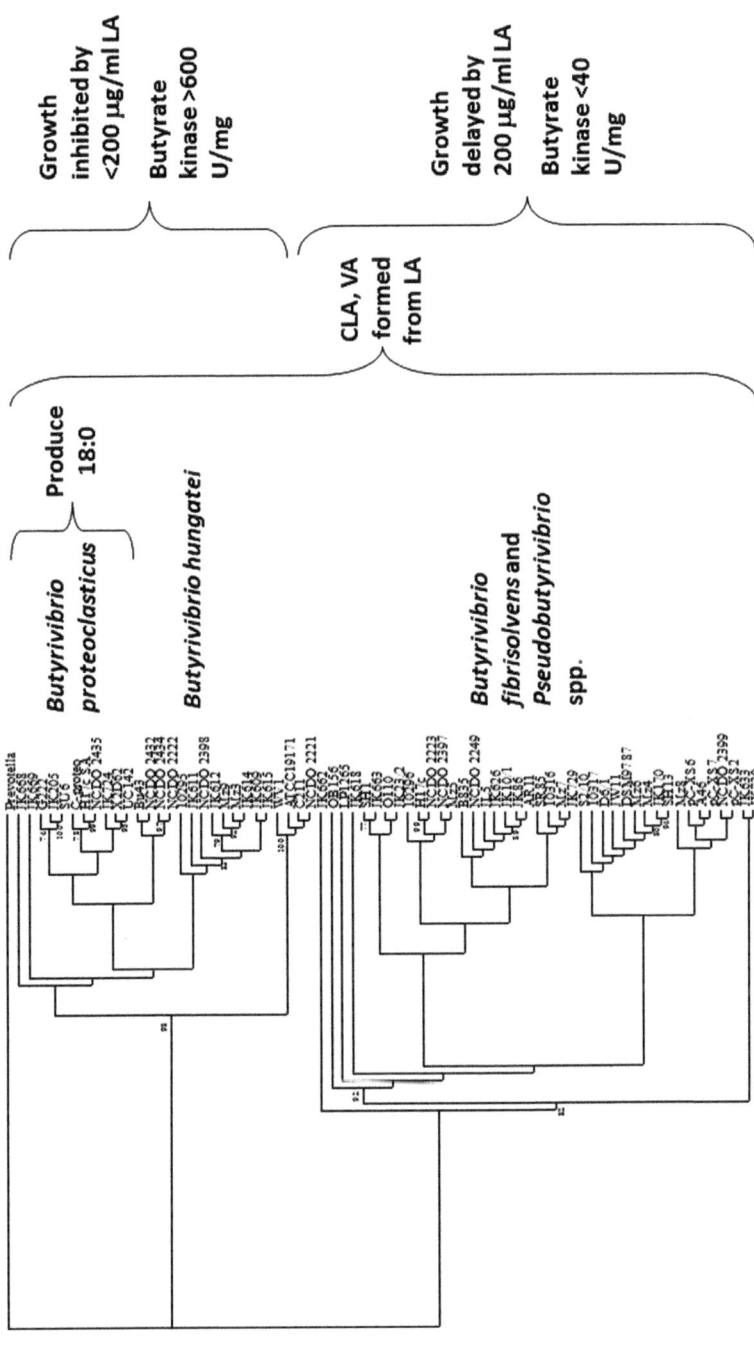

Figure 1.5 Phylogenetic tree of the *Butyrivibrio* and some key phenotypic properties. Abbreviations: CLA, conjugated linoleic acid (*cis*-9, *trans*-11 18:2); LA, linoleic acid (*cis*-9, *cis*-12 18:2); VA, vaccenic acid (*trans*-11 18:1).

favouring the production of *trans*-10 18:1 are known to occur in ruminants fed high concentrate diets containing unsaturated fatty acids[169–171] or during incubations of 18:2 n-6 with mixed ruminal bacteria containing starch[172,173] or maintained at a low pH.[127] Recently, switching from high forage to high concentrate diets was shown to be accompanied by major changes in the bacterial community in lactating cows,[174] conditions under which a shift towards the accumulation of *trans*-10 18:1 at the expense of *trans*-11 18:1 could be expected.[175,176] Non-ruminal *Lactobacillus plantarum* converts 18:2 n-6 to *cis*-9, *trans*-11 18:2 and *trans*-9, *trans*-11 18:2.[177] Of particular relevance to the *trans*-10 shift, *L. plantarum* reduces geometric 9,11 18:2 isomers to *trans*-10 18:1 rather than *trans*-11 18:1. *Lactobacillus* spp. have commonly been isolated from the rumen.[178] Their numbers increase with dietary starch content.[179] However, they have been associated with hydration rather than hydrogenation of unsaturated fatty acids.[180] A future challenge will be to associate with certainty the role of individual microbial species with alterations in rumen function and biohydrogenation pathways, such as those indicated from ribosomal intergenic spacer analysis.[174]

There are often significant differences between bacterial communities attached to solids and those free-swimming, 'planktonic' communities that inhabit the liquid phase of the fermentation mixture. The same is true for ruminal lipid metabolism, with the planktonic community converting 18:2 n-6 as far as 18:1 intermediates, whereas the solids-associated community reduced 18:2 n-6 to 18:0[181]. *B. proteoclasticus* was present only in the solids-associated community, at 12% of *Butyrivibrio*-related clones.[181]

Li *et al.*[182] carried out an interesting enrichment study by sequentially subculturing bovine ruminal digesta in the presence of 50 μg per ml *trans*-11 18:1 in order to investigate species capable of the final reduction of *trans* 18:1 to 18:0. Amounts of 18:0 were produced at progressively faster rates as the subculturing progressed. 16S rRNA amplicon analysis indicated that bacteria closely related to *B. proteoclasticus* were enriched, along with a number of previously uncultured bacteria. It was not possible to establish if any of these bacteria hydrogenated *trans*-11 18:1. Furthermore, there is no obvious reason why *trans*-11 18:1 would enrich for biohydrogenating bacteria as there would be no benefit conferred to those bacteria able to do so. More likely was that the soluble sugars present in the medium selected in favour of these bacteria.

Several studies have indicated that bacteria other than the *Butyrivibrio* spp. might be involved in biohydrogenation, particularly in the final reduction of *cis* 18:1 and *trans* 18:1 to 18:0. The long-chain unsaturated fatty acids in marine algae inhibit the biohydrogenation of 18-carbon fatty acids in the rumen and alter the ruminal microbial community.[183] These changes have been taken, by association, to indicate bacterial species involved in biohydrogenation. Inhibition of biohydrogenation by marine algae was found to induce changes in uncultivated bacteria clustering with *Butyrivibrio* and *Pseudobutyrivibrio*. Boeckaert *et al.*[184] found that this changed when biohydrogenation was inhibited. In contrast, no significant decrease in numbers of the known *Butyrivibrio*-related 18:0 producers were detected in the rumen of sheep fed marine algae.[120] However, only five animals per treatment were investigated and numerical decreases were detected as the amount of algae supplement increased. Restriction fragment length polymorphism (RFLP) of 16S rRNA amplicons indicated that *Lactospiracaea* spp. or *Quinella*-like bacteria increased, which could have occurred for other reasons. In experiments investigating the effects of fish oil supplements on the rumen bacterial community, only weak associations have been observed between digesta 18:0 concentration or ruminal 18:0 outflow and *B. proteoclasticus* group 16S rRNA concentration.[122,185] Further investigation based on multivariate analysis of data obtained from terminal RFLP (tRFLP) and denaturing gradient gel electrophoresis (DGGE) revealed many terminal restriction fragments (T-RFs) and DGGE bands that were correlated with ruminal *cis*-9, *trans*-11 18:2, *trans*-11 18:1 and 18:0 concentrations.[185] Predictive identification revealed that these linked tRFs were likely to originate from as yet uncultured bacteria, including *Prevotella*, *Lachnospiraceae* incertae sedis, and unclassified *Bacteroidales*, *Clostridiales* and *Ruminococcaceae*. DGGE bands led to similar conclusions. Consistent with other recent studies, the bacteria have not been isolated and possible isomerase or reductase activity remains to be confirmed.

1.3.2.2 Role of Ciliate Protozoa

Since up to half of the rumen microbial biomass may be protozoal in origin[186] and *ca.* 75% of the microbial fatty acids present in the rumen may be in protozoa,[125] it follows that protozoa could represent a very important source of several biohydrogenation intermediates, including isomers of CLA and *trans*-11 18:1. Early studies concluded that both protozoa and bacteria were involved in biohydrogenation.[187,188] However, the extensive ingestion of bacteria by protozoa caused Dawson and Kemp[189] to doubt this conclusion. Biohydrogenation in ruminal digesta was only slightly decreased following defaunation (*i.e.* removal of protozoa from the rumen) and the presence of protozoa was not necessary for biohydrogenation to occur.[189] Others also suggested that the minor contribution of protozoa to the biohydrogenation process was due to the activity of ingested or associated bacteria.[190,191] Yet it has been known for a long time that protozoal lipids contain proportionally more unsaturated fatty acids than the bacterial fraction.[48,192] Recently, it was established that these unsaturated fatty acids include *cis*-9, *trans*-11 18:2 and *trans*-11 18:1,[193] further highlighting the possible significance of protozoa as a means of facilitating the escape of biohydrogenation intermediates from the rumen. Protozoal species differ in fatty acid composition, with larger species like *Ophryoscolex caudatus* containing more than ten times higher concentrations of *cis*-9, *trans*-11 18:2 and *trans*-11 18:1 than small species such as *Entodinium nannelum*.[193] However, the lipid of *Isotricha prostoma*, a large species and the only holotrich examined, had low concentrations of *cis*-9, *trans*-11 18:2 and *trans*-11 18:1. Holotrichs do not ingest the large particles seen with entodinio-morphs. In incubations with fractionated ruminal digesta, 18:2 n-6 metabolism was similar between strained ruminal fluid and the bacterial enriched fraction, while the fraction containing protozoa had a much lower activity. There was no evidence of ^{14}C enrichment in *trans*-11 18:1 or *cis*-9, *trans*-11 18:2 during incubations of protozoa with ^{14}C-18:0. No genes with sequence similarity to fatty acid desaturases from other organisms were found in cDNA libraries from ruminal protozoa (E. Devillard, personal communication). Thus, the protozoa are relatively abundant in *trans*-11 18:1 or *cis*-9, *trans*-11 18:2, but do not synthesize these from 18:2 n-6 or 18:0, confirming earlier considerations.[189] It could be argued that the high un-saturated fatty acid content of protozoa arises from the ingestion of plant particles, especially chloroplasts.[187,194] A recent study showed that the engulfment of chloroplasts is a major contributor to the high 18:3 n-3 concentration of protozoa.[78] However, these observations do not explain the relatively high concentrations of *trans*-11 18:1 or *cis*-9, *trans*-11 18:2 that originate from biohydrogenation. Lourenço *et al.*[195] suggested that these findings reflect protozoa preferentially accumulating *trans*-11 18:1 or *cis*-9, *trans*-11 18:2 formed by bacteria associated with lipids in ingested plant cell organelles (Figure 1.6).

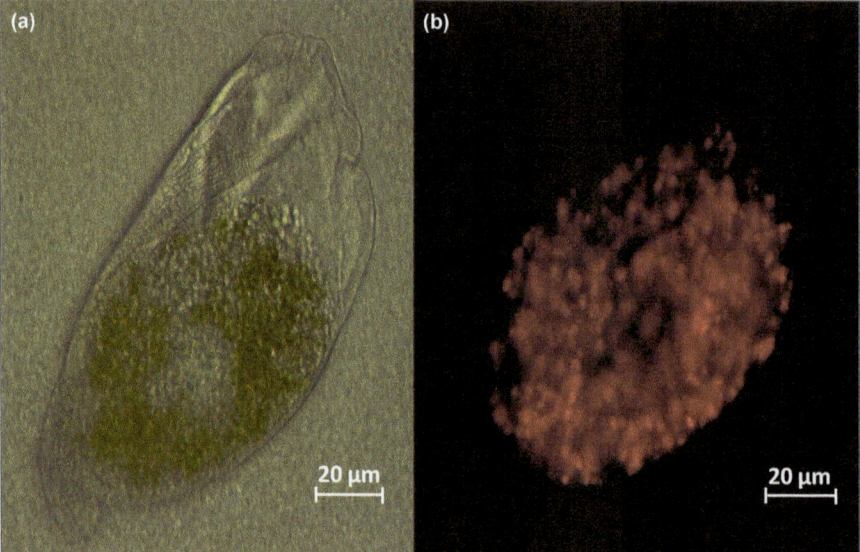

Figure 1.6　Confocal microscopy images of an *Epidinium* sp. isolated from the rumen of Hereford x Friesian steers 2 h after being offered fresh perennial ryegrass saturated with intracellular chloroplasts (adapted from Huws *et al.*[78]). Figure 1.6a is the light microscopy image and Figure 1.6b is the contrasting fluorescent image taken at a depth of 2.24 μm.

The relatively low conversion of these intermediates to 18:0 may arise because the bacteria responsible for the reduction of 18:1 intermediates are more vulnerable to protozoal digestive activities. Increased accumulation of 9,11 and 10,12 18:2 isomers in washed protozoa to antibiotic treatment[196] reflects a similar phenomenon, in as much that the enrichment of biohydrogenation intermediates is not due to endogenous protozoal activity. Until the genes encoding the enzymes involved in both bacteria and protozoa are identified, it will be difficult to resolve the issue unambiguously. Recently, *I. prostoma* was shown incapable of hydrogenating 18:2 n-6,[181] but considering its low *trans*-11 18:1 or *cis*-9, *trans*-11 18:2 CLA concentrations, this observation may not be relevant to other species, particularly entodiniomorphs.

Determinations of protozoal fatty acid composition imply that the availability of unsaturated fatty acids, including *trans*-11 18:1 or *cis*-9, *trans*-11 18:2, for absorption by the host animal could depend more on the flow of protozoa rather than bacteria from the rumen. Some ciliate protozoa are retained selectively within the rumen by a migration/sequestration mechanism that depends on chemotaxis.[197,198] As a consequence, protozoal biomass at the duodenum is proportionally less than would be the case if protozoa escaped the rumen attached to digesta particles.[199,200] The flow of microbial nitrogen at the duodenum of steers was recently shown to be 12–15% protozoal in origin. However, protozoal lipid accounted for between 30% and 43% of *cis*-9, *trans*-11 18:2 and 40% of *trans*-11 18:1 reaching the duodenum.[201] The contribution of protozoa to the flows of 16:0 and 18:0 to the duodenum was less than 20% and 10%, respectively. While there is no doubt that chloroplasts accumulate inside protozoal cells, particularly entodiomorphs, translating that capability to increase unsaturated fatty acids, including *trans*-11 18:1 or *cis*-9, *trans*-11 18:2, available for absorption may prove to be elusive.[202]

1.3.2.3   Role of Other Ruminal Microorganisms

Anaerobic fungi were first discovered in the rumen before being found in other environments.[203] It is estimated that they comprise around 7% of rumen microbial biomass, but this value is difficult to ascertain and undoubtedly highly variable.[204] Mixed ruminal fungi are capable of the isomerization of 18:2 n-6 to *cis*-9, *trans*-11 18:2,[205,206] but activity is very low in comparison with *B. fibrisolvens*.[205,207] *Orpinomyces* is the most active genus in biohydrogenation.[206]

Methanogenic archaea are also very significant members of the rumen microbial community, comprising up to perhaps 3–4% of the biomass.[208] Until recently, the mechanisms of methane formation have formed the focus for research on ruminal archaea.[209] This is changing, however, as genomic analysis reveals new, unsuspected activities.[210] Lipase genes were not noted in the *Methanobrevibacterium ruminantium* genome, but many ORFs of unknown function were present.[210] Whether these may also include enzymes involved in biohydrogenation is not clear.

1.3.3 Measuring Ruminal Synthesis

Quantitative assessment of ruminal CLA biosynthesis requires the measurement of DM flow at the omasum, abomasum or duodenum and a detailed analysis of lipid in digesta entering the sampling site. Measurement of nutrient flow in the ruminant gastro-intestinal tract is reliant on the use of indigestible marker systems to account for sampling errors that occur due to the tendency for digesta to separate during collection.[211] Obtaining representative samples of digesta for the analysis of CLA content is of particular importance owing to the heterogeneous distribution of fatty acids in digesta. It is possible to account for errors due to unrepresentative sampling using a combination of indigestible markers that specifically associate with liquid or particulate fractions of digesta that exhibit different flow characteristics.[211]

Unless a single marker is uniformly distributed across all digesta phases, unrepresentative sampling will inevitably introduce errors in estimated flows.[212] In a number of experiments, Cr_2O_3 has been used as a single marker to determine CLA at the duodenum in lactating cows,[175,213] growing cattle[214–217] and sheep.[104,218,219] In other studies multiple markers have been used for the measurement of CLA at the omasum[54,61,67,122,220] or duodenum.[68,221–229]

Analysis of CLA in sampled digesta is reliant on reliable methods for lipid extraction and the preparation of fatty acid derivatives. Conjugated fatty acids are susceptible to isomerization and may disappear during prolonged exposure to acid catalysts.[230] Under these conditions there is also a risk that conjugated products are formed from endogenous sources during methylation. Base-catalysed transesterification has been shown to be the most accurate method for determining CLA composition in range of biological samples.[231] However, NEFA are not methylated under these conditions. Virtually all of the CLA leaving the rumen is non-esterified,[54,61] and therefore this catalyst is ineffective for the analysis of digesta fatty acid composition. Isomerization and the production of artifacts with acid-based catalysts can be minimized using lower temperatures during methylation,[232] but under these conditions the methylation of PL is incomplete.[231] Preparation of FAME using methanolic sulphuric acid,[220,233] diazomethane followed by sodium methoxide,[232] or methanolic sulphuric acid followed by sodium methoxide[121] can be used while avoiding isomerization or the synthesis of allylic methoxy derivatives.

Determination of the relative distribution and abundance of specific CLA isomers also requires the use of long ($\geq$100 m) highly polar capillary columns during gas chromatography (GC) analysis in combination with silver-ion high performance liquid chromatography (HPLC).[234] During GC analysis, *cis*-9, *trans*-11 18:2 elutes with the same retention time as *trans*-8, *cis*-10 18:2 and *trans*-7, *cis*-9 18:2. The *trans*-11, *cis*-13 18:2 peak may contain minor amounts of *cis*-9, *cis*-11 18:2, while *trans*, *trans* isomers with double bonds from 7,9 to 10,12 typically elute as a single peak. The occurrence of 21 : 0 also complicates the determination of CLA by GC. Depending on the GC column used and the temperature programme applied 21 : 0 elutes anywhere between *cis*-11, *trans*-13 18 : 2 and *cis*-10, *cis*-12 18 : 2, and may therefore, be erroneously identified as an isomer of CLA.[234-236] An additional complication is that the retention of 21 : 0 relative to several minor CLA isomers may also differ between GC columns of the same type and changes over the lifetime of the GC column.[234]

1.3.4 Formation of CLA Isomers

The amount of CLA formed in the rumen has been determined for lactating cows, non–lactating cows, growing cattle and non-lactating sheep (Table 1.3). There are no estimates for lactating sheep, goats or other ruminant species. Synthesis of CLA in the rumen of sheep, growing cattle, non-lactating cows or lactating cows varies between 0.02 and 7.14, 0.14 and 6.59, 0.68 and 3.00 and 1.02 and 15.3 g per day, respectively. It should be noted that the estimates based on GC analysis alone ignore the contribution of minor isomers and are subject to possible errors due to the interference of other fatty acids. Nevertheless, it is clear that diet composition rather than the level of intake, physiological state or species is the major determinant of CLA synthesis in the rumen.

Thus far, 16 isomers of CLA have been detected in ruminal,[119,120] omasal[54,61,67,122,220] or duodenal[68,175,227,229] digesta with double bonds located at 7,9 through to 13,15. Most reports indicate that ruminal digesta contains only trace amounts of *trans*-7, *cis*-9 18 : 2. Not all 16 isomers have been isolated in a single sample of digesta, indicating that the formation of specific CLA isomers is dependent on the composition of the basal diet and the intake of fatty acid substrates. In lactating cows offered diets based on grass, grass hay, or red clover forages supplemented with cereals and solvent extracted plant protein supplements (Forage: Concentrate (F : C) ratio on a DM basis; 60 : 40), *cis*-9, *trans*-11 18 : 2 is the major isomer accounting for 48–84% of total CLA at the omasum.[54,61,67,122,220] *Cis*-9, *trans*-11 18 : 2 is the main CLA isomer formed in the rumen of cattle fed maize silage based diets[68,229] or in sheep fed diets based on lucerne hay.[119,120] However, the relative abundance of *cis*-9, *trans*-11 18 : 2 was found to be lower (27–31% of total CLA) in lactating cows fed diets (F : C ratio, 60 : 40) containing lucerne hay and maize silage.[175] Under these circumstances the amounts of *trans, trans* (7,9 through to 12,14) isomers accounted for 57–63% of total CLA at the duodenum. Reports on ruminal synthesis of specific CLA isomers in non-lactating sheep fed high concentrate diets (F : C 18 : 82) containing no additional lipid are conflicting. In one investigation, *cis*-9, *trans*-11 18 : 2 was reported to be the major CLA isomer at the duodenum,[219] whereas in a follow-up experiment *trans, trans* isomers were quantitatively more important.[104]

In ruminants fed high proportions of grass, grass silage or red clover silage, where 18 : 3 n-3 is the predominant dietary fatty acid, *trans*-11,

Table 1.3 Synthesis of CLA in the rumen of growing cattle, lactating cows and sheep.

DMI[a] (kg/d)	Forage[b]	F:C[c]	Supplement[d] (g/kg DM)	CLA[e] (g/d)	No. of isomers detected	Analytical method[f]	Reference[g]
				Non-lactating sheep			
0.86	GH	18:82-73:27	SBO (31-46)	0.35-23	3	GC	Kucuk et al.[218]
0.56	GH	18:82	SBO (0-94)	0.08-0.41	3	GC	Kucuk et al.[219]
0.80	GH	18:82	SAF (0-90)	0.02-7.14	4	GC	Atkinson et al.[104]
				Non-lactating cows			
5.90	GS	55:45	–	0.68-0.82	2	GC	Lock and Garnsworthy[240]
10.5-9.9	MS + LH	70:30	LO (0-49)	1.0-3.0	4	GC	Doreau et al.[228]
				Growing cattle			
3.60-4.78	GS/RCS	100:0	–	0.17-25	4	GC	Lee et al.[224]
4.15-8.48	GS/RCS/WCS	60:40	–	0.74-2.66	1	GC	Lee et al.[221]
6.11-6.69	GS	80:20-20:80	LO (25-33)	1.94-1.80	7	GC	Lee et al.[225]
12.0-11.4	GH	82:18	HOS (131)	0.80-0.40	3	GC	Scholljegerdes et al.[216]
12.0-11.6	GH	82:18	HLS (137)	0.80-1.30	3	GC	Scholljegerdes et al.[216]
10.4-10.7	GH	14:86	MO (0-24)	0.63-1.25	7	GC	Duckett et al.[214]
5.54-6.19	GH	12:88-36:64	SFO (20-40)	0.68-0.88	9	GC	Sackmann et al.[215]
8.33-8.77	GH	12:88	RO/MO (30-40) + FO (0-10)	0.97-2.96	9	GC	Duckett and Gillis[217]
5.18-5.20	GS	100:0	FO (0-30)	0.14-0.52	16	GC + HPLC	Lee et al.[227]
6.59-5.09	RCS	100:0	FO (0-30)	0.30-0.48	16	GC + HPLC	Lee et al.[227]

DMI[a]	Forage[b]	Ratio[c]	Lipid supplement[d]	CLA	n	Method[f]	Reference[g]
7.55–7.45	GS	50:40	FO (0–40)	2.65–6.59	4	GC	Lee et al.[222]
10.5–9.77	MS	50:40	FO (0–24)	0.39–0.52	10	GC+HPLC	Shingfield et al.[229]
8.88–8.34	MS	50:40	FO (0–30)/LO (0–30)	0.38–3.18		GC+HPLC	Shingfield et al.[68]
Lactating cows							
15.7/15.5	HSG/HNG	100:0	–	1.45/2.50	3	GC	Doreau et al.[226]
16.7–21.8	FG/GH/GS	60:40	–	3.70–5.38	12	GC+HPLC	Halmemies-Beauchet-Filleau et al.[61]
19.9–18.4	GS/RCS	60:40	–	3.94–4.12	15	GC+HPLC	Halmemies-Beauchet-Filleau et al.[54]
20.6–24.1	MS+LH	60:40-25:75	–	1.02–1.84	12	GC+HPLC	Piperova et al.[175]
19.6–20.5	GH	65:35-35:65	LO (0–30)	1.70–71	8	GC	Loor et al.[176]
15.1–15.9	GS	60:40	SFO (0–50)	3.0–15.3	13	GC+HPLC	Shingfield et al.[67]
17:1/17:2/19.3	GH	35:65	FO(25)/LO(50)/SFO(50)	4.01/6.90/8.30	8	GC	Loor et al.[223]
15.7–17.7	GS	60:40	FO (0–16)	4.44–3.46	11	GC+HPLC	Shingfield et al.[220]
18.7–15.6	GS	60:40	FO (0–19)	3.82–5.13	12	GC+HPLC	Shingfield et al.[122]
15.0–18.5	LS+GH+MS	36:64	FO (20–5) + SFO (0–20)	6.04–89	NR	GC	Qiu et al.[213]

[a]DMI, dry matter intake.
[b]Forage in the diet: GH, grass hay; GS, grass silage; FG, fresh grass; HNG, high nitrogen grass; HSG, high sugar grass; LH, lucerne haylage; MS, maize silage; RCS, red clover silage; WCS, white clover silage.
[c]Dietary forage:concentrate ratio on a dry matter basis.
[d]Lipid supplements: FO, Fish oil; LO, linseed oil; MO, Maize oil; RO, rapeseed oil; SAF, safflower oil; SBO, soyabean oil; SFO, sunflower oil.
[e]Based on sampling at the duodenum or omasum and measurements of dry matter flow at the sampling site using indigestible markers.
[f]Determination of total CLA based on gas chromatography alone or in combination with silver-ion high performance liquid chromatography.
[g]Numbers refer to citations listed in the reference section.

trans-13 18:2 or *trans*-11, *cis*-13 18:2 are the principal CLA isomers formed in the rumen.[227,228,237] *Trans*-12, *trans*-14 18:2 is also relatively abundant, whereas *cis*-9, *trans*-11 18:2 accounts for between 6.7% and 32% of total CLA synthesis.[227,228]

A number of investigations have explored the potential to increase ruminal CLA synthesis through changes in diet composition. Emphasis has been placed on alterations in the relative proportions of forages and concentrates or marine lipid supplements to manipulate rumen environment or supplementing the diet with plant oils to increase the supply of substrates for ruminal CLA synthesis. For diets containing no additional lipid supplements, increases in concentrate supplementation have variable and rather marginal effects on the amount of CLA at the duodenum (Table 1.3). Decreases in F:C ratio from 60:40 to 25:75 were shown to promote ruminal synthesis of geometric isomers of 9,11 and 10,12 18:2.[175] In contrast, decreases from 65:35 to 35:65 had no effect other than decreasing *trans*- 9, *cis*-11 18:2 and increasing formation of several *trans, trans* isomers.[176]

Studies of 18:2 n-6 metabolism *in vitro* indicate that the amount of *cis*-9, *trans*-11 18:2 leaving the rumen would be enhanced in diets that promote high passage rates, maintain rumen pH and contain high amounts of 18:2 n-6.[127,129,130,132] Increasing the intake of 18:2 n-6 on ruminal CLA synthesis has been investigated in cattle and sheep. Dietary supplements of sunflower oil, soyabean oil or safflower as a source of 18:2 n-6 has been shown to increase the amount of 8,10, 9,11 and 10,12 18:2 isomers and total CLA at the omasum or duodenum (Table 1.3). However, increases in the the formation of specific isomers to 18:2 n-6 supply is dependent on the composition of the basal diet. In lactating cows fed diets based on grass silage (F:C ratio 60:40), supplements of sunflower oil from 0 to 750 g per day progressively increased *cis*-9, *trans*-11 18:2, *trans*-9, *trans*-11 18:2, *trans*-10, *cis*-12 18:2 and *trans*-10, *trans*-12 18:2 at the omasum[67] (Table 1.4). In contrast, adding higher amounts of sunflower oil from 20 to 40 g per kg DM in high concentrate diets (average F:C ratio 24:76) stimulated ruminal *trans*-10, *cis*-12 18:2 formation (0.21–0.37 g per day) but had no effect on other measured CLA isomers at the duodenum in growing cattle.[215] Ruminal synthesis of *trans*-10,*cis*-12 CLA has been reported to be as high as 1.83 g per day in lactating cows fed high concentrate diets (F:C ratio 35:65) containing 50 g per kg DM of sunflower oil.[223] Consistent with these findings, ruminal infusion of safflower oil in non-lactating sheep fed high concentrate diets (F:C ratio 18:82) at a rate equivalent to 0 to 90 g per kg DM progressively increased the amount (g per day) of *cis*-9, *trans*-11 18:2 (<0.01–0.38), *trans*-9, *trans*-11 18:2 (0.01–2.08), *trans*-10, *cis*-12 18:2 (<0.01–4.49) and *trans*-10, *trans*-12 18:2 (0.03–0.42) recovered at the duodenum.[104] At the highest level of infusion, *trans*-10,*cis*-12 CLA was identified as the principal isomer formed in the rumen. However, an earlier experiment in non-lactating sheep fed a similar diet reported that incremental soyabean oil supplements (0-94 g per kg DM) had no influence

Table 1.4 Synthesis of CLA isomers in the rumen of growing cattle and lactating cows.

Lactating cows

DM Intake (kg/d)	Forage[a]	F:C[b]	Supplement (g/kg DM)[c]	Conjugated linoleic acid isomer (mg/d)[d] cis, trans			trans, cis						trans, trans						Reference[e]
				9,11	11,13	12,14	7,9	8,10	9,11	10,12	11,13	12,14	7,9	8,10	9,11	10,12	11,13	12,14	
20.6	MS + LH	[60:40]	–	330	21		5	9		86	9		14	38	200	109	156	88	Piperova et al.[175]
21.9	MS + LH	[60:40]	Buffer (20)[f]	276	20		7	13		54	7		13	73	130	140	179	107	
23.7	MS + LH	[25:75]	–	529	35		5	22		256	13		21	96	391	234	154	82	
24.1	MS + LH	[25:75]	Buffer (20)[f]	244	21		6	17		78	12		16	21	291	121	141	67	
17.7	GS	[60:40]	–	2858	13	52				95	460		0	9	224	47	403	193	Shingfield et al.[220]
15.7	GS	[60:40]	FO (16)	2077	11	2				21	197		46	99	552	57	89	78	
15.1	GS	[60:40]	–	1927		36				84	334			5	184	30	221	132	Shingfield et al.[67]
15.0	GS	[60:40]	SFO (17)	4768		14				144	237			26	444	131	241	139	
14.7	GS	[60:40]	SFO (34)	9228		17				182	143			37	930	226	210	121	
14.9	GS	[60:40]	SFO (50)	11571		23				396	322			36	1675	424	337	218	
18.7	GS	[60:40]	–	4246		69				104	299		95	64	286	65	559	244	Shingfield et al.[122]
18.8	GS	[60:40]	FC (4.0)	4091		26				24	256		109	76	232	31	217	140	
17.8	GS	[60:40]	FC (8.4)	3741		21				30	194		269	134	331	33	94	68	
15.6	GS	[60:40]	FC (19)	3457		17				49	55		201	86	264	58	60	4.9	
16.7	FG	[60:40]	–	2430		20				90	370			3	180	30	380	120	Halmemies-Beauchet-Filleau et al.[61]
18.4	GH	[60:40]	–	4350		20				140	110				240	30	200	80	
19.8	GH	[60:40]	–	2710		60				280	160			10	250	80	370	140	
20.2	GS	[60:40]	–	2710		30				170	360			20	210	80	760	330	
21.8	GS	[60:40]	–	2710		40				240	580			20	240	80	880	400	
19.9	GS	[60:40]	–	1880	20	20	30			160	720	70	10	30	120	60	580	180	Halmemies-Beauchet-Filleau et al.[54]
18.4	RCS	[60:40]	–	2320	10	20	30			110	510	90	0	20	90	60	600	190	

Table 1.4   (Continued)

DM Intake (kg/d)	Forage[a]	F:C[b]	Supplement (g/kg DM)[c]	Conjugated linoleic acid isomer (mg/d)[d]															Reference[e]
				cis, trans			trans, cis						trans, trans						
				9,11	11,13	12,14	7,9	8,10	9,11	10,12	11,13	12,14	7,9	8,10	9,11	10,12	11,13	12,14	
				Growing cattle															
5.18	GS	[100:0]	–	18	–	3	6	6	<1	5	5	6	1	1	4	1	78	32	Lee et al.227
5.11	GS	[100:0]	FO (10)	23	<1	–	4	2	1	6	49	2	1	0.7	6	4	40	24	
4.87	GS	[100:0]	FO (20)	66	<1	<1	16	3	<1	12	74	7	1	2	21	7	58	55	
5.20	GS	[100:0]	FO (30)	84	1	<1	15	10	2	17	126	14	8	7	25	16	74	53	
6.59	RCS	[100:0]	–	16	2	3	5	1	1	5	2	12	–	3	10	7	113	59	
6.35	RCS	[100:0]	FO (10)	37	1	–	10	8	1	6	55	17	11	7	17	12	180	81	
5.88	RCS	[100:0]	FO (20)	71	2	1	9	1	2	8	65	11	11	11	32	17	151	73	
5.09	RCS	[100:0]	FO (30)	98	<1	–	26	6	7	7	110	6	8	8	31	15	143	95	
10.5	MS	[60:40]	–	230		3		6		10	26			12	48	26	22	15	Shingfield et al.229
10.6	MS	[60:40]	FO (8)	284		3		12		17	33			13	75	29	29	25	
10.3	MS	[60:40]	FO (16)	278		3		13		7	25			15	72	20	22	18	
9.8	MS	[60:40]	FO (24)	212		4		16		8	13			19	51	16	15	14	
8.88	MS	[60:40]	–	194		1		–		17	30			8	54	27	35	13	Shingfield et al.68
8.83	MS	[60:40]	FO (30)	107		5		5		29	8			12	56	14	11	9	
8.34	MS	[60:40]	LO (30)	628		37		7		24	1194			22	163	58	690	289	
8.83	MS	[60:40]	FO+LO (30)	264		4		10		12	96			14	80	10	27	30	

[a]Forage in the diet: FG, fresh grass; GH, grass hay; GS, grass silage; LH, lucerne haylage; MS, maize silage; RCS, red clover silage.
[b]Forage:concentrate ratio of the diet on a dry matter basis.
[c]Lipid supplements: FO, Fish oil; LO, linseed oil; SFO, sunflower oil.
[d]Determined based on sampling at the duodenum or omasum and measurements of dry matter flow at the sampling site using indigestible markers. Amounts of individual CLA isomers determined by complimentary gas chromatography and silver-ion high performance liquid chromatography.
[e]Numbers refer to citations listed in the reference section.
[f]Comprised of a mixture (3:1 by weight) of NaHCO3 and MgO.

on *cis*-9, *trans*-11 18:2 and resulted in limited increases in *trans*-10, *cis*-12 18:2 from 0 to 0.22 g per day at the duodenum.[219]

Several investigations have examined the effects of dietary linseed oil supplements as a means to increase 18:2 n-6 and 18:3 n-3 intake on ruminal CLA synthesis. In non-lactating cows fed high forage diets (F:C ratio 70:30) supplements of linseed oil (49 g per kg DM) increased duodenal flows of unresolved *trans*-8, *cis*-10 18:2 and *cis*-9, *trans*-11 18:2, *trans*-11, *cis*-13 18:2 and *trans*-11, *trans*-13 18:2 from 0.22, 0.12 and 0.13 to 0.64, 0.46 and 0.64 g per day, respectively.[228] In growing cattle fed maize silage based diets, linseed oil (30 g per kg DM) stimulated ruminal *cis*-9, *trans*-11 18:2, *trans*-8, *cis*-10 18:2, *trans*-11, *cis*-13 18:2, *trans*-13, *cis*-15 18:2 and *trans*, *trans* (9,11 to 13,15) 18:2 synthesis[68] (Table 1.4). Consistent with reports on the influence of 18:2 n-6 supply, the composition of the basal diet is also a major determinant of the changes in ruminal CLA synthesis to linseed oil. In lactating cows fed high or low forage diets (F:C 65:35 vs. 35:65), linseed oil resulted in similar increases in ruminal *cis*-9, *trans*-11 18:2, *trans*-11, *cis*-13 18:2 and *trans*-11, *trans*-13 18:2 synthesis, irrespective of diet composition.[176] However, *trans*-8, *cis*-10 18:2, *trans*-9, *cis*-11 18:2 and *cis*-11, *trans*-13 18:2 formation was increased on the low forage diet, but remained unchanged on the high forage diet.

It is well established that the concentrations of *cis*-9, *trans*-11 18:2 and total CLA are higher in milk in grazing ruminants compared with that on diets based on conserved forages.[45,47,49] Relatively few studies have compared the synthesis of CLA isomers in the rumen of grazing ruminants compared with diets containing conserved forages.[61,237] Even though the number of observations that have been made is limited, there is little evidence that the effects on milk fat CLA concentrations are related to differences in the synthesis of CLA isomers in the rumen. Similarly, replacing grass silage with grass hay[61] or red clover silage[54,227] has limited influence on total CLA synthesis, but may result in differences in the amounts of minor isomers formed in the rumen (Table 1.4).

The effect of dietary marine lipid supplements on ruminal CLA synthesis has also been investigated. Even though the highly unsaturated fatty acids in fish oil[119,122,220] or marine algae[120,184,238] inhibit the complete biohydrogenation of unsaturated 18-, 20-, 21- and 22-carbon unsaturated fatty acids in the rumen, the influence on the accumulation of total CLA is rather small. In some experiments, supplementing grass silage, red clover silage or grass silage based diets with fish oil has been demonstrated to increase *trans*-7, *trans*-9 18:2, *trans*-8, *trans*-10 18:2 and *trans*-9, *trans*-11 18:2 formation and decrease *cis*-12, *trans*-14 18:2 synthesis,[220,227] but not in all cases[122] (Table 1.4). Studies in growing cattle fed maize silage-based diets indicated that fish oil supplementation up to 16 g per kg diet DM resulted in marginal increases in *cis*-9, *trans*-11 18:2 and *trans*-9, *trans*-11 18:2 at the duodenum, changes that were not evident when the amount of fish oil in the diet was increased above 24 g per kg diet DM.[68,229]

1.3.5   Formation of CLA Precursors

Secretion of *trans*-7, *cis* 11 18:2 and *cis*-9, *trans*-11 18:2 in milk is known to be several-fold higher compared with the amounts of these isomers at the duodenum in lactating cows.[45,239,240] These differences are explained by endogenous synthesis in ruminant tissues. It now known that the majority of *cis*-9, *trans*-11 18:2 in milk is formed from the desaturation of *trans*-11 18:1 in the mammary glands. Most, if not all, of the *trans*-7, *cis*-11 18:2 found in ruminant milk also originates from the desaturation using *trans*-7 18:1 as a substrate. The potential to increase the synthesis of CLA in the rumen is relatively limited compared with projections of the enrichment in ruminant foods required to confer benefits to human health. For this reason, considerable emphasis has been placed on understanding the role of diet on the synthesis of CLA precursors, *trans*-11 18:1 in particular, in the rumen. The reduction of *trans* 18:1 intermediates to 18:0 is considered rate limiting for the complete biohydrogenation of 18-carbon unsaturated fatty acids in the rumen.[125] Studies *in vitro* have demonstrated that increases in the amount of 18:2 n-6 or 18:3 n-3 incubated with mixed ruminal bacteria results in the accumulation of *trans* 18:1 (Table 1.2). *In vivo* changes in diet composition tend to have a much greater influence on the amount and relative proportions of *trans* 18:1 leaving the rumen (Table 1.5) compared with ruminal CLA synthesis (Table 1.3). Therefore, formulating diets that increase ruminal formation of *trans*-11 18:1 rather than specifically targeting *cis*-9, *trans*-11 18:2 synthesis is more effective for increasing the CLA content of ruminant foods.

In ruminants fed diets containing forages and concentrate ingredients, *trans*-11 18:1 is typically the major biohydrogenation intermediate leaving the rumen. Increases in concentrate supplementation have been shown to cause the accumulation of *trans* 18:1 intermediates in the rumen.[176,218,241] Switching from grass hay based diets containing 35% to 65% of concentrates increased the flow of *trans* 18:1 (Δ4 to 16) at the duodenum in lactating cows.[176] Changes in *trans*-11 18:1 in response to concentrate supplementation were marginal, with most of the increase being associated with the *trans*-10 isomer (Table 1.5). More extreme increases in the amount of concentrates offered to lactating cows, from 60% to 75%, caused all *trans* 18:1 isomers to accumulate, other than *trans*-11 and *trans*-16.[176] Under these conditions, the most marked influence was on *trans*-10 18:1 at the duodenum, being increased four-fold compared with the high forage diets (Table 1.5). Switching from a diet based on dehydrated grass to concentrate ingredients was found to result in the almost complete disappearance of *trans*-11 18:1 and a substantial increase in *trans*-10 18:1 concentration of abomasal digesta in growing lambs.[242]

Examination of the effects of pH (6.4 vs. 5.6) or dietary F:C ratio (70:30 *vs.* 30:70) on the production of biohydrogenation intermediates in dual-flow continuous culture suggested that decreases in pH rather that increases in concentrate *per se* was the principal factor responsible for the shift from *trans*-11 18 : 1 to *trans*-10 18 : 1 accumulation.[127] However, this interpretation has been challenged based on recent reports of 18 : 2 n-6 metabolism by ruminal bacteria[243] and measurements of ruminal fatty acid composition in cows fed diets containing low or high amounts of starch.[173] The findings from both investigations led to the conclusion that the influence of starch on *trans*-10 18 : 1 formation was independent of the decreases in pH that occur during starch fermentation.

Under most situations, metabolism of 18 : 2 n-6 and 18 : 3 n-3 in the rumen leads to the formation of *trans*-11 18 : 1 during the complete hydrogenation to 18 : 0 (Figures 1.3 and 1.4). It is perhaps unsurprising that dietary supplements of oils enriched in 18 : 2 n-6 or 18 : 3 n-3 have used to increase the amounts of *trans*-11 18 : 1 formed in the rumen. Including plant oils in the diet have been shown to result in dose dependent increases in the amount of *trans* 18 : 1 intermediates leaving the rumen in sheep[104,218,219], growing cattle,[68,214–217] lactating[67,176] and non-lactating cows.[228] However, the profile of *trans* 18 : 1 intermediates that accumulate is dependent on a complex interaction between the relative proportions of forages and concentrates in the diet and the amount and source of oil supplement (Table 1.5). On high forage diets, supplements of sunflower oil[67] or linseed oil[176] increase the amount of *trans*-11 18 : 1 escaping the rumen. In marked contrast, inclusion of the same oils in high concentrate diets has minimal influence on *trans*-11 18 : 1, but results in *trans*-10 18 : 1 accumulating several-fold.[176,215] Increases in the concentration of *trans*-10 18 : 1 are also accompanied by decreases in the synthesis of fatty acids in the mammary glands of cows and sheep, but not in goats.[107] Changes in milk fatty acid composition also serve as a proxy of the relative amounts of *trans* 18 : 1 escaping the rumen, indicating that the major pathways of biohydrogenation are altered under these conditions causing *trans*-10 18 : 1 to displace *trans*-11 18 : 1 as the principal intermediate formed in the rumen. The most recent evidence has indicated that a low rumen pH and high dietary concentrations of starch and oil are prerequisite in promoting the '*trans*-10 shift'.[172,173]

Palmquist *et al.*[45] proposed that ruminal *trans*-11 18 : 1 formation is dependent on three inter-dependent processes; 1) substrate supply, 2) inhibition of the reduction of *trans* 18 : 1 to 18 : 0 and 3) prevention of the *trans*-10 shift. It was argued that substrate supply has a typically permissive role in determining the extent of *trans* 18 : 1 fatty acid accumulation in the rumen in response to the other two processes, such that the balance between the inhibition of *trans* 18 : 1 reduction and induction of the shift towards *trans*-10 18 : 1 at the expense of *trans*-11 18 : 1 regulates the magnitude of the overall response to changes in diet composition.

Fish oil and marine algae contain relatively high amounts of long chain (≥ 20-carbon) polyunsaturated fatty acids. When added to the diet of ruminants, both lipid sources lead to an inhibition of the complete biohydrogenation of 18-carbon unsaturated fatty acids causing *trans* 18 : 1 and *trans* 18 : 2 intermediates to accumulate.[68,120,122] Fish oil is an effective means to increase the amount of *trans*-11 18 : 1 leaving the rumen

Table 1.5 Synthesis of *trans* 18:1 isomers in the rumen of growing cattle, lactating cows and sheep.

Forage[a]	F:C[b]	Supplement[c] (g/kg DM)	Δ4	Δ5	Δ6–8[e]	Δ9	Δ10	Δ11	Δ12	Δ13/14[e]	Δ15	Δ16	Reference[f]
						Non-lactating sheep							
GH	18:82	Control (0)				0.23		2.62	0.22				Atkinson et al.[104]
		SAF (90)				3.07		40.8	1.92				
						Growing cattle							
HSG	100:0	–			0.45	0.23	0.33	7.85	0.48	0.69	0.88		Lee et al.[224]
GS	100:0	–			0.26	0.14	0.25	4.60	0.29	0.51	0.55		
RCS	100:0	–			0.26	0.15	0.26	3.83	0.49	0.74	0.92		
GS	80:20	–	0.08	0.05	2.24	1.18	1.57	24.3	2.66	1.84	4.14	0.23	Lee et al.[225]
GS	20:80	–	0.15	0.10	2.92	1.28	4.68	29.2	3.57	2.98	5.56	1.22	
GH	12:88	30				0.70	41.4	3.48	1.29				Sackmann et al.[2–5]
GH	36:64	30				2.14	15.5	11.8	2.53				
GS	100:0	Control	<0.01	0.15	0.1	0.17	2.28	0.29	0.26	0.5	0.92	<0.01	Lee et al.[227]
GS	100:0	FO (30)	0.06	0.03	0.77	0.67	1.08	7.87	1.49	0.69	1.28	1.22	
RCS	100:0	Control	0.02	0.01	0.34	0.23	0.47	3.02	0.94	0.62	1.7	2.29	
RCS	100:0	FO (30)	0.09	0.04	1.26	1.11	1.8	11.2	2.47	0.7	1.94	1.75	
MS	60:40	Control	0.23	0.16	1.58	1.09	2.05	13.8	1.47	1.01	1.4	1.63	Shingfield et al.[68]
		LO (30)	0.69	0.42	4.67	2.84	3.91	51.2	4.79	4.47	5.76	5.63	
		FO (30)	0.15	0.11	5.08	4.45	19.2	66.7	5.85	2.68	3.25	0.98	
						Non-lactating cows							
HNG	100:0	–	0.36	1.63	1.97	1.03	2.85	62.0	2.75	13.9	0.30	5.65	Doreau et al.[226]
HSG	100:0	–	0.27	1.26	1.51	0.78	2.05	41.6	2.25	9.89	0.20	4.07	

Forage[a]	Forage:concentrate[b]	Treatment[c]	Lactating cows										Reference[f]
FG	60:40		0.24	0.19	1.37	0.96	2.26	20.1	1.95	4.47	2.02	2.59	Halmemies-Beauchet-Filleau et al.[61]
GH	60:40		0.12	0.15	1.14	0.75	1.89	12.5	1.83	2.93	1.66	2.06	
GH	60:40		0.20	0.19	1.44	0.85	2.72	15.0	2.15	4.16	1.84	2.27	
GS	60:40		0.24	0.25	1.97	1.3	2.84	18.9	3.53	8.74	4.54	5.45	
GS	60:40		0.29	0.29	2.16	1.43	3.53	22.9	4.12	10.3	4.81	5.81	
GS	60:40		0.27	0.20	2.08	1.57	2.93	25.8	3.56	9.04	4.95	5.09	Halmemies-Beauchet-Filleau et al.[54]
RCS	60:40		0.32	0.23	2.36	1.71	2.92	25.8	4.10	10.8	5.80	5.69	
MS+LH	60:40	-			1.02	2.41	5.73	20.76	5.68	14.11	5.74	5.45	Piperova et al.[175]
	25:75	-			3.72	4.55	29.13	33.61	9.52	22.86	8.53	7.98	
GS	60:40	Control	0.3	0.3	1.3	0.8	1.3	14.9	1.7	4.5	2.3	2.9	Shingfield et al.[67]
		SFO (50)	1.9	1.3	10.5	7.0	20.6	126.2	12.9	22.6	10.7	12.3	
GH	65:35	Control	0.37	1.29	1.83	1.38	1.46	21.4	1.93	4.17	1.95	2.34	Loor et al.[176]
		LO (30)	1.06	3.12	6.75	3.89	6.61	61.7	8.32	29.6	12.1	11.5	
GH	35:65	Control	0.88	1.81	5.98	2.96	20.2	26.0	3.78	10.3	4.84	3.98	
		LO (30)	1.87	3.36	16.2	13.1	50.6	139	9.57	42.9	16.8	10.7	
GS	60:40	Control	0.62	0.44	3.46	2.02	4.25	22	4.31	11.3	5.24	5.99	Shingfield et al.[122]
		FO (8.4)	0.74	0.62	7.94	6.37	9.95	80.8	11.8	21	9.2	7.59	
		FO (19)	0.4	0.4	6.83	5.72	56.4	66.5	9.41	15.9	6.68	2.84	

[a]Forage in the diet: FG, fresh grass; GH, grass hay; GS, grass silage; HNG, high nitrogen grass; HSG, high sugar grass; LH, lucerne haylage; MS, maize silage; RCS, red clover silage.

[b]Forage:concentrate ratio of the diet (on a dry matter basis).

[c]Lipid supplements; FO, Fish oil; LO, linseed oil; SAF, safflower oil; SFO, sunflower oil.

[d]Determined based on sampling at the duodenum or omasum and measurements of dry matter flow at the sampling site using indigestible markers.

[e]Isomers not resolved during analysis.

[f]Numbers refer to citations listed in the reference section.

(Table 1.5), but in high amounts can induce a shift towards *trans*-10 18:1 formation.[122] Incubations with mixed ruminal bacteria have shown that both 20:5 n-3 and 22:6 n-3 cause *trans* 18:1 to accumulate.[244-246] It is possible that other 20-, 22- or 24-carbon unsaturated fatty acids contained in fish oil and marine algae may also inhibit the reduction of *trans* 18:1 to 18:0 in the rumen.

Relatively few studies have examined the effects of forage conservation or forage species on ruminal biohydrogenation. The most recent investigations indicate that the relative abundance of *trans*-11 18:1 in the rumen of lactating cows is higher on pasture compared with zero-grazing or grass silage.[237] Flows of *trans*-11 18:1 at the omasum have also been shown to be higher in lactating cows during zero-grazing compared with diets based on hay or silage prepared at the same time from the same grass swards.[61] Conservation of grass by drying or ensiling had no effect on the amount of *trans*-11 18:1 leaving the rumen.[61] Replacing grass silage with red clover silage[54,227] also has little influence on *trans*-11 18:1 or total *trans*-18:1 formation in the rumen (Table 1.5).

Studies involving the incubation of fatty acids with mixed ruminal bacteria have demonstrated that the biohydrogenation of *cis*-9 18:1, 18:2 n-6 and 18:3 n-3 results in the formation of *trans*-7 18:1 as an intermediate (Table 1.2). Because *trans*-6, -7 and -8 elute as a single peak during GC analysis there are no definitive reports on the effect of diet on *trans*-7 18:1 formation in the rumen. Direct comparisons of milk fat *trans*-7, *cis*-9 18:2 concentrations in ruminants offered plant oil or oilseed supplements have shown a consistently higher enrichment when rapeseed lipids are fed.[46,47] By implication, biohydrogenation of *cis*-9 18:1 appears to be the main source of *trans*-7 18:1 in the rumen.

Recent studies have also provided evidence that *trans*-9 16:1 (palmitelaidic acid) may also serve as a substrate for endogenous *cis*-9, *trans*-11 18:2 synthesis in ruminant tissues.[247] The metabolic origins of *trans* 16:1 formed in the rumen have not been extensively investigated. A range of *trans* 16:1 with ethylenic bonds between positions Δ3 to 14 are known to be formed in the rumen and incorporated into milk fat.[54,248] Fish oil is known to cause *trans*-9 16:1 to accumulate in the rumen.[68,122,227,229] Under these situations, it seems plausible that *trans*-9 16:1 originates from the incomplete biohydrogenation of *cis*-9, *cis*-12 16:2 or *cis*-6, *cis*-9, *cis*-12, *cis*-15 16:4. On conventional diets, *trans* 16:1 isomers are thought to originate from the incomplete biohydrogenation of *cis*-6, *cis*-9, *cis*-12 16:3 present in trace amounts in the ruminant diet.[248]

1.4   Endogenous CLA Synthesis in Ruminants

Early reports indicated that the concentrations of *cis*-9, *trans*-11 18:2 in milk were increased on pasture[249,250] or on diets supplemented with linseed oil[251] or fish oil.[252] Such findings were perplexing. The convention at the time was that ruminal biohydrogenation of 18:2 n-6 was responsible for *cis*-9, *trans*-11

18 : 2 formation in ruminants, and yet pasture and linseed oil contain high proportions of 18 : 3 n-3 and fish oil contains few 18-carbon unsaturated fatty acids. Much earlier studies had demonstrated that several *trans* 18 : 1 isomers could be converted to *cis*-9 containing 18 : 2 products during incubations with rat liver microsomal preparations.[253,254]

These seminal investigations confirmed that *trans* 18 : 1 fatty acids with double bonds at Δ4–7 and Δ11–13 were substrates for the enzyme stearoyl-CoA desaturase (SCD; E.C. 1.14.99.5). Based on the observed changes in milk fat composition to diet and the substrate specificity of the SCD enzyme, Griinari and Bauman[255] proposed that the rumen was not the sole source of *cis*-9, *trans*-11 18 : 2 and that endogenous synthesis due to the desaturation of *trans*-11 18 : 1 also occurred in ruminant tissues. An overview of the metabolic pathways of endogenous CLA synthesis in ruminants is described in Figure 1.7.

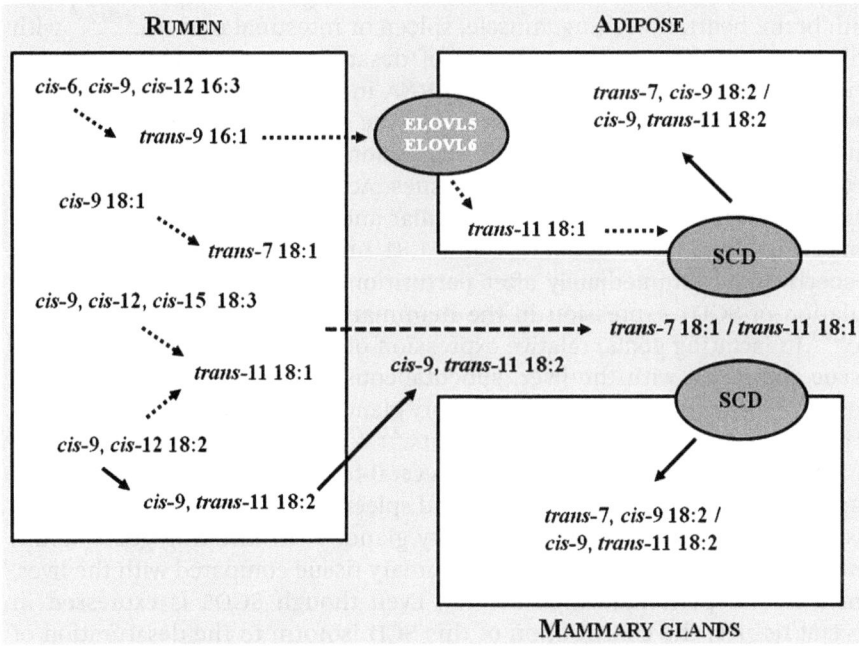

Figure 1.7 Metabolic pathways of endogenous conjugated linoleic acid synthesis in ruminants. Most of the *cis*-9, *trans*-11 18 : 2 incorporated into tissue lipids and milk fat is synthesized by the action of stearoyl-CoA desaturase (SCD) on *trans*-11 18 : 1 in adipose and the mammary glands. *Trans*-11 18 : 1 is formed as the penultimate intermediate of 18 : 2 n-6 and 18 : 3 n-3 metabolism in the rumen. *Cis*-9, *trans*-11 18 : 2 in adipose may also be synthesized from the elongation and desaturation of *trans*-9 16 : 1 formed in the rumen during the biohydrogenation of 16 : 3 n-4. Ruminal synthesis of *trans*-7, *cis*-9 18 : 2 is negligible. The activity of SCD on *trans*-7 18 : 1 formed during the biohydrogenation of *cis*-9 18 : 1 in the rumen is the main source of *trans*-7, *cis*-9 18 : 2 in ruminant tissues and milk fat.

1.4.1   Stearoyl CoA Desaturase

1.4.1.1   *Tissue Specific Expression*

The SCD gene encodes a protein of 359 amino acid residues located in the endoplasmic reticulum (ER) that catalyzes the desaturation of 10- to 19-carbon fatty acyl-CoA substrates, resulting in the formation of a *cis* double bond located between carbons 9 and 10 of the fatty acid moiety.[256,257] Homologues of the SCD gene have been characterized in several mammalian species, including the mouse, rat, hamster, pig and human. Some mammalian genomes have been shown to contain multiple SCD isoforms. In mice, four isoforms (SCD1, SCD2, SCD3 and SCD4) have been documented.[258–261] Until recently only one isoform of the SCD gene, now referred to as SCD1, had been characterized in sheep,[262] cows[263] and goats.[264] A second SCD isoform (SCD5) was recently identified in cattle[265] and subsequently found to be expressed in the goat.[265]

Expression of the SCD1 gene is several-fold higher in adipose compared with brain, heart, liver, lung, muscle, spleen or intestinal mucosa,[265,267] with adipose tissue being the major site of desaturase activity in growing ruminants.[268,269] Abundance of SCD1 mRNA in subcutaneous adipose tissue increases after weaning[270] with activity of the desaturase enzyme increasing during growth even after SCD gene expression starts to decline.[271] Catalytic activity of SCD also varies between tissues. Activity of SCD was found to be higher in intestinal mucosa, interfasicular and subcutaneous adipose compared with liver (18.1, 10.6, 5.24 and 1.81 nmol per mg protein per min, respectively).[272] Immediately after parturition, there is a substantial upregulation of SCD1 expression in the mammary glands of sheep[262] and cattle.[273] In lactating goats, relative expression of SCD1 is higher in mammary tissue compared with the liver, subcutaneous, omental and perirenal adipose[274–277] consistent with the mammary glands being the major site of fatty acid desaturation in lactating ruminants.[278,279]

Transcript abundance of SCD5 is several-fold higher in the brain compared with heart, liver, lung, muscle and spleen in growing cattle.[265] SCD5 is also expressed in the bovine mammary gland.[280] In lactating goats, abundance of SCD5 mRNA is higher in mammary tissue compared with the liver, omental and perirenal adipose.[277,281] Even though SCD5 is expressed in several tissues, the contribution of this SCD isoform to the desaturation of fatty acids in ruminants remains unclear.

1.4.1.2   *Catalytic Activity*

The SCD enzyme is the rate-limiting step in the synthesis of monoenoic fatty acids that are utilized as substrates for TAG, PL and cholesterol ester (CE) formation.[282] Stearoyl and palmitoyl- CoA are the preferred substrates but a range of 10- to 19-carbon fatty acyl-CoA are desaturated in various ruminant tissues.[46,256] *Cis*-9 18 : 1 is the major product of the SCD enzyme in mammals, the incorporation of which into PL contributes to the maintenance of membrane fluidity.[282] Activity of SCD in the ruminant mammary gland is thought to occur as a mechanism to maintain and regulate the fluidity of milk.[283]

Fatty acid desaturases are nonheme iron-containing enzymes that introduce a double bond between defined carbons of fatty acyl chains. In common with other desaturases, SCD catalyses a highly regio- and stereoselective reaction on long-chain fatty acids that consist of essentially equivalent methylene chains in the absence of a distinguishing feature close to the site of desaturation. The Δ9 desaturase homologous to SCD isolated from *Ricinus communis* (castor) is a homodimeric protein with each monomer folded into a compact single domain comprised of nine helices.[284] A diiron active site is located within a core four-helix bundle that is positioned alongside a deep, bent, narrow hydrophobic channel in which the substrate is bound during catalysis.[285] It is thought that SCD has 4 transmembrane domains in the ER membrane and that both the -NH_2 and -COOH terminal groups are oriented toward the cytosol.[286]

Oxidation of fatty acyl-CoA substrates catalysed by SCD in mammals is an aerobic process with an absolute requirement for molecular oxygen, cytochrome b5 reductase and the electron acceptor cytochrome b5.[287] Desaturation is initiated by the energy demanding abstraction of a hydrogen atom from the methylene group at the Δ9 position followed by the abstraction of a second hydrogen atom from the Δ10 position.[288] This is made possible by the recruitment and activation of molecular oxygen with the use of an active-site diiron cluster.[289] Electrons are transferred sequentially from NAD(P)H, via NADH-cytochrome b_5 reductase and cytochrome b_5, to the terminal desaturase SCD, and finally to active oxygen, which is reduced to H_2O[287] (Figure 1.8).

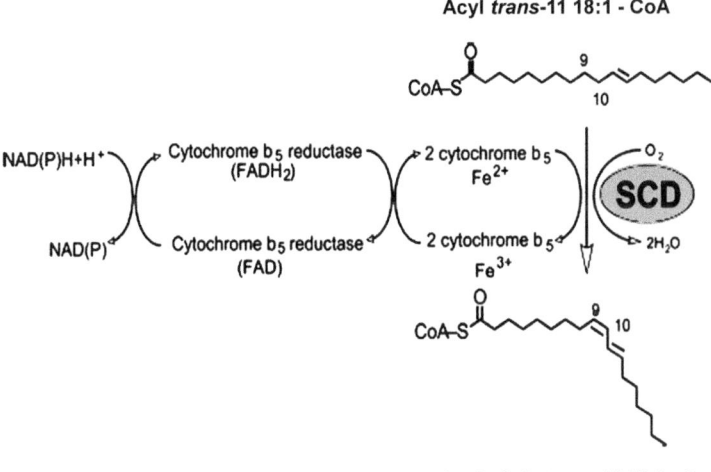

Figure 1.8 Scheme for the transfer of electrons during the oxidation of acyl *trans*-11 18:1-CoA yielding acyl *cis*-9, *trans*-11 18:2 CoA by stearoyl-CoA desaturase (SCD) adapted from Paton and Ntambi.[288] Adipose is the major site of desaturation of fatty acids in growing ruminants, whereas the mammary glands are the main site in lactating ruminants.

For membrane bound desaturases the initial breaking of the C–H bond is rate-limiting.[290] Electron transfer is facilitated by a catalytic di iron protein complex of the terminal desaturase.[291] Insertion of Fe to the di-iron centre is thought to occur spontaneously and anchored to the active site by binding with carboxyl and nitrogen groups of glutamine and histidine residues, with evidence that the affinity of the two iron atoms comprising the di-iron center differ.[289]

The promoter region of the bovine SCD1 gene was shown to contain a 36-bp sequence of critical importance, designated the stearoyl-CoA desaturase transcriptional enhancer element (STE) that contains three putative transcription factor binding complexes.[292] In the bovine MAC-T cell, the SCD gene promoter was shown to be up-regulated by insulin and down-regulated by *cis*-9 18 : 1, with the STE region shown to play a key role in the inhibitory effect of *trans*-10, *cis*-12 18 : 2 on SCD gene transcription. There is no known allosteric or feedback inhibition by substrates or products but the SCD enzyme is regulated by several dietary factors including glucose and PUFA, hormones such as insulin, glucagon and thyroid hormone and the sterol responsive element binding protein-1c transcription factor.[257,282,288] Transcriptional regulation is made possible due to the relatively short half-life (3–4 h) of the mammalian SCD protein.[293] In ruminants, SCD mRNA abundance and activity in the mammary glands has in some, but not all studies, been reported to be decreased on diets containing plant oils, oilseeds or marine lipid supplements, whereas corresponding effects on the transcription and activity of the SCD enzyme in adipose are marginal.[274–277] In growing cattle, plant oils and fish oil have been shown to lower SCD gene and protein expression in adipose but not the amount of substrate desaturated per unit SCD protein.[47] Abomasal infusion of *trans*-10, *cis*-12 18 : 2[280,294] or oral dosing or ruminal administration of cobalt EDTA or cobalt acetate[295,296] have been shown to decrease fatty acid desaturation in the bovine mammary glands.

1.4.2   Endogenous Synthesis in the Mammary Glands

It is well established that fatty acids are extensively desaturated in the mammary glands of ruminants. The activity of the SCD enzyme contributes to *ca.* 56%, 60% and 90% of *cis*-9 18 : 1, *cis*-9 16 : 1 and *cis*-9 14 : 1, respectively, in bovine milk.[239,297–299] Several approaches have been used to estimate the contribution of endogenous *cis*-9, *trans*-11 18 : 2 synthesis in lactating ruminants. These have involved: 1) postruminal infusion of sterculic oil to inhibit SCD activity in the mammary glands and measuring subsequent changes in milk fatty acid composition, 2) postruminal infusions of *trans*-11 18 : 1 and measuring the output of *cis*-9, *trans*-11 18 : 2 in milk, 3) comparison of ruminal synthesis and output of secretion of CLA isomers in milk, 4) measurements of enrichment of *cis*-9, *trans*-11 18 : 2 in milk following the administration of ^{13}C-labelled *trans*-11 18 : 1 and 5) measuring arterio-venous differences of circulating NEFA and TAG across the mammary glands and the output of fatty acids in milk fat (Table 1.6).

Table 1.6 Endogenous *cis*-9, *trans*-11 18 : 2 synthesis in the mammary glands of lactating cows, goats and sheep. Values determined based on 1) post-ruminal infusion of sterculic oil to inhibit stearoyl-CoA desaturase activity and measuring the subsequent changes in milk fatty acid composition, 2) post-ruminal infusion of *trans*-11 18 : 1 and measuring changes in *cis*-9, *trans*-11 18 : 2 in milk, 3) administration of ^{13}C-labelled *trans*-11 18 : 1 and determination of ^{13}C enrichment in *cis*-9, *trans*-11 18 : 2, 4) measuring arterio-venous differences across the mammary glands and secretion of fatty acids in milk and 5) comparison of ruminal synthesis and output of *trans*-11 18 : 1 and *cis*-9, *trans*-11 18 : 2 in milk.

Species	Forage[a]	F:C[b]	Supplement[c] (g/kg DM)	Methodology	Desaturation of *trans*-11 18 : 1, %	Endogenous *cis*-9, *trans*-11 18 : 2 synthesis, %	Reference[d]
Bovine	LH	45 : 55	–	Abomasal infusion of sterculic oil		≥64	Grünari *et al.*[300]
	LH	45 : 55	–	Abomasal infusion of sterculic oil		78	Corl *et al.*[301]
	Pasture	100 : 0	–	Abomasal infusion of sterculic oil	24	≥91	Kay *et al.*[306]
	Pasture	100 : 0	SFO (28)	Abomasal infusion of sterculic oil	20	≥91	Kay *et al.*[306]
	LH	47 : 53	–	Abomasal infusion of 12.5 g *trans*-11 18:1/d	–	–	Grünari *et al.*[300]
	LH	45 : 55	–	Abomasal infusion of 23.5 g *trans*-11 18:1/d	23.4		Corl *et al.*[301]
	GS	75 : 25	–	Abomasal infusion of 7.5–30 g *trans*-11 18:1/d	28.9	84.4	Shingfield *et al.*[302]
	LS/BS/LH	53 : 47	–	Abomasal administration of 1.5 g 1-[13]C *trans*-11 18:1	25.7	83.1	Mosley *et al.*[279]
	FG/GH/GS	60 : 40	–	Arterio-venous differences across the mammary glands	18.6–34.1	78.0	Halmemies-Beauchet-Filleau *et al.*[299]
	Multiple	35 : 65 100 : 0	–	Meta-analysis of fatty acid flow at the duodenum and secretion in milk	21.0	94.7	Glasser *et al.*[239]
Caprine	GH	66 : 34	SFO (46)	i.v. administration of 1.5 g of 1-[13]C *trans*-11 18:1	31.7	73.1	Bernard *et al.*[266]
	GH	68 : 3 2	SFO (30) + FO (15)	i.v. administration of 1.5 g of 1-[13]C *trans*-11 18:1	31.6	62.9	Bernard *et al.*[266]
Ovine	Pasture	100 : 0	–	i.v. administration of sterculic oil		74	Bichi *et al.*[307]

[a]Forage in the diet: BS, barley silage; FG, fresh grass; GH, grass hay; GS, grass silage; LH, lucerne haylage; LS, lucerne silage.
[b]Forage:concentrate ratio of the diet (on a dry matter basis).
[c]Lipid supplements: FO, Fish oil; SFO, sunflower oil.
[d]Numbers refer to citations listed in the reference section.

The first demonstration that endogenous synthesis of CLA occurred in ruminants was reported in lactating cows. Abomasal infusion of a mixture of fatty acids in lactating cows providing 12.5 g per day of *trans*-11 18:1 were shown to increase milk *cis*-9, *trans*-11 18:2 concentrations by 31%.[300] Subsequent studies demonstrated that infusions of higher amounts of *trans*-11 18:1 of up to 30 g per day enriched *cis*-9, *trans*-11 18:2 in milk with no evidence that the extent of desaturation varied according to substrate supply.[301,302] In goats, intravenous injections of *trans*-11 18:1 were found to elevate *cis*-9, *trans*-11 18:2 concentrations in milk and upregulate SCD gene expression in mammary tissue.[303] Administration of *trans*-11 18:1 was also associated with an upregulation of proteasome (prosome, macropain) subunit α type 5 (PSMA5) and downregulation of peroxiredoxin-1 and translationally controlled tumor protein 1 in mammary tissue. The PSMA5 complex is involved in the proteolytic degradation via the ubiquitin-proteasome pathway. Given that SCD is a short-lived protein and constitutively degraded within the ER by the ubiquitin-proteasome system,[304] it has been suggested that PSMA5 is involved in SCD degradation.[303]

As an alternative to increasing substrate supply, several experiments have used sterculic oil containing 7-2-octyl-1-cyclopropenyl heptanoic acid and 8-2-octyl-1-cyclopropenyl octanoic acid to inhibit SCD activity and measured the changes in *trans*-11 18:1 and *cis*-9, *trans*-11 18:2 concentrations. Using *cis*-9 14:1/14:0 concentration ratios as a proxy of the extent of SCD inhibition, the contribution of endogenous synthesis to *cis*-9, *trans*-11 18:2 in milk was estimated to vary between 64% and 91% in lactating cows.[300,301,305,306] Using the same methodology, between 85% and 100% of *trans*-7, *cis*-9 18:2 in bovine milk was estimated to originate from the desaturation of *trans*-7 18:1.[305] Similarly intravenous infusion of sterculic oil and measurements of milk fat composition in sheep estimated that endogenous synthesis contributed to 74% of *cis*-9, *trans*-11 18:2 milk.[307] Quantitative estimates of endogenous CLA synthesis in lactating ruminants have been made using 1-^{13}C labelled *trans*-11 18:1. Modelling of the exponential decay of ^{13}C enrichment of fatty acids in milk indicated that 26% of *trans*-11 18:1 taken up by the mammary glands was converted to *cis*-9, *trans*-11 18:2.[279] Desaturation of *trans*-11 18:1 was determined as 32% in the caprine mammary glands.[266] Endogenous *cis*-9, *trans*-11 18:2 synthesis in lactating cows has also been determined from measurements of arteriovenous differences across the mammary glands and the output of fatty acids in milk.[299] Such experiments indicated that between 18.6% and 34.1% of *trans*-11 18:1 extracted from circulating NEFA and TAG and taken up by the mammary glands was converted to *cis*-9, *trans*-11 18:2. Overall, these experiments have confirmed that endogenous synthesis is the major source of *cis*-9, *trans*-11 18:2 in ruminant milk. Numerous investigations have demonstrated that the concentrations of *trans*-11 18:1 and *cis*-9, *trans*-11 18:2 in bovine,[255,308] caprine[309,310] and ovine[311,312] milk fat are highly correlated over a wide range of diets. Regression analysis indicate that the relationship between *cis*-9, *trans*-11 18:2 and *trans*-11 18:1 is linear, and that the slopes are similar (*ca.* 0.40), irrespective of the source of milk analysed. Such evidence indicate that increases in *trans*-11 18:1 supply does not inhibit desaturation in the mammary glands and that the conversion to *cis*-9, *trans*-11 18:2 is very similar in the cow, goat and sheep.

1.4.3 Endogenous Synthesis in Other Tissues

Estimating the extent of endogenous CLA synthesis in growing ruminants is particularly challenging given that CLA isomers accumulate in adipose over the life time of the animal. Several studies have attempted to estimate endogenous synthesis of the major CLA isomer in growing cattle based on comparisons of *trans*-11 18:1 and *cis*-9, *trans*-11 18:2 abundance in abomasal digesta and muscle. Such comparisons suggest that *ca.* 86% of *cis*-9, *trans*-11 18:2 in *Longissimus dorsi* originates from *trans*-11 18:1 and more than 80% in adipose tissue.[313,314] Modelling of the relative proportions of *trans*-11 18:1 and *cis*-9, *trans*-11 18:2 in mesenteric adipose, subcutaneous adipose and longissimus muscle in lambs have also been used to predict endogenous synthesis. Between 45% and 95% of *cis*-9, *trans*-11 18:2 deposited in muscle and adipose was estimated to originate from *trans*-11 18:1 with the extent of conversion ranging between 11% and 22%.[315]

Recent studies in lactating goats have attempted to quantify endogenous CLA synthesis in several tissues of lactating goats using 1-^{13}C *trans*-11 18:1 as a chemical tracer.[277] Goats were slaughtered 4 d after jugular injection and ^{13}C enrichment of *trans*-11 18:1 and *cis*-9, *trans*-11 18:2 was determined in mammary secretory tissue, liver, omental and perirenal adipose. From these measurements, it was calculated that 27% of *trans*-11 18:1 was desaturated to *cis*-9, *trans*-11 18:2 in perirenal adipose but conversion was only 2% in omental fat. The same experiment also demonstrated a higher enrichment of ^{13}C *trans*-11 18:1 in the liver compared with mammary or adipose tissues. Due to hepatic uptake and incorporation of ^{13}C *cis*-9, *trans*-11 18:2 synthesized in other tissues, no conclusions can be drawn on the role of the liver in endogenous CLA synthesis in ruminants. Even though mRNA encoding for SCD has been identified in the liver of ruminants, reports on the relative importance of hepatic fatty acid desaturation are inconsistent.[45] Studies *in vitro* indicate that *trans*-11 18:1 is not converted to *cis*-9, *trans*-11 18:2 in the bovine liver.[316]

A recent experiment also provided the first evidence that *trans*-9 16:1 may also serve as a substrate for endogenous *cis*-9, *trans*-11 18:2 synthesis in ruminant tissues. Incubations of bovine primary stromal vascular cells with incremental amounts of *trans*-9 16:1 resulted in a progressive increase in adipocyte *trans*-11 18:1 and *cis*-9, *trans*-11 18:2 concentrations.[247] Measurements indicated that *ca.* 50% of *trans*-9 16:1 incorporated into adipocytes was elongated to *trans*-11 18:1, with about 8% being desaturated to *cis*-9, *trans*-11 18:2. It was proposed that *trans*-9 16:1 served as a substrate for the fatty acid elongases 5 (*ELOVL5*) and 6 (*ELOVL6*), which catalysed the addition of two carbon atoms to yield *trans*-11 18:1 that could subsequently be converted to *cis*-9, *trans*-11 18:2.

1.5   Synthesis of CLA in Humans

1.5.1   Endogenous Synthesis

Synthesis of *cis*-9, *trans*-11 18:2 via the action of SCD on *trans*-11 18:1 has been documented in mice,[317–319] rats,[320–322] hamsters[323] and pigs.[324] Two isoforms of the SCD gene, SCD1[325] and SCD5[326] have been characterized in humans. The SCD1 gene is expressed in most tissues, whereas that of SCD5 is essentially confined to the brain and pancreas.[288] Several key observations indicated that endogenous CLA synthesis via the activity of the SCD enzyme also occurs in humans. Firstly, the concentration of *cis*-9, *trans*-11 18:2 in plasma was found to be elevated in volunteers consuming diets high in *trans* fatty acids.[327,328] Secondly, ^{13}C enrichment of *cis*-9, *trans*-11 18:2 was detected in the blood of a single male subject offered 1-[13]C *trans*-11 18:1.[329] Thirdly, *trans*-11 18:1 was shown to be converted to *cis*-9, *trans*-11 18:2 during incubations with human mammary (MCF-7) and colon (SW480) cancer cell lines.[330]

The first quantitative estimates of endogenous *cis*-9, *trans*-11 18:2 synthesis were derived from an intervention involving 30 healthy subjects.[331] Volunteers were offered a diet rich in oleic acid for two weeks, followed by diets providing 1.5, 3.0 or 4.5 g of *trans*-11 18:1 per day for nine days. Plasma concentrations of *trans*-11 18:1 were elevated 194%, 407% and 620% above baseline, respectively. These changes were also accompanied by 50%, 169% and 198% increases in *cis*-9, *trans*-11 18:2 concentrations. Regression of the change in *cis*-9, *trans*-11 18:2 against the change in *trans*-11 18:1 plus the change in *cis*-9, *trans*-11 18:2 in circulating TAG of very low density lipoproteins indicated that 19% of *trans*-11 18:1 was desaturated. These estimates were supported by a longer intervention in which 12 volunteers were offered 3 g per day of *trans*-11 18:1 over a 42 d interval.[332] Concentrations of *trans*-11 18:1 and *cis*-9, *trans*-11 18:2 were increased eight- and two-fold relative to baseline. Enrichment of *trans*-11 18:1 and *cis*-9, *trans*-11 18:2 estimated that 24% and 19% of *trans*-11 18:1 was desaturated in plasma and red blood cell membranes, respectively. These observations led the authors to propose that *ca.* 25% of *trans*-11 18:1 supplied from the diet was converted to *cis*-9, *trans*-11 18:2 in humans.[332] Such estimates do not, however, account for the disappearance of *trans*-11 18:1-derived *cis*-9, *trans*-11 18:2 into organs or tissues from the circulation, and are not an indication of the extent of desaturation, but simply a reflection of the net sum of end-products surviving metabolism. Conversion of *trans*-11 18:1 in humans was confirmed in an intervention involving four lactating women offered 2.5 mg of ^{13}C-labelled *trans*-11 18:1 per kg body weight.[333] Enrichment of ^{13}C was detected for *cis*-9, *trans*-11 18:incorporated into TAG, CE and PL fractions in plasma and in milk lipids. Up to 10% of *cis*-9, *trans*-11 18:2 in milk was found to originate from *trans*-11 18:1, indicating that endogenous synthesis of CLA in the mammary glands of humans is considerably lower than for ruminants (Table 1.6).

In all human studies to date, the extent of endogenous *cis*-9, *trans*-11 18:2 synthesis has been shown to vary considerably between individuals. These differences have been attributed to variations in diet composition, physiological state or genetics. In rats, the conversion of *trans*-11 18:1 to *cis*-9, *trans*-11 18:2 has been shown to differ between tissues, ranging from 36.2% to 4.2% (testes and kidneys > adipose > ovaries > muscle > liver > heart).[334] In addition to tissue specific expression, genetic variations may also influence the SCD enzyme activity and endogenous *cis*-9, *trans*-11 18:2 synthesis. A recent study provided the first indications of an association between single nucleotide polymorphisms of the SCD gene and concentration ratios of *cis*-9, *trans*-11 18:2/*trans*-11 18:1 in the plasma of Caucasian and Asian adults.[335]

Recent studies *in vitro* have also raised the possibility that *trans*-9 16:1 is elongated and desaturated to *cis*-9, *trans*-11 18:2 in humans.[247] There are no reports on the metabolic fate of *trans*-9 16:1 in humans. Evidence emerging from clinical prospective studies have reported that higher *trans*-9 16:1 concentrations in blood are associated with a more favourable metabolic profile and lower incidence of diabetes.[336,337] Thus far, no cause and effect has been established.

Milk and dairy products are the major source of CLA in the human diet.[31,35-39] However, these foods provide more CLA than would be indicated based on proximal analysis of CLA composition. Irrespective of species, diet and production system, the relative concentrations of *trans*-11 18:1 and *cis*-9, *trans*-11 18:2 in ruminant milk are relatively constant in a ratio of 2.5:1.[255,308-312] The relationship between *trans*-11 18:1 and *cis*-9, *trans*-11 18:2 in ruminant muscle, is however, much more variable, which reflects both differences in the relative proportions of TAG and PL in intramuscular lipid and the differential incorporation of *trans*-11 18:1 and *cis*-9, *trans*-11 18:2 in these lipid fractions. In growing cattle, the relative abundance of *trans*-11 18:1 and *cis*-9, *trans*-11 18:2 varies between fat deposits in a ratio ranging between 0.15 and 0.23 compared with 0.17, 0.26 and 0.85 in heart, liver and erythrocyte PL, respectively.[338-340] Relative concentrations of *cis*-9, *trans*-11 18:2 to *trans*-11 18:1 also differ between subcutaneous (0.19), omental (0.15), perirenal adipose (0.12) and liver (0.30) in growing lambs.[243] However, ruminant meat is not the principal source of CLA in the human diet.[34-36,38,44] Assuming a mean conversion of *trans*-11 18:1 to *cis*-9, *trans*-11 18:2 in humans of 20%, it has been proposed that the effective physiological dose supplied by ruminant-derived foods is 1.4–1.5 times the measured CLA content.[39,42]

1.5.2 Synthesis in the Hindgut

Evidence of CLA formation by human intestinal bacteria was first obtained indirectly in the experiments of Chin *et al.*,[341] who noted that germ-free rats had a lower incorporation of CLA in liver, lung, kidney, skeletal muscle and abdominal adipose tissue than conventional animals. Subsequently, mixed human intestinal flora were shown to convert 18 : 2 n-6 to isomers of CLA,[342] thus offering the possibility that the human gut microbiota could contribute to the CLA status in humans. This prospect was diminished when it was discovered that little or none of the CLA formed in the rat intestine was absorbed and incorporated into host tissues. Absorption of CLA by the colonic epithelium has not been demonstrated, and it is therefore difficult to understand the mechanisms by which CLA synthesis could elicit a systemic effect. Nevertheless, more recent studies suggest a possible important role *in situ* for CLA produced in the intestine. Anti-inflammatory activity of *cis*-9, *trans*-11 18 : 2 and *trans*-10, *cis*-12 18 : 2 has been reported in the mouse model of inflammatory bowel disease (IBD).[343] Moreover, in animal studies, *cis*-9, *trans*-11 18 : 2 has proved to be a potent anti-carcinogen by lowering the incidence of chemically induced mouse aberrant crypt foci in the rat colon.[11] Isomers of CLA have also been found to exhibit anti-proliferative properties *in vitro*, inhibiting the growth of human colon cancer cells.[344] The possibility therefore arises that, if 18 : 2 n-6 can be delivered to the intestine, and the correct bacterial population is present, possibly through the simultaneous delivery of a CLA-producing 'probiotic', the susceptibility to IBD and colorectal cancer in humans might be decreased.

A survey of 30 representative strains of human Gram-positive intestinal bacteria indicated that several indigenous human intestinal bacteria could form isomers of CLA from 18 : 2 n-6.[345] *Roseburia* species were among the most active. Different *Roseburia* spp. formed either *trans*-11 18 : 1 or a 10-OH 18 : 1 intermediate, both of which served as precursors for *cis*-9, *trans*-11 18 : 2 synthesis. Bacteria from other ecosystems and from food products, but that are also found in the human gut, including strains of *Lactobacillus, Propionibacterium* and *Bifidobacterium*, have been known for some time to be capable of 18 : 2 n-6 isomerisation to *cis*-9, *trans*-11 18 : 2.[156,346–348]

However, the more abundant bacterial species belonging to clostridial clusters IV and XIVa also metabolized 18 : 2 n-6 at among the fastest rates of all bacteria investigated, forming products that can serve as precursors of CLA synthesis (Figure 1.9). Given the greater abundance of *Clostridium*-like bacteria present in the human intestinal microbiota,[349,350] it may be deduced that 18 : 2 n-6 metabolism by this major group will be quantitatively more important than that of the *Lactobacillus, Propionibacterium* and *Bifidobacterium* groups. Nonetheless, given that the *Lactobacillus* and *Bifidobacterium* genera provide many probiotics in common use, and because, unlike the *Roseburia*, they are not strict anaerobes, these two genera may be of more practical significance.[347] *Bifidobacterium breve* has been shown to have multiple effects on tissue concentrations of polyunsaturated fatty acids in rats used as a model for IBS,[351] although the ability to hydrogenate fatty acids, if any, in causing these benefits was unclear.

Thus, there appear to be two potential routes of *cis*-9, *trans*-11 18 : 2 formation in the human intestine (Figure 1.9). When faecal bacteria from four human donors and six species of human intestinal bacteria capable of 18 : 2 n-6 metabolism were incubated with 18 : 2 n-6 in deuterium oxide-enriched

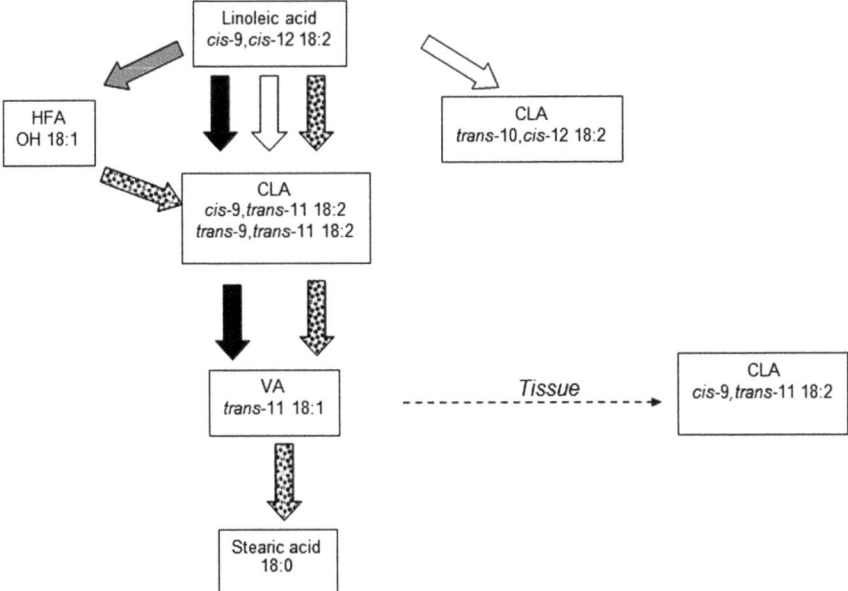

Figure 1.9 Proposed pathways of *cis*-9, *cis*-12 18:2 metabolism by bacterial species isolated from the human gut adapted from Devillard *et al.*[352] The open arrows represent the bacterial activity of *Lactobacillus, Propionibacterium* and *Bifidobacterium* species leading to the formation of 9,11 and 10,12 isomers of conjugated linoleic acid (CLA). The shaded arrows represent the bacterial activity of some *Lactobacillus, Propionibacterium* and *Bifidobacterium* species and some *Clostridium*-like bacteria belonging to clusters IV (e.g. *Eubacterium siraeum*) and XIVa (e.g. *R. intestinalis* and *Roseburia faecis*) leading to the formation of hydroxyl fatty acids (HFA). The solid arrows represent the bacterial activity of *Clostridium*-like bacteria belonging to cluster XIVa leading to the formation of *trans*-11 18:1 (VA) (e.g. *Roseburia hominis* and *R. inulinivorans*). The dotted arrows represent activities observed in faecal microbiota for which the responsible bacterial species are still unknown.

medium, the main products in faecal suspensions were *cis*-9, *trans*-11 18:2 and *trans*-9, *trans*-11 18:2, which were labelled at Δ13, as were the other 9,11 geometric isomers formed.[352] Traces of *trans*-10, *cis*-12 18:2 were formed, but this product was labelled to a much lower extent than *cis*-9, *trans*-11 18:2. In pure culture, *Bifidobacterium breve* formed labelled *cis*-9, *trans*-11 18:2 and *trans*-9, *trans*-11 18:2, while a human faeces-derived *B. fibrisolvens, Roseburia hominis, Roseburia inulinivorans* and *Ruminococcus obeum*-like strain A2-162 converted 18:2 n-6 to *trans*-11 18:1, labelled in a manner indicating that *trans*-11 18:1 was formed via C-13-labelled *cis*-9, *trans*-11 18:2.[352] *Propionibacterium freudenreichii* subsp. *shermanii*, a possible probiotic strain, formed mainly *cis*-9, *trans*-11 18:2 with smaller amounts of *trans*-10, *cis*-12 18:2 and *trans*-9, *trans*-11 18:2, labelled in the same manner as in the mixed microbiota. Ricinoleic acid (12-OH-*cis*-9 18:1) was not converted to one or more CLA isomers in the mixed microbiota, in contrast to that described for *Lactobacillus plantarum*.[352] These results were similar to those reported for the mixed microbiota of the rumen. Thus, although the bacterial genera and species responsible for biohydrogenation in the human intestine differs from the rumen, and a second route of *cis*-9, *trans*-11 18:2 formation via a 10-OH 18:1 is present in the intestine, the overall labelling patterns of the different CLA isomers formed are common to both gut eco-systems. A hydrogen-abstraction enzyme mechanism was proposed to explain the role of a 10-OH 18:1 intermediate in the formation of *cis*-9, *trans*-11 18:2 during incubations of 18:2 n-6 with pure and mixed cultures.[352]

1.6 Conclusions

Following the discovery that isomers of CLA exhibit anti-mutagenic properties in the rodent model of cancer, subsequent investigations have demonstrated that *cis*-9, *trans*-11 18:2 and *trans*-10, *cis*-12 18:2 elicit a diverse range of biological activities in cell culture and animal models. These findings have served as a catalyst to explore the biochemical, microbial and physiological mechanisms regulating the synthesis of CLA in ruminants, and more specifically, the transformations of dietary lipid that occur in the rumen. Even though *cis*-9, *trans*-11 18:2 is the principal isomer in ruminant tissues and milk, numerous isomers are formed in the rumen. Thus far, sixteen isomers have been detected in ruminal digesta that originate from the metabolism of 18:2 n-6 and 18:3 n-3. Bacteria rather than protozoa or fungi are the primary microorganisms responsible for catalysing the *cis-trans* isomerization yielding conjugated intermediates. Several ruminal bacteria capable of converting 18:2 n-6 to 9,11 or 10,12 geometric 18:2 isomers have been identified. Much less is known about the metabolic origins of other CLA isomers formed in the rumen. Diets rich in 18:3 n-3 promote the synthesis of 11,13 18:2 and 12,14 18:2 isomers in the rumen, but the metabolic pathways explaining their formation are not well characterized. Synthesis of CLA isomers can be increased several fold by supplementing the diet of ruminants with oils enriched in 18:2 n-6 and 18:3 n-3. Secretion of *trans*-7, *cis*-9 18:2 and *cis*-9, *trans*-11 18:2 in milk or the appearance of these isomers in host tissues is many fold higher compared with their synthesis in the rumen, due to endogenous synthesis catalysed by the stearoyl CoA desaturase (SCD) enzyme. Virtually all of the *trans*-7, *cis*-9 18:2 in ruminants is synthesized via the action of SCD on *trans*-7 18:1 formed as an intermediate of *cis*-9 18:1 biohydrogenation in the rumen. Endogenous synthesis using *trans*-11 18:1 as a substrate accounts for at least 60% of *cis*-9, *trans*-11 18:2 in ruminant milk. Between 45% and 95% of *cis*-9, *trans*-11 18:2 in ruminant tissues originates from *trans*-11 18:1. *Trans*-11 18:1 is the penultimate intermediate formed during the biohydrogenation of 18:2 n-6 and 18:3 n-3 in the rumen. Reduction of *trans* 18:1 intermediates is thought to be rate limiting for the complete biohydrogenation of 18-carbon unsaturated fatty acids to 18:0. Changes in diet composition and increases in 18:2 n-6 and 18:3 n-3 supply have a profound influence on the formation of the *trans* 18:1 precursors for endogenous synthesis, more so than on the formation of CLA isomers in the rumen. For this reason, nutritional strategies for enhancing the CLA content of ruminant foods have focused on increasing the amount of *trans*-11 18:1 escaping the rumen and preventing shifts in biohydrogenation pathways leading to the formation of other *trans* 18:1 isomers. Recent investigations suggest that *trans*-9 16:1 escaping the rumen may also serve as a substrate for endogenous *cis*-9, *trans*-11 18:2 synthesis via a mechanism that involves elongation to *trans*-11 18:1 followed by SCD catalysed desaturation. Formation of *trans*-9 16:1 has not been extensively investigated, but is thought to be an intermediate of 16:2 n-4, 16:3 n-4 and 16:4 n-1 metabolism in the rumen. Appearance of other CLA isomers is directly related to their synthesis in the rumen. Endogenous synthesis of *cis*-9, *trans*-11 18:2 via the action of SCD on *trans*-11 18:1 has also been demonstrated in humans. Desaturation of *trans*-11 18:1 in the mammary glands is much lower in women compared with cows, goats or sheep. Conversion of *trans*-11 18:1 to *cis*-9, *trans*-11 18:2 implies that the effective physiological dose of CLA supplied from ruminant-derived foods is 1.4–1.5 times the measured content.

References

1. *World Health Organ, Tech. Rep. Ser.*, 2003, **916**, 1–149.
2. P. A. Heidenreich, J. G. Trogdon, O. A. Khavjou, J. Butler, K. Dracup, M. D. Ezekowitz, E. A. Finkelstein, Y. Hong, S. C. Johnston, A. Khera, D. M. Lloyd-Jones, S. A. Nelson, G. Nichol, D. Orenstein, P. W. F. Wilson and Y. J. Woo, *Circulation*, 2011, **123**, 933.
3. S. Allender, P. Scarborough, V. Peto, M. Raynor, J. Leal, R. Luengo-Fernández and A. Gray, *Eur Heart Net.*, 2008, http://www.ehnheart.org.
4. D. E. Bloom, E. T. Cafiero, E. Jané-Llopis, S. Abrahams-Gessel, L. R. Bloom, S. Fathima, A. B. Feigl, T. Gaziano, M. Mowafi, A. Pandya, K. Prettner, L. Rosenberg, B. Seligman, A. Z. Stein, and C. Weinstein, *World Economic Forum*, 2011, http://www3.weforum.org/docs/.
5. D. I. Givens, *Animal*, 2010, **4**, 1941.
6. M. W. Pariza, S. H. Ashoor, F. S. Chu and D. B. Lund, *Cancer Lett.*, 1979, 7, 63.
7. M. W. Pariza and W. A. Hargraves, *Carcinogenesis*, 1985, **6**, 591.
8. Y. L. Ha, N. K. Grimm and M. W. Pariza, *Carcinogenesis*, 1987, **8**, 1881.
9. G. Jahreis, J. Kraft, F. Tischendorf, F. Schöne and C. von Loeffelholz, *Eur. J. Lipid Sci. Technol.*, 2000, **102**, 695.
10. L. D. Whigham, M. E. Cook and R. L. Atkinson, *Pharmacol. Res.*, 2000, **42**, 503.
11. D. Kritchevsky, *Br. J. Nutr.*, 2000, **83**, 459.
12. M. A. Belury, *Ann. Rev. Nutr.*, 2002, **22**, 505.
13. T. M. Larsen, S. Toubro and A. Astrup, *J. Lipid Res.*, 2003, **44**, 2234.

14. M. W. Pariza, *Am. J. Clin. Nutr.*, 2004, **79**, 1132S.
15. C. G. Taylor and P. Zahradka, *Am. J. Clin. Nutr.*, 2004, **79**, 1164S.
16. K. W. Wahle, S. D. Heys and D. Rotondo, *Prog. Lipid Res.*, 2004, **43**, 553.
17. Y. Wang and P. J. Jones, *Am. J. Clin. Nutr.*, 2004, **79**, 1153S.
18. A. Bhattacharya, J. Banu, M. Rahman, J. Causey and G. Fernandes, *J. Nutr. Biochem.*, 2006, **17**, 789.
19. P. Yaqoob, S. Tricon, G. C. Burdge and P. C. Calder, *Improving the Fat Content of Foods*, ed. C Williams and J Buttriss, Woodhead Publishing Limited, Cambridge, UK, 2006, pp. 182–209.
20. N. S. Kelley, N. E. Hubbard and K. L. Erickson, *J. Nutr.*, 2007, **137**, 2599.
21. P. L. Mitchell and R. S. McLeod, *Biochem. Cell. Biol.*, 2008, **86**, 293.
22. J. Salas-Salvadó, F. Márquez-Sandoval and M. Bulló, *Crit. Rev. Food Sci. Nutr.*, 2006, **46**, 479.
23. L. D. Whigham, A. C. Watras and D. A. Schoeller, *Am. J. Clin. Nutr.*, 2007, **85**, 1203.
24. S. Benjamin and F. Spener, *Nutr. Metab.*, 2009, **6**, 36.
25. S. W. Ing and M. A. Belury, *Nutr. Rev.*, 2011, **69**, 123.
26. R. P. Jutzeler van Wijlen, *Eur. J. Lipid Sci. Technol.*, 2011, **113**, 1077.
27. T. A. McCrorie, E. M. Keaveney, J. M. Wallace, N. Binns and M. B. Livingstone, *Nutr. Res. Rev.*, 2011, **24**, 206.
28. A. Dilzer and Y. Park, *Crit. Rev. Food Sci. Nutr.*, 2012, **52**, 488.
29. F. C. Parrish, B. R. Wiegand, D. C. Beitz, D. U. Ahn, M. Du and A. H. Trenkle, *Advances in Conjugated Linoleic Acid Research*, ed. J.-L. Sébédio, W. W. Christie and R. Adlof, AOCS Press, Champaign, Il, USA, Vol. 2, pp. 189–217.
30. M. Collomb, A. Schmid, R. Sieber, D. Wechsler and E.-L. Ryhänen, *Int. Dairy J.*, 2006, **16**, 1347.
31. P. W. Parodi, *Aust. J .Dairy Technol.*, 1994, **49**, 93.
32. J. Fritsche and H. Steinhart, *Z. Lebensm. Unters. Forsch*, 1998, **206**, 77.
33. A. Aro, S. Männistö, I. Salminen, M.-L. Ovaskainen, V. Kataja and M. Uusitupa, *Nutr. Cancer*, 2000, **38**, 151.
34. J. G. Ens, D. W. Ma, K. S. Cole, C. J. Field and M. T. Clandinin, *Nutr. Res.*, 2001, **21**, 955.
35. R. E. Lawson, A. R. Moss and D. I. Givens, *Nutr. Res. Rev*, 2001, **14**, 153.
36. K. L. Ritzenthaler, M. K. McGuire, R. Falen, T. D. Shultz, N. Dasgupta and M. A. McGuire, *J. Nutr.*, 2001, **131**, 1548.
37. L. Laloux, L. du Chaffaut, L. Razanamahefa and L. Lafay, *Eur. J. Lipid Sci. Technol.*, 2007, **109**, 918.
38. S. V. Martins, P. A. Lopes, C. M. Alfaia, V. S. Ribeiro, T. V. Guerreiro, C. M. G. A. Fontes, M. F. Castro, G. Soveral and J. A. M. Prates, *Br. J. Nutr.*, 2007, **98**, 1206.
39. R. P. Jutzeler van Wijlen and P. C. Colombani, *Int. Dairy J.*, 2010, **20**, 433.
40. S. Mushtaq, E. H. Mangiapane and K. A. Hunter, *Br. J. Nutr.*, 2010, **103**, 1366.
41. S. F. Chin, W. Liu, J. M. Storkson, Y. L. Ha and M. W. Pariza, *J. Food Comp. Anal.*, 1992, **5**, 185.

42. P. W. Parodi 2003. *Advances in Conjugated Linoleic Acid Research*, ed. J. L. Sébédio, W. W. Christie and R. Adlof, AOAC Press, Champaign, IL, USA, Vol 2, pp. 101–122.
43. R. C. Khanal and K. C. Olson, *Pak. J. Nutr.*, 2004, **3**, 82.
44. A. Schmid, M. Collomb, R. Sieber and G. Bee, *Meat Sci.*, 2006, **73**, 29.
45. D. L. Palmquist, A. L. Lock, K. J. Shingfield and D. E. Bauman, *Adv. Food Nutr. Res.*, **50**, 179.
46. K. J. Shingfield, Y. Chilliard, V. Toivonen, P. Kairenius and D. I. Givens, *Adv. Expt. Med. Biol.*, 2008, **606**, 3.
47. K. J. Shingfield, M. Bonnet and N. D. Scollan, *Animal*, 2013, 7(S1), 132.
48. C. G. Harfoot and G. P. Hazlewood, *The Rumen Microbial Ecosystem*, ed. P. N. Hobson and C. S. Stewart, Chapman and Hall, London, 1997, pp. 382–426.
49. R. J. Dewhurst, K. J. Shingfield, M. R. F. Lee and N. D. Scollan, *Anim. Feed Sci. Technol.*, 2006, **131**, 168.
50. A. Buccioni, M. Decandia, S. Minieri, G. Molle and A. Cabiddu, *Anim. Feed Sci. Technol.*, 2012, **174**, 1.
51. M. R. F. Lee, A. L. Winters, N. D. Scollan, R. J. Dewhurst, M. K. Theodorou and F. R. Minchin, *J. Sci. Food Agric.*, 2004, **84**, 1639.
52. G. Van Ranst, V. Fievez, M. Vandewalle, J. De Riek and E. Van Bockstaele, *Grass For. Sci.*, 2009, **64**, 196.
53. G. Van Ranst, V. Fievez, M. Vandewalle, C. Van Waes, J. De Riek and E. Van Bockstaele, *Animal*, 2010, **4**, 1528.
54. A. Halmemies-Beauchet-Filleau, A. Vanhatalo, V. Toivonen, T. Heikkilä, M. R. F. Lee and K. J. Shingfield, *J. Dairy Sci.*, 2013, **96**, 5882.
55. R. J. Dewhurst, N. D. Scollan, S. J. Youell, J. K. S. Tweed and M. O. Humphreys, *Grass For. Sci.*, 2001, **56**, 68.
56. A. Vanhatalo, K. Kuoppala, V. Toivonen and K. J. Shingfield, *Eur. J. Lipid Sci. Technol.*, 2007, **109**, 856.
57. K. J. Shingfield, C. K. Reynolds, B. Lupoli, V. Toivonen, M. P. Yurawecz, P. Delmonte, J. M. Griinari, A. S. Grandison and D. E. Beever, *Anim. Sci.*, 2005, **80**, 225.
58. K. J. Shingfield, P. Salo-Väänänen, E. Pahkala, V. Toivonen, S. Jaakkola, V. Piironen and P. Huhtanen, *J. Dairy Res.*, 2005, 72, 349.
59. A. A. AbuGhazaleh, D. J. Schingoethe and A. R. Hippen, *J. Dairy Sci.*, 2001, **84**, 1845.
60. H. Boufaïed, P. Y. Chouinard, G. F. Tremblay, H. V. Petit, R. Michaud and G. Bélanger, *Can. J. Anim. Sci.*, 2003, **83**, 501.
61. A. Halmemies-Beauchet-Filleau, P. Kairenius, S. Ahvenjärvi, L. K. Crosley, S. Muetzel, P. Huhtanen, A. Vanhatalo, V. Toivonen, R. J. Wallace and K. J. Shingfield, *J. Dairy Sci.*, 2013, **96**, 2428.
62. I. Feussner and C. Wasternack, *Ann. Rev. Plant Biol.*, 2002, **53**, 275.
63. A. Mosblech, I. Feussner and I. Heilmann, *Plant Physiol. Biochem.*, 2009, **47**, 511.
64. F. Noci, P. O'Kiely, F. J. Monahan, C. Stanton and A. P. Moloney, *Meat Sci.*, 2005, **69**, 509.

65. V. B. Woods and A. M. Fearon, *Lives. Sci.*, 2009, **126**, 1.
66. D. I. Givens, K. E. Kliem, D. J. Humphries, K. J. Shingfield and R. Morgan, *Animal*, 2009, **3**, 1067.
67. K. J. Shingfield, S. Ahvenjärvi, V. Toivonen, A Vanhatalo, P. Huhtanen and J. M. Griinari, *Br. J. Nutr.*, 2008, **99**, 971.
68. K. J. Shingfield, M. R. F. Lee, D. J. Humphries, N. D. Scollan, V. Toivonen, D. E. Beever and C. K. Reynolds, *J. Anim. Sci.*, 2011, **89**, 3728.
69. C. Hurtaud and J. L. Peyraud, *J. Dairy Sci.*, 2007, **90**, 5134.
70. A. Halmemies-Beauchet-Filleau, T. Kokkonen, A.-M. Lampi, V. Toivonen, K. J. Shingfield and A. Vanhatalo, *J. Dairy Sci.*, 2011, **94**, 4413.
71. C. Boeckaert, B. Vlaeminck, J. Dijkstra, A. Issa-Zacharia, T. Van Nespen, W. Van Straalen and V. Fievez, *J. Dairy Sci.*, 2008, **91**, 4714.
72. G. A. Garton, P. N. Hobson and A. K. Lough, *Nature*, 1958, **182**, 1511.
73. R. M. C. Dawson and N. Hemington, *Br. J. Nutr.*, 1974, **32**, 327.
74. R. M. C. Dawson, N. Hemington and G. D. Hazlewood, *Br. J. Nutr.*, 1977, **38**, 225.
75. J. M. Omar Faruque, B. D. W. Jarvis and J. C. Hawke, *J. Sci. Food Agric.*, 1974, **25**, 1313.
76. M. R. F. Lee, M. K. Theodorou, T. T. Chow, M. Enser and N. D. Scollan, *Proc. Nutr. Soc.*, 2002, **61**, 103A.
77. G. Van Ranst, V. Fievez, M. Vandewalle, J. De Riek and E. Van Bockstaele, *J. Agric. Food Chem.*, 2009, **57**, 6611.
78. S. A. Huws, E. J. Kim, A. H. Kingston-Smith, M. R. F. Lee, S. M. Muetzel, A. R. Cookson, C. J. Newbold, R. J. Wallace and N. D. Scollan, *FEMS Microbiol. Ecol.*, 2009, **69**, 461.
79. P. N. Hobson and S. O. Mann, *J. Gen. Microbiol.*, 1961, **25**, 227.
80. G. N. Jarvis, C. Strompl, E. R. B. Moore and J. H. Thiele, *Sys. Appl. Microbiol.*, 1998, **21**, 135.
81. D. G. Cirne, O. D. Delgado, S. Marichamy and B. Mattiasson, *Int. J. Sys. Evol. Microbiol.*, 2006, **56**, 625.
82. D. Paillard, N. McKain, L. C. Chaudhary, N. D. Walker, F. Pizette, I. Koppova, N. R. McEwan, J. Kopecny, P. E. Vercoe, P. Louis and R. J. Wallace, *Antonie Van Leeuwenhoek*, 2007, **91**, 417.
83. D. E. Wright, *NZ J. Agric. Res.*, 1961, **4**, 216.
84. R. W. Bailey and B. H. Howard, *Biochem. J.*, 1963, **87**, 146.
85. G. S. Coleman, P. Kemp and R. M. Dawson, *Biochem. J.*, 1971, **123**, 97.
86. J. L. Arpigny and K. E. Jaeger, *Biochem. J.*, 1999, **343**, 177.
87. C. Henderson, *J. Gen. Microbiol.*, 1971, **65**, 81.
88. C. Henderson and W. Hodgkiss, *J. Gen. Microbiol.*, 1973, **76**, 389.
89. F. Privé, N. N. Kaderbhai, S. Girdwood, H. J. Worgan, E. Pinloche, N. D. Scollan, S. A. Huws and C. J. Newbold, *Plos One*, 2013, **8**, e69076.
90. J. Liu, D. Wang, S. Bu, C. Zhao, P. McSweeney, Yu and D. Li, *Biochem. Biophys. Res. Commun.*, 2009, **385**, 605.
91. J. C. Hawke and W. R. Silcock, *Biochim. Biophys. Acta.*, 1970, **218**, 201.

92. J. H. Moore, R. C. Noble, W. Steele and J. W. Czerkawski, *Br. J. Nutr.*, 1969, **23**, 869.
93. R. C. Noble, J. H. Moore and C. G. Harfoot, *Br. J. Nutr.*, 1974, **31**, 99.
94. T. Gerson, A. John and B. R. Sinclair, *J. Agric. Sci., Camb.*, 1983, **101**, 97.
95. T. Gerson, A. John and A. S. D. King, *J. Agric. Sci., Camb.*, 1985, **105**, 27.
96. T. Gerson, A. John and A. S. D. King, *J. Agric. Sci., Camb.*, 1986, **106**, 445.
97. T. M. Beam, T. C. Jenkins, P. J. Moate, R. A. Kohn and D. L. Palmquist, *J. Dairy Sci.*, 2000, **83**, 2564.
98. T. T. Chow, V. Fievez, A. P. Moloney, K. Raes, D. Demeyer and S. De Smet, *Anim. Feed Sci. Technol.*, 2004, **117**, 1.
99. M. J. Latham, J. E. Storry and M. E. Sharpe, *Appl. Microbiol.*, 1972, **24**, 871.
100. C. J. Van Nevel and D. I. Demeyer, *Reprod. Nutr. Dev.*, 1996, **36**, 53.
101. M. R. F. Lee, J. de, J. O. Colmenero, A. L. Winters, N. D. Scollan and F. R. Minchin, *J. Sci. Food Agric.*, 2006, **86**, 1503.
102. M. R. F. Lee, M. B. Scott, J. K. S. Tweed, F. R. Minchin and D. R. Davies, *Anim. Feed Sci. Technol.*, 2008, **144**, 125.
103. R. Bickerstaffe, E. F. Annison and D. E. Noakes, *Biochem. J.*, 1972, **130**, 607.
104. R. L. Atkinson, E. J. Scholljegerdes, S. L. Lake, V. Nayigihugu, B. W. Hess and D. C. Rule, *J. Anim. Sci.*, 2006, **84**, 387.
105. F. B. Shorland, R. O. Weenink and A. T. Johns, *Nature*, 1955, **175**, 1129.
106. T. C. Jenkins, R. J. Wallace, P. J. Moate and E. E. Mosley, *J. Anim. Sci.*, 2008, **86**, 397.
107. K. J. Shingfield, L. Bernard, C. Leroux and Y. Chilliard, *Animal*, 2010, **4**, 1140.
108. R. Reiser, *Fed. Proc.*, 1951, **10**, 236.
109. P. F. Wilde and R. M. C. Dawson, *Biochem. J.*, 1966, **98**, 469.
110. C. E. Polan, J. J. McNeill and S. B. Tove, *J. Bact.*, 1964, **88**, 1056.
111. C. R. Kepler, K. P. Hirons, J. J. McNeill and S. B. Tove, *J. Biol. Chem.*, 1966, **241**, 1350.
112. F. V. Ward, T. W. Scott and R. M. C. Dawson, *Biochem. J.*, 1964, **92**, 60.
113. C. R. Kepler and S. B. Tove, *J. Biol. Chem.*, 1967, **242**, 5686.
114. R. W. White, P. Kemp and R. M. C. Dawson, *Biochem. J.*, 1970, **116**, 767.
115. P. Kemp, R. W. White and D. J. Lander, *J. Gen. Microbiol.*, 1975, **90**, 100.
116. P. Kemp and D. J. Lander, *Biochem. J.*, 1983, **216**, 519.
117. M. R. G. Maia, C. A. S. Correia, S. P. Alves, A. J. M. Fonseca and A. R. J. Cabrita, *J. Anim. Sci.*, **90**, 900.
118. S. P. Alves, M. R. G. Maia, R. J. B. Bessa, A. J. M. Fonseca and A. R. J. Cabrita, *Lipids*, 2012, **47**, 171.
119. P. G. Toral, K. J. Shingfield, G. Hervás, V. Toivonen and P. Frutos, *J. Dairy Sci.*, 2010, **93**, 4804.
120. P. G. Toral, A. Belenguer, K. J. Shingfield, G. Hervás, V. Toivonen and P. Frutos, *J. Dairy Sci.*, 2012, **95**, 794.
121. P. Kairenius, V. Toivonen and K. J. Shingfield, *Lipids*, 2011, **46**, 587.

122. K. J. Shingfield, P. Kairenius, A. Arölä, D. Paillard, S. Muetzel, S. Ahvenjärvi, A. Vanhatalo, P. Huhtanen, V. Toivonen, J. M. Griinari and R. J. Wallace, *J. Nutr.*, 2012, **142**, 1437.
123. M. Doreau and A. Ferlay, *Anim. Feed Sci. Technol.*, 1994, **45**, 379.
124. F. Glasser, P. Schmidely, D. Sauvant and M. Doreau, *Animal*, 2008, **2**, 691.
125. M. Keeney, *Physiology of Digestion and Metabolism in the Ruminant*, ed. A. T. Phillipson, Oriel Press, Newcastle-upon-Tyne, UK, 1970, pp. 489–503.
126. A. A. AbuGhazaleh, M. B. Riley, E. E. Thies and T. C. Jenkins, *J. Dairy Sci.*, 2005, **88**, 4334.
127. M. C. Fuentes, S. Calsamiglia, P. W. Cardozo and B. Vlaeminck, *J. Dairy Sci.*, 2009, **92**, 4456.
128. S. A. Martin and T. C. Jenkins, *J. Anim. Sci.*, 2002, **80**, 3347.
129. A. Troegeler-Meynadier, M. C. Nicot, C. Bayourthe, R. Moncoulon and F. Enjalbert, *J. Dairy Sci.*, 2003, **86**, 4054.
130. X. Qiu, M. L. Eastridge, K. E. Griswold and J. L. Firkins, *J. Dairy Sci.*, 2004, **87**, 3473.
131. N. J. Choi, J. Y. Imm, S. Oh, B. C. Kim, H. J. Hwang and Y. J. Kim, *Anim. Feed Sci. Technol.*, 2005, **123–124**, 643.
132. A. Troegeler-Meynadier, L. Bret-Bennis and F. Enjalbert, *Reprod. Nutr. Dev.*, 2006, **46**, 713.
133. T. C. Jenkins, A. A. AbuGhazaleh, S. Freeman and E. J. Thies, *J. Nutr.*, 2006, **136**, 926.
134. N. McKain, K. J. Shingfield and R. J. Wallace, *Microbiol.*, **156**, 579.
135. P. Kemp and D. J. Lander, *Br. J. Nutr.*, 1984, **52**, 165.
136. J. A. Hudson, C. A. M. Mackenzie and K. N. Joblin, *Appl. Microbiol. Biotechnol.*, 1995, **44**, 1.
137. J. A. Hudson, Y. Cai, R. J. Corner, B. Morvan and K. N. Joblin, *J. Appl. Microbiol.*, 2000, **88**, 286.
138. E. E. Mosley, G. L. Powell, M. B. Riley and T. C. Jenkins, *J. Lipid Res.*, 2002, **43**, 290.
139. J. M. Proell, E. E. Mosley, G. L. Powell and T. C. Jenkins, *J. Lipid Res.*, 2002, **43**, 2072.
140. J.-P. Jouany, B. Lassalas, M. Doreau and F. Glasser, *Lipids*, 2007, **42**, 351.
141. A. M. Honkanen, J. M. Griinari, A. Vanhatalo, S. Ahvenjärvi, V. Toivonen and K. J. Shingfield, *J. Dairy Sci.*, 2012, **95**, 1376.
142. R. J. Wallace, N. McKain, K. J. Shingfield and E. Devillard, *J. Lipid Res.*, 2007, **48**, 2247.
143. M. M. Or-Rashid, O. AlZahal and B. W. McBride, *Appl. Microbiol. Biotechnol.*, 2011, **89**, 387.
144. J. A. Hudson, B. Morvan and K. N. Joblin, *FEMS Microbiol. Lett.*, 1998, **169**, 277.
145. M. J. Kellens, H. L. Goderis and P. P. Tobback, *Biotechnol. Bioeng.*, 1986, **28**, 1268.

146. I. Wąsowska, M. Maia, K. M. Niedźwiedzka, M. Czauderna, J. M. C. Ramalho Ribeiro, E. Devillard, K. J. Shingfield and R. J. Wallace, *Br. J. Nutr.*, 2006, **95**, 1199.
147. S. Fukuda, Y. Nakanishi, E. Chikayama, H. Ohno, T. Hino and J. Kikuchi, *PLoS ONE*, 2009, **4**, e4893.
148. Y. J. Lee and T. C. Jenkins, *J. Nutr.*, 2011, **141**, 1445.
149. F. Destaillats, J. P. Trottier, J. M. G. Galvez and P. Angers, *J. Dairy Sci.*, 2005, **88**, 3231.
150. M. Plourde, F. Destaillats, P. Y. Chouinard and P. Angers, *J. Dairy Sci.*, 2007, **90**, 5269.
151. S. Lerch, K. J. Shingfield, A. Ferlay, A. Vanhatalo and Y. Chilliard, *J. Dairy Sci.*, 2012, **95**, 7269.
152. C. R. Kepler, W. P. Tucker and S. B. Tove, *J. Biol. Chem.*, 1970, **245**, 3612.
153. C. R. Kepler, W. P. Tucker and S. B. Tove, *J. Biol. Chem.*, 1971, **246**, 2765.
154. A. Liavonchanka, E. Hornung, I. Feussner and M. G. Rudolph, *Proc. Nat. Acad. Sci. U. S. A.*, 2006, **103**, 2576.
155. A. Liavonchanka and I. Feussner, *Chembiochem*, 2008, **9**, 1867.
156. J. Ogawa, S. Kishino, A. Ando, S. Sugimoto, K. Mihara and S. Shimizu, *J. Biosci. Bioeng.*, 2005, **100**, 355.
157. S. Kishino, S. B. Park, M. Takeuchi, K. Yokozeki, S. Shimizu and J. Ogawa, *Biochem. Biophys. Res. Commun.*, 2011, **416**, 188.
158. S. J. Park, K. A. Park, C. W. Park, W. S. Park, J. O. Kim and Y. L. Ha, *J. Food Sci. Nutr.*, 1996, **1**, 244.
159. R. A. Rosson, M.-D. Deng, A. D. Grund and S. S. Peng, Patent WO 01/00846, 2001.
160. S. Fukuda, H. Furuya, Y. Suzuki, N. Asanuma and T. Hino, *J. Gen. Appl. Microbiol.*, 2005, **51**, 105.
161. S. Fukuda, Y. Suzuki, T. Komori, K. Kawamura, N. Asanuma and T. Hino, *J. Appl. Microbiol.*, 2007, **103**, 365.
162. C. E. Polan, J. J. McNeill and S. B. Tove, *J. Bact.*, 1964, **88**, 1056.
163. G. P. Hazlewood, P. Kemp, D. Lauder and R. M. C. Dawson, *Br. J. Nutr.*, 1976, **35**, 293.
164. J. L. van de Vossenberg and K. N. Joblin, *Lett. Appl. Microbiol.*, 2003, **37**, 424.
165. R. J. Wallace, L. C. Chaudhary, N. McKain, N. R. McEwan, A. J. Richardson, P. E. Vercoe, N. D. Walker and D. Paillard, *FEMS Microbiol. Lett.*, 2006, **265**, 195.
166. D. Moon, D. M. Pacheco, W. J. Kelly, S. C. Leahy, D. Li, J. Kopecny and G. T. Attwood, *Int. J. System. Evol. Microbiol.*, 2008, **58**, 2041.
167. M. R. G. Maia, L. C. Chaudhary, C. S. Bestwick, A. J. Richardson, N. McKain, T. R. Larson, I. A. Graham and R. J. Wallace, *BMC Microbiol.*, 2010, **10**, 521.
168. Y. J. Kim, R. H. Liu, J. L. Rychlik and J. B. Russell, *J. Appl. Microbiol.*, 2002, **92**, 976.
169. D. E. Bauman and J. M. Griinari, *Lives. Prod. Sci.*, 2001, **70**, 15.
170. D. E. Bauman and J. M. Griinari, *Ann. Rev. Nutr.*, 2003, **23**, 203.

171. K. J. Shingfield and J. M. Griinari, *Eur. J. Lipid Sci. Technol.*, 2007, **109**, 799.

172. A. Zened, F. Enjalbert, M. C. Nicot and A. Troegeler-Meynadier, *Animal*, 2012, **6**, 459.

173. A. Zened, F. Enjalbert, M. C. Nicot and A. Troegeler-Meynadier, *J. Dairy Sci.*, 2013, **96**, 451.

174. P. J. Weimer, D. M. Stevenson and D. R. Mertens, *J. Dairy Sci.*, 2010, **93**, 265.

175. L. S. Piperova, J. Sampugna, B. B. Teter, K. F. Kalscheur, M. P. Yurawecz, Y. Ku, K. M. Morehouse and R. A. Erdman, *J. Nutr.*, 2002, **132**, 1235.

176. J. J. Loor, K. Ueda, A. Ferlay, Y. Chilliard and M. Doreau, *J. Dairy Sci.*, 2004, **87**, 2472.

177. S. Kishino, J. Ogawa, K. Yokozeki and S. Shimizu, *Appl. Microbiol. Biotechnol.*, 2009, **84**, 87.

178. M. E. Sharpe, M. J. Latham, E. I. Garvie, J. Zirugibl and O. Kaudler, *J. Gen. Microbiol.*, 1973, 77, 37.

179. L. Slyter, *J. Anim. Sci.*, 1976, **43**, 910.

180. B. Morvan and K. N. Joblin, *Anaerobe*, 1999, **5**, 605.

181. C. Boeckaert, D. P. Morgavi, J. P. Jouany, L. Maignien, N. Boon and V. Fievez, *Animal*, 2009, **3**, 961.

182. D. Li, J. Q. Wang and D. P. Bu, *BMC Res. Notes*, 2012, **5**, 97.

183. C. Boeckaert, B. Vlaeminck, J. Mestdagh and V. Fievez, *Anim. Feed Sci. Technol.*, 2007, **136**, 63.

184. C. Boeckaert, B. Vlaeminck, V. Fievez, L. Maignien, J. Dijkstra and N. Boon, *Appl. Environ. Microbiol.*, 2008, **74**, 6923.

185. S. A. Huws, E. J. Kim, M. R. F. Lee, M. B. Scott, J. K. Tweed, E. Pinloche, R. J. Wallace and N. D. Scollan, *Environ. Microbiol.*, 2011, **13**, 1500.

186. A. G. Williams and G. S. Coleman, *The Rumen Protozoa*, Springer-Verlag, New York, USA, First Edition, 1992, pp. 441.

187. D. E. Wright, *Nature*, 1959, **184**, 875.

188. D. E. Wright, *Nature*, 1960, **185**, 546.

189. R. M. C. Dawson and P. Kemp, *Biochem. J.*, 1969, **115**, 351.

190. V. Girard and J. C. Hawke, *Biochim. Biophys. Acta*, 1978, **528**, 17.

191. S. Singh and J. C. Hawke, *J. Sci. Food Agric.*, 1979, **30**, 603.

192. I. Katz and M. Keeney, *J. Dairy Sci.*, 1966, **49**, 962.

193. E. Devillard, F. M. McIntosh, C. J. Newbold and R. J. Wallace, *Br. J. Nutr.*, 2006, **96**, 697.

194. M. D. Stern, W. H. Hoover and J. B. Leonard, *J. Dairy Sci.*, 1977, **60**, 911.

195. M. Lourenço, E. Ramos-Morales and R. J. Wallace, *Animal*, 2010, **4**, 1008.

196. M. M. Or-Rashid, O. AlZahal and B. W. McBride, *Appl. Microbiol. Biotechnol.*, 2008, **81**, 533.

197. M. Abe, T. Iriki, N. Tobe and H. Shibui, *Appl. Environ. Microbiol.*, 1981, **41**, 758.

198. P. Ankrah, S. C. Loerch and B. A. Dehority, *J. Gen. Microbiol.*, 1990, **136**, 1869.

199. R. E. Hungate, J. Reichl and R. Prins, *Appl. Microbiol.*, 1971, **22**, 1104.
200. R. A. Weller and A. F. Pilgrim, *Br. J. Nutr.*, 1974, **32**, 341.
201. D. R. Yáñez-Ruiz, N. D. Scollan, R. J. Merry and C. J. Newbold, *Br. J. Nutr.*, 2006, **96**, 861.
202. S. A. Huws, M. R. F. Lee, A. H. Kingston-Smith, E. J. Kim, M. B. Scott, J. K. S. Tweed and N. D. Scollan, *Br. J. Nutr.*, 2012, **108**, 2207.
203. C. G. Orpin, *Anaerobic Fungi. Biology, Ecology, and Function*, ed. D. O. Mountfort and C. G. Orpin, Marcel Dekker, Inc., New York, 1994, pp. 1–45.
204. D. O. Mountfort, *Anaerobic Fungi. Biology, Ecology, and Function*, ed. D. O. Mountfort and C. G. Orpin, Marcel Dekker, Inc., New York, 1994, pp. 271–279.
205. I. S. Nam and P. C. Garnsworthy, *J. Appl. Microbiol.*, 2007, **103**, 551.
206. I. S. Nam and P. C. Garnsworthy, *Asian-Austr. J. Anim. Sci.*, 2007, **20**, 1694.
207. M. R. G. Maia, L. C. Chaudhary, L. Figueres and R. J. Wallace, *Antonie van Leeuwenhoek*, 2007, **91**, 303.
208. P. H. Janssen and M. Kirs, *Appl. Environ. Microbiol.*, 2008, **74**, 3619.
209. D. P. Morgavi, E. Forano, C. Martin and C. J. Newbold, *Animal*, 2010, **4**, 1024.
210. S. C. Leahy, W. J. Kelly, E. Altermann, R. S. Ronimus, C. J. Yeoman, D. M. Pacheco, D. Li, Z. H. Kong, S. McTavish, C. Sang, S. C. Lambie, P. H. Janssen, D. Dey and G. T. Attwood, *Plos One*, 2010, **5**, e8926.
211. G. J. Faichney, *Digestion and Metabolism in the Ruminant*, ed. I. W. McDonald and A. C. I. Warner, The University of New England Publishing Unit, Sydney, Australia, 1975, pp. 277–291.
212. S. Ahvenjärvi, A. Vanhatalo, K. J. Shingfield and P. Huhtanen, *Br. J. Nutr.*, 2003, **90**, 41.
213. X. Qiu, M. L. Eastridge and J. L. Firkins, *J. Dairy Sci.*, 2004, **87**, 4278.
214. S. K. Duckett, J. G. Andrae and F. N. Owens, *J. Anim. Sci.*, 2002, **80**, 3353.
215. J. R. Sackmann, S. K. Duckett, M. H. Gillis, C. E. Realini, A. H. Parks and R. B. Eggelston, *J. Anim. Sci.*, 2003, **81**, 3174.
216. E. J. Schelljegerdes, B. W. Hess, G. E. Moss, D. L. Hixon and D. C. Rule, *J. Anim. Sci.*, 2004, **82**, 3577.
217. S. K. Duckett and M. H. Gillis, *J. Anim. Sci.*, 2010, **88**, 2684.
218. O. Kucuk, B. W. Hess, P. A. Ludden and D. C. Rule, *J. Anim. Sci.*, 2001, **79**, 2233.
219. O. Kucuk, B. W. Hess and D. C. Rule, *J. Anim. Sci.*, 2004, **82**, 2985.
220. K. J. Shingfield, S. Ahvenjärvi, V. Toivonen, A. Arölä, K. V. V. Nurmela, P. Huhtanen and J. M. Griinari, *Anim. Sci.*, 2003, **77**, 165.
221. M. R. F. Lee, L. J. Harris, R. J Dewhurst, R. J. Merry and N. D. Scollan, *Anim. Sci.*, 2003, **76**, 491.
222. M. R. F. Lee, J. K. S. Tweed, A. P. Moloney and N. D. Scollan, *Anim. Sci.*, **80**, 361.
223. J. J. Loor, K. Ueda, A. Ferlay, Y. Chilliard and M. Doreau, *Anim. Feed Sci. Technol.*, 2005, **119**, 203.

224. M. R. F. Lee, P. L. Connelly, J. K. S. Tweed, R. J. Dewhurst, R. J. Merry and N. D. Scollan, *J. Anim. Sci.*, 2006, **84**, 3061.

225. M. R. F. Lee, J. K. S. Tweed, R. J. Dewhurst and N. D. Scollan, *Anim. Sci.*, **82**, 31.

226. M. Doreau, D. Rearte, J. Portelli and J. L. Peyraud, *Eur. J. Lipid Sci. Technol.*, 2007, **109**, 790.

227. M. R. F. Lee, K. J. Shingfield, J. K. S. Tweed, V. Toivonen, S. A. Huws and N. D. Scollan, *Animal*, 2008, **2**, 1859.

228. M. Doreau, S. Laverroux, J. Normand, G. Chesneau and F. Glasser, *Lipids*, 2009, **44**, 53.

229. K. J. Shingfield, M. R. F. Lee, D. J. Humphries, N. D. Scollan, V. Toivonen, C. K. Reynolds and D. E. Beever, *Br. J. Nutr.*, 2010, **104**, 56.

230. J. K. G. Kramer, V. Fellner, M. E. R. Dugan, F. D. Sauer, M. M. Mossoba and M. P. Yurawecz, *Lipids*, 1997, **32**, 1219.

231. M. P. Yurawecz, J. K. G. Kramer and Y. Ku, *Advances in Conjugated Linoleic Acid Research*, ed. M. P. Yurawecz, M. M. Mossoba, J. K. G. Kramer, M. W. Pariza and G. J. Nelson, AOCS Press, Champaign, IL, USA, Vol 1., 1999, pp. 64–82.

232. S. F. W. Chin, J. M. Liu, Y. Storkson, L. Ha and M. W. Pariza, *J. Food Comp. Anal.*, 1992, **5**, 185.

233. J. K. G. Kramer and J. Zhou, *Eur. J. Lipid Sci. Technol.*, 2001, **103**, 594.

234. J. K. G. Kramer, C. Cruz-Hernandez, Z. Deng, J. Zhou, G. Jahreis and M. E. R. Dugan, *Am. J. Clin. Nutr.*, 2004, **79**(S), 1137S.

235. J. K. G. Kramer, N. Sehat, M. E. Dugan, M. M. Mossoba, M. P. Yurawecz, J. A. Roach, K. Eulitz, J. L. Aalhus, A. L. Schaefer and Y. Ku, *Lipids*, 1998, **33**, 549.

236. J. A. Roach, M. P. Yurawecz, J. K. G. Kramer, M. M. Mossoba, K. Eulitz and Y. Ku, *Lipids*, 2000, **35**, 797.

237. R. Mohammed, C. S. Stanton, J. J. Kennelly, J. K. G. Kramer, J. F. Mee, D. R. Glimm, M. O'Donovan and J. J. Murphy, *J. Dairy Sci.*, **92**, 3874.

238. C. Boeckaert, V. Fievez, D. Van Hecke, W. Verstraete and N. Boon, *Eur. J. Lipid Sci. Technol.*, 2007, **109**, 767.

239. F. Glasser, A. Ferlay, M. Doreau, P. Schmidely, D. Sauvant and Y. Chilliard, *J. Dairy Sci.*, 2008, **91**, 2771.

240. A. L. Lock and P. C. Garnsworthy, *Anim. Sci.*, 2002, **74**, 163.

241. K. F. Kalscheur, B. B. Teter, L. S. Piperova and R. A. Erdman, *J. Dairy Sci.*, 1997, **80**, 2115.

242. Z. C. T. R. Daniel, R. J. Wynn, A. M. Salter and P. J. Buttery, *J. Anim. Sci.*, 2004, **82**, 747.

243. M. R. G. Maia, R. J. Bessa and R. J. Wallace, *Proceedings of the XIth International Symposium of Ruminant Physiology*, ed. Y. Chilliard, F. Glasser, Y. Faulconnier, F. Bocquier, I. Veissier, and M. Doreau, Wageningen Academic Publishers, Wageningen, the Netherlands, 2009, pp. 276–277.

244. A. A. AbuGhazaleh and T. C. Jenkins, *J. Dairy Sci.*, 2004, **87**, 645.

245. A. A. AbuGhazaleh and T. C. Jenkins, *J. Dairy Sci.*, 2004, **87**, 1047.

246. C. M. Klein and T. C. Jenkins, *J. Dairy Sci.*, 2011, **94**, 4676.
247. K. G. Kadegowda, T. A. Burns, M. C. Miller and S. K. Duckett, *J Anim. Sci.*, 2013, **91**, 1614.
248. F. Destaillats, R. L. Wolff, D. Precht and J. Molkentin, *Lipids*, 2000, **35**, 1027.
249. S. Banni, C. Carta, M. S. Contini, E. Angioni, M. Deiana, M. A. Dessi, M. P. Melis and F. P. Corongiu, *J. Nutr. Biochem.*, 1996, 7, 150.
250. M. L. Kelly, E. S. Kolver, D. E. Bauman, M. E. Van Amburgh and L. D. Muller, *J. Dairy Sci.*, 1998, **81**, 1630.
251. M. L. Kelly, J. R. Berry, D. A. Dwyer, J. M. Griinari, P. Y. Chouinard, M. E. Van Amburgh and D. E. Bauman, *J. Nutr.*, 1998, **128**, 881.
252. N. W. Offer, M. Marsden, J. Dixon, B. K. Speake and F. E. Thacker, *Anim. Sci.*, 1999, **69**, 613.
253. M. M. Mahfouz, A. J. Valicenti and R. T. Holman, *Biochim. Biophys. Acta*, 1980, **618**, 1.
254. M. R. Pollard, F. D. Gunstone, A. T. James and L. J. Morris, *Lipids*, 1980, **15**, 306.
255. J. M. Griinari and D. E. Bauman, *Advances in Conjugated Linoleic Acid Research*, ed. M. P. Yurawecz, M. M. Mossoba, J. K. G. Kramer, M. W. Pariza and G. J. Nelson, AOCS Press, Champaign, IL, USA, Vol. I ., 1999, pp. 180–200.
256. D. E. Bauman and C. L. Davis, *Lactation: A Comprehensive Treatise*, ed. B. L. Larson and V. R. Smith, Academic Press, New York, USA, Vol. 2, 1974, pp. 31–75.
257. J. M. Ntambi, *J. Lipid Res.*, 1999, **40**, 1549.
258. K. H. Kaestner, J. M. Ntambi, T. J. Kelly and M. D. Lane, *J. Biol. Chem.*, 1989, **264**, 14755.
259. M. Miyazaki, M. J. Jacobson, W. C. Man, P. Cohen, E. Asilmaz, J. M. Friedman and J. M. Ntambi, *J. Biol. Chem.*, 2003, **278**, 33904.
260. J. M. Ntambi, S. A. Buhrow, K. H. Kaestner, R. J. Christy, E. Sibley, T. J. Kelly and M. D. Lane, *J. Biol. Chem.*, 1988, **263**, 17291.
261. Y. Zheng, S. M. Prouty, A. Harmon, J. P. Sundberg, K. S. Stenn and S. Parimoo, *Genomics*, 2001, **71**, 182.
262. R. J. Ward, M. T. Travers, S. E. Richards, R. G. Vernon, A. M. Salter, P. J. Buttery and M. C. Barber, *Biochim. Biophys. Acta.*, 1998, **1391**, 145.
263. M. Chung, S. Ha, S. Jeong, J. Bok, K. Cho, M. Baik and Y. Choi, *Biosci. Biotechnol. Biochem.*, 2000, **64**, 1526.
264. L. Bernard, C. Leroux, H. Hayes, M. Gautier, Y. Chilliard and P. Martin, *Gene*, 2001, **281**, 53.
265. A. J. Lengi and B. A. Corl, *Lipids*, 2007, **42**, 499.
266. L. Bernard, J. Mouriot, J. Rouel, F. Glasser, P. Capitan, E. Pujos-Guillot, J. M. Chardigny and Y. Chilliard, *Br. J. Nutr.*, 2010, **104**, 346.
267. P. J. Cameron, M. Rogers, J. Oman, S. G. May, D. K. Lunt and S. B. Smith, *J. Anim. Sci.*, 1994, 72, 2624.
268. L. C. St John, D. K. Lunt and S. B. Smith, *J. Anim. Sci.*, 1991, **69**, 1064.

269. J. H. P. Chang, D. K. Lunt and S. B. Smith, *J. Nutr.*, 1992, **122**, 2074.

270. G. S. Martin, D. K. Lunt, K. G. Britain and S. B. Smith, *J. Anim. Sci.*, 1999, **77**, 630.

271. K. Y. Chung, D. K. Lunt, H. Kawachi, H. Yano and S. B. Smith, *J. Anim. Sci.*, 2007, **85**, 380.

272. S. L. Archibeque, D. K. Lunt, C. D. Gilbert, R. K. Tume and S. B. Smith, *J. Anim. Sci.*, 2005, **83**, 1153.

273. M. Bionaz and J. J. Loor, *BMC Genomics*, 2008, **9**, 366.

274. L. Bernard, J. Rouel, C. Leroux, A. Ferlay, Y. Faulconnier and P. Legrand, *J. Dairy Sci.*, 2005, **88**, 1478.

275. L. Bernard, M. Bonnet, C. Leroux, K. J. Shingfield and Y. Chilliard, *J. Dairy Sci.*,**92**, 6083.

276. L. Bernard, C. Leroux, Y. Faulconnier, D. Durand, K. J. Shingfield and Y. Chilliard, *J. Dairy Res.*, 2009, **76**, 241.

277. P. G. Toral, Y. Chilliard and L. Bernard, *J. Dairy Sci.*, 2012, **95**, 6755.

278. J. E. Kinsella, *Lipids*, 1972, 7, 349.

279. E. E. Mosley, B. Shafii, P. J. Moate and M. A. McGuire, *J. Nutr.*, **136**, 570.

280. R. Gervais, J. W. McFadden, A. J. Lengi, B. A. Corl and P. Y. Chouinard, *J. Dairy Sci.*, 2009, **92**, 5167.

281. P. G. Toral, L. Bernard, C. Delavaud, D. Gruffat, C. Leroux and Y. Chilliard, *Animal*, 2013, 7, 948.

282. M. T. Nakamura and T. Y. Nara, *Annu. Rev. Nutr.*, 2004, **24**, 345.

283. H. Timmen and S. Patton, *Lipids*, 1988, **23**, 685.

284. M. Moche, J. Shanklin, A. Ghoshal and Y. Lindqvist, *J. Biol. Chem.*, 2003, **278**, 25072.

285. J. Shanklin, J. E. Guy, G. Mishra and Y. Lindqvist, *J. Biol. Chem.*, 2009, **284**, 18559.

286. W. C. Man, M. Miyazaki, K. Chu and J. M. Ntambi, *J. Biol. Chem.*, 2006, **281**, 1251.

287. J. M. Ntambi, *Prog. Lipid Res.*, 1995, **34**, 139.

288. C. M. Paton and J. M. Ntambi, *Am. J. Physiol. Endocrinol. Metab.*, 2009, **297**, E28.

289. B. G. Fox, J. Shanklin, C. Somerville and E. Münck, *Proc. Natl. Acad. Sci. U. S. A.*, 1993, **90**, 2486.

290. P. H. Buist, *Nat. Prod. Rep.*, 2004, **21**, 249.

291. Y. Lindqvist, W. Huang, G. Schneider and J. Shanklin, *The EMBO J.*, 1996, **15**, 4081.

292. A. F. Keating, J. J. Kennelly and F. Q. Zhao, *Biochem. Biophys. Res. Comm.*, 2006, **344**, 233.

293. F. S. Heinemann and J. Ozols, *Mol. Biol. Cell.*, 1998, **9**, 3445.

294. L. H. Baumgard, E. Matitashvili, B. A. Corl, D. A. Dwyer and D. E. Bauman, *J. Dairy Sci.*, 2002, **85**, 2155.

295. K. J. Shingfield, A. Äröla, S. Ahvenjärvi, A. Vanhatalo, V. Toivonen, J. M. Griinari and P. Huhtanen, *J. Nutr.*, 2008, **138**, 710.

296. I. J. Karlengen, O. M. Harstad, O. Taugbøl, I. Berget, A. H. Aastveit and D. I. Våge, *J. Anim. Physiol. Anim. Nutr.*, 2012, **96**, 1065.

297. F. Enjalbert, M. C. Nicot, C. Bayourthe and R. Moncoulon, *J. Nutr.*, 1998, **128**, 1525.
298. E. E. Mosley and M. A. McGuire, *Lipids*, 2007, **42**, 939.
299. A. Halmemies-Beauchet-Filleau, P. Kairenius, S. Ahvenjärvi, V. Toivonen, P. Huhtanen, A. Vanhatalo, D. I. Givens and K. J. Shingfield, *J. Dairy Sci.*, 2013, **96**, 5267.
300. J. M. Griinari, B. A. Corl, S. H. Lacy, P. Y. Chouinard, K. V. V. Nurmela and D. E. Bauman, *J. Nutr.*, 2000, **130**, 2285.
301. B. A. Corl, L. H. Baumgard, D. A. Dwyer, J. M. Griinari, B. S. Phillips and D. E. Bauman, *J. Nutr. Biochem.*, 2001, **12**, 622.
302. K. J. Shingfield, S. Ahvenjärvi, V. Toivonen, A. Vanhatalo and P. Huhtanen, *J. Nutr.*, 2007, **137**, 1154.
303. Y. C. Jin, Z. H. Li, Z. S. Hong, C. X. Xu, J. A. Han, S. H. Choi, J. L. Yin, Q. K. Zhang, K. B. Lee, S. K. Kang, M. K. Song, Y. J. Kim, H. S. Kang, Y. J. Choi and H. G. Lee, *J. Dairy Sci.*, **95**, 4286.
304. H. Kato, K. Sakaki and K. Mihara, *J. Cell Sci.*, 2006, **119**, 2342.
305. B. A. Corl, L. H. Baumgard, J. M. Griinari, P. Delmonte, K. M. Morehouse, M. P. Yuraweczc and D. E. Bauman, *Lipids*, 2002, **37**, 681.
306. J. K. Kay, T. R Mackle, M. J. Auldist, N. A. Thomson and D. E. Bauman, *J. Dairy Sci.*, 2004, **87**, 369.
307. E. Bichi, P. G. Toral, G. Hervás, P. Frutos, P. Gómez-Cortés, M. Juárez and M. A. de la Fuente, *J. Dairy Sci.*, 2012, **95**, 5242.
308. Y. Chilliard, A. Ferlay and M. Doreau, *Lives. Prod. Sci.*, 2001, **70**, 31.
309. Y. Chilliard, A. Ferlay, J. Rouel and G. Lamberet, *J. Dairy Sci.*, 2003, **86**, 1751.
310. A. Nudda, G. Battacone, M. G. Usai, S. Fancellu and G. Pulina, *J. Dairy Sci.*, 2006, **89**, 277.
311. A. Cabiddu, M. Decandia, M. Addis, G. Piredda, A. Pirisi and G. Molle, *Small Rum. Res.*, 2005, **59**, 169.
312. A. Nudda, M. A. McGuire, G. Battacone and G. Pulina, *J. Dairy Sci.*, 2005, **88**, 1311.
313. M. H. Gillis, S. K. Duckett, J. R. Sackmann and D. H. Keisler, *J. Anim. Sci.*, 2003, **81**(S2), 12.
314. E. Pavan and S. K. Duckett, *J. Anim. Sci.*, 2007, **85**, 1731.
315. D. L. Palmquist, N. St-Pierre and K. E. McClure, *J. Nutr.*, 2004, **134**, 2407.
316. D. Gruffat, A. De La Torre, J. Chardigny, D. Durand, O. Loreau and D. Bauchart, *Lipids*, 2005, **40**, 295.
317. J. E. Santora, D. L. Palmquist and K. L. Roehrig, *J. Nutr.*, 2000, **130**, 208.
318. J. J. Loor, X. Lin and J. H. Herbein, *Reprod. Nutr. Dev.*, 2002, **42**, 85.
319. X. Lin, J. J. Loor and J. H. Herbein, *J. Nutr.*, **134**, 1362.
320. C. Ip, S. Banni, E. Angioni, G. Carta, J. McGinley, H. J. Thompson, D. Barbano and D. E. Bauman, *J. Nutr.*, 1999, **129**, 2135.
321. S. Banni, E. Angioni, E. Murru, G. Carta, M. P. Melis, D. E. Bauman, Y. Dong and C. Ip, *Nutr. Cancer*, 2001, **41**, 91.

322. B. A. Corl, D. M. Barbano, D. E. Bauman and C. Ip, *J. Nutr.*, 2003, **133**, 2893.

323. A. L. Lock, C. A. M. Horne, D. E. Bauman and A. M. Salter, *J. Nutr.*, 2005, **135**, 1934.

324. K. R. Gläser, M. R. L. Scheeder and C. Wenk, *Eur. J. Lipid Sci. Technol.*, 2000, **102**, 684.

325. L. Zhang, L. Ge, S. Parimoo, K. Stenn and S. M. Prouty, *Biochem. J.*, 1999, **340**, 255.

326. J. Wang, L. Yu, R. E. Schmidt, C. Su, X. Huang, K. Gould and G. Cao, *Biochem. Biophys. Res. Commun.*, 2005, **332**, 735.

327. I. Salminen, M. Mutanen, M. Jauhiainen and A. Aro, *J. Nutr. Biochem.*, 1998, **9**, 93.

328. P. Weill, B. Schmitt, G. Chesneau, N. Daniel, F. Safraou and P. Legrand, *Ann. Nutr. Metab.*, 2002, **46**, 182.

329. R. O. Adlof, S. Duval and E. A. Emken, *Lipids*, 2000, **35**, 131.

330. A. Miller, E. McGrath, C. Stanton and R. Devery, *Lipids*, 2003, **38**, 623.

331. A. M. Turpeinen, M. Mutanen, A. Aro, I. Salminen, S. Basu, D. L. Palmquist and J. M. Griinari, *Am. J. Clin. Nutr.*, 2002, **76**, 504.

332. K. Kuhnt, J. Kraft, P. Moeckel and G. Jahreis, *Br. J. Nutr.*, 2006, **95**, 752.

333. E. E. Mosley, M. K. McGuire, J. E. Williams and M. A. McGuire, *J. Nutr*, 2006, **136**, 2297.

334. J. Kraft, L. Hanske, P. Möckel, S. Zimmermann, A. Härtl, J. K. G. Kramer and G. Jahreis, *J. Nutr.*, 2006, **136**, 1209.

335. S. A. Abdelmagid, S. E Clarke, J. Wong, K. Roke, D. Nielsen, A. Badawi, A. El-Sohemy, D. M. Mutch and D. W. L. Ma, *Nutr. Metab.*, 2013, **10**, 50.

336. D. Mozaffarian, H. Cao, I. B. King, R. N. Lemaitre, X. Song, D. S. Siscovick and G. S. Hotamisligil, *Ann. Intern. Med.*, 2010, **153**, 790.

337. D. Mozaffarian, M. C. de Oliveira Otto, R. N. Lemaitre, A. M. Fretts, G. Hotamisligil, M. Y. Tsai, DS Siscovick and J. A. Nettleton, *Am. J. Clin. Nutr.*, 2013, **97**, 854.

338. F. Noci, F. J. Monahan, P. French and A. P. Moloney, *J. Anim. Sci.*, 2005, **83**, 1167.

339. X. Shen, K. Nuernberg, G. Nuernberg, R. Zhao, N. D. Scollan, K. Ender and D. Dannenberger, *Lipids*, 2007, **42**, 1093.

340. T. Moreno, M. G. Keane, F. Noci and A. P. Moloney, *Meat Sci.*, 2008, **78**, 157.

341. S. F. Chin, J. M. Storkson, K. J. Albright and M. W. Pariza, *J. Nutr.*, 1994, **124**, 694.

342. B. Kamlage, L. Hartmann, B. Gruhl and M. Blaut, *J.Nutr.*, 1999, **129**, 2212.

343. J. Bassaganya-Riera, K. Reynolds, S. Martino-Catt, Y. Z. Cui, L. Hennighausen, F. Gonzalez, J. Rohrer, A. U. Benninghoff and R. Hontecillas, *Gastroenterology*, 2004, **127**, 777.

344. M. Q. Kemp, B. D. Jeffy and D. F. Romagnolo, *J. Nutr.*, 2003, **133**, 3670.

345. E. Devillard, F. M. McIntosh, S. H. Duncan and R. J. Wallace, *J. Bact.*, 2007, **189**, 2566.

346. L. Alonso, E. P. Cuesta and S. E. Gilliland, *J. Dairy Sci.*, 2003, **86**, 1941.

347. M. Coakley, R. P. Ross, M. Nordgren, G. Fitzgerald, R. Devery and C. Stanton, *J. Appl. Microbiol.*, 2003, **94**, 138.
348. E. Rosberg-Cody, M. C. Johnson, G. F. Fitzgerald, P. R. Ross and C. Stanton, *Microbiol.*, 2007, **153**, 2483.
349. C. Lay, L. Rigottier-Gois, K. Holmstrøm, M. Rajilic, E. E. Vaughan, W. M. de Vos, M. D. Collins, R. Thiel, P. Namsolleck, M. Blaut and J. Doré, *Appl. Environ. Microbiol.*, 2005, **71**, 4153.
350. R. I. Aminov, A. W. Walker, S. H. Duncan, H. J. M. Harmsen, G. W. Welling and H. J. Flint, *Appl. Environ. Microbiol.*, 2006, **72**, 6371.
351. E. Barrett, P. Fitzgerald, T. G. Dinan, J. F. Cryan, R. P. Ross, E. M. Quigley, F. Shanahan, B. Kiely, G. F. Fitzgerald, P. W. O'Toole and C. Stanton, *Plos One*, 2012, **7**, e48159.
352. F. M. McIntosh, K. J. Shingfield, E. Devillard and R. J. Wallace, *Microbiol.*, 2009, **155**, 285.
353. M. Doreau, V. Fievez, A. Troegeler-Maynadier and F. Glasser, *INRA Productions Animales*, 2012, **25**, 361.

CHAPTER 2

Use of CLA in Animal Feed

N. EVERAERT,[a] A. KOPPENOL[b,c] AND J. BUYSE*[c]

[a] University of Liège, Gembloux Agro-Bio Tech, Animal Science Unit, Passage des Déportés 2, 5030, Gembloux, Belgium; [b] ILVO Animal Sciences Unit, Scheldeweg 68, 9090, Melle, Belgium; [c] KU Leuven, Department of Biosystems, Laboratory of Livestock Physiology, Kasteelpark Arenberg 30 – box 2456, 3001, Leuven, Belgium
*Email: Johan.buyse@biw.kuleuven.be

2.1 Introduction

Conjugated linoleic acids (CLA) were first identified in rumen fluid as an intermediate of the biohydrogenation process.[1] In 1966, the primary microorganism responsible for the formation of CLA was identified, *Butyrivibrio fibrisolvens*.[2] In the intestinal contents of ruminants, a dozen or more CLA isomers have yet been characterized.[3] Most of the CLA found in tissues of ruminant animals are probably derived from the desaturation of *trans*-vaccenic acid. It was demonstrated that *trans*-vaccenic acid absorbed from the digestive tract, is a substrate for stearoyl desaturase and can be converted into the *cis*-9, *trans*-11 isomer of CLA.[4,5] Whereas the major CLA found in ruminants is largely the *cis*-9, *trans*-11 isomer, synthetic CLA preparations usually contain about the equal amounts of this isomer and the *trans*-10, *cis*-12 isomer.[6]

As anti-carcinogenic and anti-obesity effects of CLA were demonstrated in rodents *in vivo* and *in vitro*,[7] interest in the applications of CLA in both animal and human nutrition has grown exponentially ever since.[6] There are three main purposes of livestock studies. Firstly, an attempt to increase the CLA content in tissues destined for human consumption as a means to

RSC Catalysis Series No. 19
Conjugated Linoleic Acids and Conjugated Vegetable Oils
Edited by Bert Sels and An Philippaerts
© The Royal Society of Chemistry 2014
Published by the Royal Society of Chemistry, www.rsc.org

increase CLA intake of humans for health benefits.[6] Whereas *cis*-9, *trans*-11 CLA is supposed to be responsible for the anti-inflammatory effect,[8] *trans*-10, *cis*-12 CLA has been demonstrated to promote inflammation in primary cultures of differentiated human adipocytes.[9] The level of CLA in various food products were first reported in the early 1990s.[10–12] In general, CLA levels in various tissues of monogastric animals are much lower than those of ruminants. Secondly the performance of the animals is aimed to be ameliorated, *e.g.* reduction of body fat, improvement of growth and feed efficiency and better processing of the meat. Lastly, CLA might have a direct health benefit based on its anti-inflammatory and/or immune-ameliorating effects in farm animals.[13]

The present chapter will first discuss the effects of CLA supplementation in ruminants, followed by pigs, chicken and fish. The focus lies on changes due to CLA supplementation in zootechnical performance, body composition, fatty acid profiles of milk, fat, yolk or muscle. In addition, the effect of CLA on reproduction, and on the immune system is discussed.

2.2 Ruminants

2.2.1 Dairy Cows

2.2.1.1 Milk Yield

Many studies have shown that abomasally infused CLA causes milk fat depression[14–16] in lactating dairy cows and *trans*-10, *cis*-12 linoleic acid was identified as a responsible CLA isomer.[17,18] Moreover, research with rumen-inert CLA (RI-CLA) supplements (Ca^{2+} salts) indicated that RI-CLA also decrease milk fat when fed to cows either receiving a total mixed ration or rotationally grazed.[19–23] The *trans*-10, *cis*-12 CLA isomer has been shown to inhibit lipid synthesis in different animal models, with a dose-dependent decrease in milk fat synthesis in lactating dairy cows.[24,25] The biohydrogenation hypothesis of dietary CLA induced milk fat depression, proposes that under certain dietary situations, the pathways of rumenal biohydrogenation are altered, resulting in the production of unique fatty acid intermediates that act at the mammary gland as potent inhibitors of milk fat synthesis.[26] Sæbo *et al.* (2005) demonstrated that *cis*-10, *trans*-12 CLA also caused a reduction in milk fat synthesis[27] and Perfield *et al.* (2007) found that *trans*-9, *cis*-11 CLA abomasal infusion reduced milk fat yield.[28] For an overview of the possible mechanisms of CLA inducing milk fat depression, the reader is referred to Bauman *et al.* (2008).[25] However, results of CLA on the milk fat depression are not always consistent. Pappritz *et al.* (2011) fed 5 or 10 g of *trans*-10, *cis*-12 CLA per day from 1 d postpartum to the 26th week of lactation and observed a reduction in milk fat content of 0.7% but no significant decrease in milk fat yield.[29] They found a decreased dry matter intake (DMI) and consequently a compromised calculated energy balance in early lactation. Petzold *et al.* (2013) fed CLA (8 g/d of each *trans*-10, *cis*-12 CLA and *cis*-9, *trans*-11 CLA) 21 days antepartum until 60 days post-partum, and observed no effect on milk yield, milk protein or lactose and milk fat composition.[30] A possible explanation might be an insufficient protection of CLA against microbial degradation in the rumen, or a low milk yield.

In some studies, an increase in milk yield was observed,[16,31,32] which differs from the findings of Castañeda-Gutierrez *et al.* (2005) and Hutchinson *et al.* (2011).[33,34] Also Hutchinson *et al.* (2012) reported a decrease in milk fat concentration and yield, an increased milk yield and no effect on milk energy output by CLA supplementation from 0 to 60 days in milk.[32] Mackle *et al.* (2003) suggested that energy spared from a decrease in milk fat synthesis possibly has a greater positive effect on milk production in pasture-fed cows than cows fed a total mixed ration diet that could more closely meet energy requirements.[16] The studies of Mackle *et al.* (2003), Kay *et al.* (2006), Medeiros *et al.* (2010) and Hutchinson *et al.* (2012) support this hypothesis.[16,32,35,36] A clear positive relationship exists between moderate milk fat depression and milk yield response in early lactation that does not exist in later lactation, which is probably related to the animals' energetic status.[37,38] Immediately postpartum, a larger (about three times) amount of CLA is required compared with that needed in established lactation to obtain milk fat depression.[22,35] Therefore, Odens *et al.* (2007), used high RI-CLA (600 g/d) supplementations about 9 days before the expected calving date and ceased at 40 days of milk production.[39] As expected, CLA treatments decreased overall milk fat content, which became significant by day 8.[39] CLA treatment increased milk yield after the second week of lactation, and improved overall net energy balance. Indeed, supplemented cows were producing similar or even higher volumes of milk while consuming comparable quantities of feed throughout the entire trial. An improved energy balance at 2 weeks postpartum was also observed by Petzold *et al.* (2013) after supplementing CLA from 21 days antepartum until 60 days postpartum together with a high concentrate, resulting in an increased DMI.[30] Other studies agree with an improved energy status[34,40] and body condition score[34] with CLA supplementation although an absence of an effect on calculated energy balance has also been reported.[31,33] Hutchinson *et al.* (2012) suggested that the energy spared by CLA-induced milk fat depression, is partitioned primarily toward milk production rather than body reserves. In addition, they were the first to report a decrease in milk lactose concentration.[32]

The milk fat depression, and therefore reduction of the energy require-ment, may be used as an energy source to improve the synthesis of add-itional milk, to alter milk composition or to use the energy for a different physiological state (*i.e.* reproduction).[41] Moore *et al.* (2005) used heat stress, a time when nutrient availability may be limiting due to decreased dry matter intake, in combination with RI-CLA supplementation to test whether the decreased milk fat would allow for spared energy to be partitioned for increased production of milk or milk components.[41] Milk fat content and yield was decreased, confirming other trials in mid- and late lactating cows, but did however not affect protein or total milk synthesis.[19,20,42] On tropical pastures, several nutritional and climatic challenges may constrain milk production. Therefore, CLA supplementation might have a positive response in milk and milk protein production, due to the inhibition of milk fat secretion, resulting from CLA supplementation. Medeiros *et al.* (2010) supplemented Holstein x Zebu cows with a mixture of Ca salts of CLA from 28 to 84 days during the production period.[36] An increased milk production and volume, reduced milk fat concentration, fat production and milk energy concentration were observed. Interestingly, milk protein concentration and total milk protein increased, which contradicts several studies reporting no change in milk yield or milk protein content. During the first week after the CLA treatment, the milk fat content was still reduced in the previously CLA-supplemented cows, but disappeared thereafter. Indeed, restoration of normal milk fat synthesis took a few days to occur. Milk production and milk protein concentration were greater during the first week of the post-treatment period (85 to 112 days in milk), an episode dependent of the fat reduction.

2.2.1.2 Milk Fatty Acid Composition

Duodenal infusion of *trans*-10, *cis*-12 CLA (1.85 g/d) during 14 days decreased the secretion of all milk fatty acids (FA), including the odd-chain FA.[43] This decrease was proportionally greater for those FA synthesized *de novo*, resulting in a decrease in short and medium-chain FA percentages and an increase in preformed long-chain FA percentages. These results were in agreement with the studies of Shingfield and Griinari (2007) and Harvatine *et al.* (2009) concerning the inhibiting role of this isomer on *de novo* milk FA synthesis.[44,45] Generally, CLA supplementation trials reported a reduction in the production of FA of all chain lengths, but the major effect is seen on short-chain FA, resulting in a shift toward a greater content of longer chain FA.[16,20,32–34,36] Furthermore, the use of rumen intact-CLA increases the milk fat content of total CLA.[19,20,22,36,41] Moreover, supplementation with *trans*-10, *cis*-12 CLA results in a higher proportion of this isomer in milk fat.[22,30,31]

2.2.1.3 Body Composition

Von Soosten *et al.* (2011, 2012) supplemented primiparous lactating German Holstein cows from 1 until 105 days in milk production (DIM) with 10 g/d of *trans*-10, *cis*-12 CLA and 10 g/d *cis*-9, *trans*-11 CLA.[46,47] The decrease in retroperitoneal adipose depot from the first day in milk until 105 DIM tended to proceed slower in the CLA supplemented group, suggesting a CLA-induced deceleration of mobilization of this fat depot.[46] No effects on whole body composition (dry matter, ether extract, crude protein, ash, energy content of several body parts) were observed.[47] There was however a trend for a decreased body mass (fat and protein) mobilization suggesting a protective effect of CLA supplementation against excessive use of body reserves within 42 DIM. Continuous CLA supplementation until 105 DIM increased protein accretion and decreased heat production. These results suggested a more efficient utilization of metabolizable energy in early lactating dairy cows due to CLA supplementation.[47] In another similar study, but with lower doses of *cis*-9, *trans*-11 CLA (5.7 g/d) and *trans*-10, *cis*-12 (6.0 g/d), von Soosten *et al.* (2013) reported that the transfer of supplemented CLA isomers into the dairy cow's body was only marginal during the first 105 DIM, as the concentration of *trans*-10, *cis*-12 CLA due to CLA supplementation increased only in the retroperitoneal, mesenteric, subcutaneous adipose tissue, offal and mammary gland.[48]

2.2.1.4   Reproduction

Prolonged days to first ovulation, acyclicity, and subsequent reproductive failure, metabolic problems, and decreased production are associated with duration, magnitude and days of negative net energy balance and can be exacerbated by transition period disorders.[38,49,50] De Veth *et al.* (2009) performed a meta-analysis of five published studies on CLA and reproduction in early lactating cows and evaluated the association of CLA with time of first ovulation and time to conception.[51] The probability of cows becoming pregnant increased in a non-linear manner when *trans*-10, *cis*-12 CLA dose increased, with the optimal dose predicted to be 10.1 g/d. At the optimal dose, the probability of pregnancy increased by 26% compared with those animals receiving no CLA. Similarly, the time to conception was decreased in a nonlinear manner with increasing dietary *trans*-10, *cis*-12 CLA dose. The predicted optimal dose was 10.5 g/d of *trans*-10, *cis*-12 CLA. At this dose, the median time to conception was decreased by 34 days when compared with those cows receiving no CLA. A herd-scale evaluation of the effect of protected (lipid-encapsulated) CLA on reproductive performance in lactating dairy cattle revealed no effect on any estrous cycle characteristics or measures of reproductive performance.[32]

2.2.1.5   Immunomodulatory Effect

Studies in ruminants were only recently initiated to investigate the possible immunomodulatory effects of CLA. CLA supplementation did not affect serum haptoglobin (Hp) concentrations, a major acute phase protein in cattle, nor the mRNA abundance of Hp in liver and in four of the six measured fat depots. The observed decrease in Hp mRNA abundance due to CLA supplementation in two out of six different fat depots was considered to be of marginal importance.[52] There was no effect of CLA supplementation on the stimulation index (SI) of peripheral blood mononuclear cells (PBMC) obtained from primiparous cows *ex vivo* 42 and 105 days postpartum. However, the SI of splenocytes from the same animals were decreased following CLA supplementation.[53] In a long term CLA study with primiparous and pluriparous cows, the proportion of *trans*-10, *cis*-12 CLA of total fatty acids in PBMC increased, but there was no effect on the proportion of the *cis*-9, *trans*-11 isomer. These changes did not lead to differences in the mitogen induced activation of the cells. If the altered fatty acid profile impacts other immunological parameters remains to be investigated.[54] Dänicke *et al.* (2012) exposed dairy fetuses to CLA during 19 to 102 days of early fetal development through the diet of their mothers receiving 50 or 100 g/d of a CLA-containing fat supplement.[55] The stimulation ability of the PBMC of the calves originating from the CLA50 group was decreased, and that of the corresponding cows was also compromised. The FA pattern of erythrocytes of the calves from the CLA100 group was affected: the proportion of C16:0 was significantly increased whereas the proportions of 16:1 and C18:1 tended to decrease.

Altogether, the effects of CLA supplementation to dairy cows on the immune function are low and several effects seen in other species[56,57] could not be observed.

2.2.2 Beef Cattle

2.2.2.1 Zootechnical Performance

Park *et al.* (1997) and Pariza *et al.* (2001) suggested that muscle mass may be preserved or enhanced as a result of CLA-induced changes in the regulation of some cytokines that profoundly affect skeletal muscle catabolism and immune function.[58,59] Therefore, Schlegel *et al.* (2012) supplemented young Simmental heifers with CLA, and observed no influence on body weight gain, feed efficiency, carcass conformation or fatness.[60] These results were further confirmed by Alberti *et al.* (2013) in young Holstein bulls.[61] These findings are in contrast with observations in mice, rats and pigs where CLA supplementation caused a marked improvement of feed efficiency, increased whole-body protein, and profoundly reduced body fat accretion,[58,62,63] but in agreement with another study in older heifers of heavy body weight.[64]

Schiavon *et al.* (2010) tested the hypothesis that CLA might have protein-sparing effects in double-muscled Piemontese young bulls, animals with a greater potential for lean growth, by providing a low-protein (LP) or control diet, and supplemented or not with rumen-protected CLA (80g/d).[65] The addition of CLA exerted a significant reduction in the *in vivo* fatness score, but only during the middle part of the growth period and not during the finishing period. CLA addition did not have any influence on dry matter intake, in agreement with Gillis *et al.* (2004a).[66] The CLA supply had a positive effect on leg and hoof health conditions. CLA administration decreased feed efficiency of animals fed the control diets, but increased feed efficiency of those reared on the LP diets. Therefore, under conditions of a reduced intake of dietary protein, these results support the hypothesis of Park *et al.* (1997) and Pariza *et al.* (2001) that muscle mass could be preserved or enhanced by CLA.[58,59]

2.2.2.2   Fatty Acid Composition

Increased tissue concentrations of *trans*-10, *cis* 12 CLA have been observed in several studies after CLA supplementation, whereas no effect was observed on *cis*-9, *trans*-11 CLA concentrations.[60,66–68] The latter is in contrast to findings in non-ruminants.[69–71] The *cis*-9, *trans*-11 CLA can be synthesized endogenously by Δ^9-desaturation of *trans*-vaccenic acid that is catalysed by Δ^9-desaturatase. In beef, CLA feeding was reported to inhibit Δ^9-desaturation, probably resulting in a reduced endogenous formation of *cis*-9, *trans*-11 CLA from *trans*-vaccenic acid by Δ^9-desaturatase. Possibly, the increased uptake of *cis*-9, *trans*-11 CLA from the CLA supplement and its incorporation into the tissue lipids, compensated for the reduced endogenous formation of this isomer. This is further supported as the proportions of C18 : 1 and total MUFA (that are also formed by Δ^9-desaturatase) were reduced, whereas those of C18 : 0 and total SFA were increased in muscle and adipose tissue in CLA-fed heifers.[60]

2.3   Pigs

Pigs are monogastric animals, and their acid stomach content contains few microorganisms relative to ruminant animals. In pigs, only a small amount of CLA is produced by way of bacterial biohydrogenation and pork meat usually contains a limited amount of CLA (0.1–0.2 mg per g fatty acids).[72]

2.3.1   Zootechnical Performance

The effects of CLA on growth performance and carcass fat deposition of pigs have been reviewed by Azain (2003) and Corino *et al.* (2005).[6,73] Inconsistencies between studies have been attributed to the breed of pigs, dietary factors such as the source of CLA and the dietary fat content, the duration of feeding or the gender. Azain (2003) concluded that CLA supplementation to pigs with >23 mm subcutaneous fat thickness at 100 kg body weight reduced carcass fat, but not when fat thickness was <20 mm.[6] In addition, barrows seem to respond more than gilts,[74] and a greater response was obtained with low-energy diets compared to diets with added fat.[75] In some studies, an improvement in feed efficiency of pigs fed dietary CLA was observed,[76–80] whereas other studies did not report a difference compared to control animals,[81] even with a high inclusion (4%) of CLA.[82] The increase in feed efficiency is likely due to a decrease in fat deposition with a concomitant increase in muscle protein accretion.[80]

2.3.2   Body Composition

Backfat depth was reduced when CLA was supplemented,[75,79,83,84] or not affected.[82,86] An increase in lean content in pigs fed CLA was observed in some studies,[75,79,80,82–84] but not in other studies.[82,85] Inconsistencies also occur for intramuscular fat (IMF): no effect of CLA supplementation,[74,77,82,86] or an increase of IMF in pigs supplemented with CLA was found.[79,87–94]

Apart from the CLA concentration and supplementation time, the slaughter weight might also contribute to these discrepancies.[95] They observed an increase in IMF content in *Longissimus dorsi*, when CLA was supplemented to heavy pigs (>125 kg) with doses up to 2% CLA while, so far, no effect on IMF concentration have been found in pigs slaughtered at heavier weights, including less than 1% CLA.[96–99]

2.3.3 Fatty Acid Composition

The effect of CLA on FA composition differed between muscle types,[82] when finishing pigs were fed a high dose (4%) of CLA, confirming the study of Intarapichet *et al.* (2008).[100] Indeed, Tous *et al.* (2013) reported that the percentage of SFA increased in *longissimus thoracis* muscle (LT), *semimembranosus* muscle (SM), liver and LT subcutaneous fat, in agreement with Joo *et al.* (2002), Bee (2000a), Eggert *et al.* (2001) and Weber *et al.* (2006), whereas MUFA were only reduced in LT and LT subcutaneous fat, and PUFA in SM, liver and LT subcutaneous fat.[69,80,82,91,101] The reduction of PUFA in LT subcutaneous fat is in agreement with the findings of Wiegand *et al.* (2002) and Sun *et al.* (2004).[80,89] Although the CLA supplement contained similar quantities of *cis*-9, *trans*-11 CLA and *trans*-10, *cis*-12 CLA, the concentration deposited in all tissues was higher for the *cis*-9, *trans*-11 isomer as was previously observed by Ostrowska *et al.* (2003), Lo Fiego *et al.* (2005), Bee *et al.* (2008) and Cordero *et al.* (2010).[95,98,102,103]

2.3.4 Immunomodulatory Effects

The immunomodulatory capacity of CLA has been inferred from the increase in immunoglobulin production,[104–106] suggesting a beneficial effect on the humoral immune system. However, dietary CLA did not reverse the stress caused by LPS, resulting in a decreased growth performance.[106,107] Corino *et al.* (2009) reported that piglets weaned from sows fed a diet supplemented with 0.5% CLA had greater BW and increased total IgG production than pigs weaned from sows that did not receive dietary CLA.[108] In addition, CLA influenced CD4+ and CD8+ lymphocytes percentages in swine,[105,109–111] and decreased proinflammatory cytokines in response to immune challenge,[105] thus influencing the cellular immune response. In addition, Hontecillas *et al.* (2002) supplemented CLA in the diet before the induction of colitis and observed a decreased mucosal damage, a maintainance of cytokine profiles (interferon-γ (IFN-γ), interleukin-10 (IL-10)) and of lymphocyte subset populations, resembling those of noninfected pigs, and attenuated growth reduction.[112] Moreover, Bassaganya-Riera *et al.* (2003) observed a decrease in IFN-γ production by CD4+ Th1 cells in pigs infected with circovirus that also received 1.33% dietary CLA.[111] They concluded that isomers *cis*-9, *trans*-11 and *trans*-10, *cis*-12 CLA have important immunomodulatory properties. Lai *et al.* (2005b) found inhibitory actions of CLA on IL-1ß, IL-6, tumor necrotic factor-α (TNF-α) and increased IL-10 levels after *in vivo* lipopolysaccharide (LPS) stimulation in pigs supplemented with CLA for only 14 days.[113] They attributed the effects mainly to *trans*-10, *cis*-9 CLA. However, in the study of Moraes *et al.* (2012), CLA supplementation did not affect lymphocyte proliferation nor the percentage of CD4+ and CD8+ cells after an LPS challenge.[106] In agreement, Wiegand *et al.* (2011) could not observe an altered lymphocyte subpopulation cell distribution in CLA supplemented growing-finishing pigs, without an immune challenge.[114] The reported effects of CLA on the immune system range from stimulation to inhibition and lack of influence.[110] Discrepancies between the studies can be attributed to factors such as cell culture condition and cell type, and for *in vivo* experiments diet composition, species and age.[115] In addition, the prevalent CLA isomer and the dose of supplementation may also affect the biological activity.[59]

2.3.5 Reproduction and Transfer to the Progeny

CLA has been demonstrated to induce milk fat depression as occurs in cows, and CLA supplementation altered the FA composition of sow colostrum.[116] In agreement, Peng *et al.* (2010) found that CLA supplementation of sows during the last 50 days of gestation and throughout a 26 day lactation period, increased the concentrations of total SFA, but decreased total MUFA in colostrum.[117] However, it increased the concentrations of total SFA but had no effect on the total PUFA in milk, which is consistent with previous studies.[101,118] The CLA isomer amounts were increased in colostrum and milk, but did not alter the birth weight and body weight gain of piglets,[117] which was in agreement with findings of Poulos *et al.* (2004).[119] However, Patterson *et al.* (2007) observed that piglets from sows fed diets supplemented with 2% CLA were lighter than piglets from control sows.[120] In suckling piglets from sows supplemented with CLA, umbilical cord blood, plasma, backfat, and the *Longissimus* muscle CLA concentrations of piglets were also increased,[117] which is consistent with other reports.[101,118–121] However, *trans*-10, *cis*-12 CLA rather than *cis*-9, *trans*-11 CLA was measured in the umbilical cord blood, suggesting that CLA may be transported from the sow to the fetus in an isomer-specific manner.[117] On the other hand, Krogh *et al.* (2012) observed that dietary CLA supplementation reduced colostrum yield and piglet performance, but stimulated milk yield, without affecting the time of initiation of lactation.[122] CLA supplementation to gestating sows did not affect sow performance characteristics such as body weight and daily feed intake.[108,119,123]

2.4 Poultry

2.4.1 Broilers

2.4.1.1 *Zootechnical Performance*

In general CLA is known to reduce feed intake, however results in the literature are contradicting. A level of 0.1% CLA supplementation did not influence feed intake, whereas 0.2% CLA supplementation in the first 3 weeks enhanced daily feed intake.[124] However, Javadi *et al.* (2007) observed a depressed feed intake after feeding 1% CLA diet during 3 weeks.[125] Suksombat *et al.* (2007) noted no effect of CLA (0.5, 1.0 and 1.5%) on daily feed intake (42 days).[70] Kawahara *et al.* (2009) also recorded no differences in feed intake when feeding CLA for 28 days.[126]

CLA is a potent inhibitor of body fat accumulation in mice as reported by Pariza *et al.* (1996) and hence may act as a fat to lean repartitioning agent.[127] The reduction in the ratio of fat to lean should be reflected in an improved feed conversion ratio. Lower feed conversion ratios have been reported in broilers fed CLA.[128,129] 0.2% CLA supplementation improved the feed to gain ratio according to Halle *et al.* (2012).[124] Suksombat *et al.* (2007); Zhang *et al.* (2007) and Bölükbasi *et al.* (2006) observed an increased feed to gain ratio with increasing CLA, whereas Javadi *et al.* (2007) did not observe any significant differences.[70,125,130,131]

A level of 0.2% CLA supplementation did not significantly influence body weight.[124] Javadi *et al.* (2007) neither found differences in body weight gain in a 3 week study with 1% CLA supplementation.[125] Zhang *et al.* (2007) also observed that CLA supplementation for 126 days (0.25, 0.5, 1.0 and 2.0%) did not significantly influence body weight of chickens, which was in agreement with Kawahara *et al.* (2009) and Du and Ahn (2002).[126,130,132] In the trial of Suksombat *et al.* (2007), however, daily weight gain was significantly and linearly decreased when supplementing 0.5, 1.0 or 1.5% CLA, in agreement with the study of Szymczyk *et al.* (2001).[70,129] In contrast, Tanai *et al.* (2011) measured higher body weights at the age of 42 days after feeding 0.5 and 1.0% CLA diets but yet decreased body weights after feeding 2% CLA.[133] Birds given 1.0, 2.0 or 3.0% CLA had greater body weights and body weight gains than the control group in the study of Bölükbasi *et al.* (2006).[131] The discrepancy of effects of CLA on animal growth performance could be ascribed to differences in dietary CLA concentration, type of isomers of CLA, feeding periods, and nutritional status of the animals.

2.4.1.2 Body Composition

No effects of up to 1.5% CLA on body weight, carcass percentage, and percentages of thigh, boneless thigh, pectoralis major and pectoralis minor weight were observed in the study of Suksombat *et al.* (2007).[70] However, percentages of liver weight were increased and percentages of drumstick and boneless drumstick were decreased by dietary CLA. Bölükbasi (2006) found no difference in heart liver and leg weight, but a decreased wing weight, whereas breast weight responded differently to increasing levels of dietary CLA.[131] Du and Ahn (2003) measured significantly increased liver weights of broilers after CLA feeding but there was no difference in liver fat content among the different CLA treatments.[134]

As a result of increased consumer demand for leaner meat products, there is considerable interest in including CLA in poultry feeds to improve lean production efficiency and quality. However, the effect of CLA on body fat in the avian species has not been as consistent as in mammals. Javadi *et al.* (2007) reported a higher proportion of fat in broilers fed 1% CLA for 21 days, in contradiction with Du and Ahn (2002), who reported that the whole body fat content decreased from 14.2% in the control to 11.9 and 12.2% in the 2 and 3% CLA groups, respectively.[125,132] Zhang *et al.* (2007) showed that feeding 1 to 2% CLA for 126 days to Beijingyou chickens resulted in decreased intramuscular fat in breast and thigh as well as decreased abdominal fat weight.[130] Similar results on abdominal fat weight or percentage were shown by Szymczyk *et al.* (2000; 2001), and Suksombat *et al.* (2007).[70,129,135] In the trial of Halle *et al.* (2012), however, the supplementation of 0.1 and 0.2% CLA diet did not lead to changes in the abdominal and visceral fat percentages compared to the control animals.[124] Moreover, no effect was observed in the protein and fat content of the breast muscle.[124] The latter was in agreement with the study of Suksombat *et al.* (2007) and Bölükbasi (2006), who concluded that supplementation of dietary CLA up to 3% did not affect the lipid content of muscles.[70,131] Kawahara *et al.* (2009) stated that total lipid and triglyceride concentrations in breast meat tended to be decreased in 2% CLA-fed broilers.[126] These opposing results with regard to body fat could possibly be explained by the onset of CLA feeding during the growth phase of the chickens and/or length of CLA feeding.

2.4.1.3   Fatty Acid Composition

It has been well demonstrated that dietary CLA is readily incorporated in tissue lipids in broilers.[70,129,131,132] Supplementing diets with CLA modified the fatty acid composition of breast muscle as the proportion of CLA was increased.[124,130,136,137] Feeding incremental levels of CLA up to 2% resulted in linear increases in the concentration of CLA isomers in breast- and drumstick meat of broilers.[124,126,138] Similarly, Aletor *et al.* (2003), Du and Ahn (2002) and Tanai *et al.* (2011) investigated the effects of increasing CLA concentrations in the feed and found a linear increase of CLA in tissue samples.[132,133,139] The concentration of individual CLA isomers in muscle lipid, however, does not completely reflect the CLA isomer content of the diet. Some isomers are more effectively mobilized compared to others.[140] It was found that incorporation of individual CLA isomers into body lipids differed as indicated by preferential incorporation of *cis*-9, *trans*-11 CLA at the expense of *trans*-10, *cis*-12 CLA.[70,126,129] In the study of Halle *et al.* (2012), the *cis*-9, *trans*-11 CLA isomer accounted for 56–59% of the total CLA in breast muscle lipids, confirming the results of Szymczyk *et al.* (2001).[124,129]

   Several reports indicated that CLA supplementation increased the amount of SFA and decreased the MUFA fraction in tissues of broilers.[126,129,130,132,138,141] The decrease of proportion of MUFA results from a reduced delta-9 desaturase activity in consequence of feeding CLA.[142,143] Studies showed that the *cis*-9, *trans*-11 CLA isomer does not inhibit the activity of the delta-9 desaturase whereas the *trans*-10, *cis*-12 CLA isomer is biologically highly active.[143,144] Zhang *et al.* (2007) measured increased PUFA proportion of breast muscle after CLA supplementation, as confirmed by Kawahara *et al.* (2009).[126,130] In particular, the composition of DHA (22 : 6 n-3) and EPA (20 : 4 n-6) significantly increased, although that of AA (20 : 4 n-6) did not change.[126] Javadi *et al.* (2007) and Szymczyk *et al.* (2001) however found a decreased PUFA proportion in the abdominal fat and muscles.[125,129]

2.4.1.4 Immunomodulatory Effect

Research in nutritional immunology revealed an important role for dietary CLA in diminishing inflammation-associated diseases.[145–147] Data about the effects of dietary CLA in chicken immunity are scarce. It has been suggested that CLA protected the catabolic responses against endotoxin in chicks.[148,149] Although some studies in mammals demonstrate that dietary CLA enhances immunoglobulin production,[150,151] the effect of CLA on antibody production of chicks is still unclear. Plasma Ig G concentrations in chicks fed CLA were higher than those in chicks fed a basal diet, showing dietary CLA enhanced antibody production in broilers.[149] Cook *et al.* (1993) showed that antibody production in chicks against sheep red blood cell (SRBC) was not affected by feeding 5% CLA, whereas Takahashi *et al.* (2003) reported that 1% dietary CLA enhanced anti-SRBC antibody production of broilers.[148,152] CLA has been shown to alleviate immunosuppression. CLA, predominantly the *cis*-9, *trans*-11 isomer, enhanced immune function under the cyclosporine A-immunosuppressive status in chickens, which prevents the synthesis of T-cell cytokines by blocking a late-stage signaling pathway initiated by the T-cell receptor, which in turn affects the production of IL-2.[153] Hence, T-cell proliferation is affected.[154,155] Furthermore, CLA have been shown to have anti-inflammatory activities. CLA decreased the production of inflammatory cytokines such as TNF-alpha and IL-6 in both human and animal models.[156] The immunoregulatory actions of CLA to viral disease pathogenesis and immune responses in chicken were investigated by Long *et al.* (2011).[157] They created a model of infectious bursal disease virus (IBDV) in chickens and determined the effects of dietary CLA on bursal lymphocyte proliferation, antibody titers to IBDV, and mRNA expression of proinflammatory cytokine profiles (*i.e.*, IL-6 and IFN-gamma). Compared with the 1% CLA diet, lymphocytes depletion was more accentuated in chickens fed the control diet, whereas IFN-gamma and IL-6 mRNA relative expression were upregulated. Additionally, histopathological examination of the bursa revealed that the pathological changes tended to be more severe in infected chickens fed the control diet, which were also characterized by a decrease in lymphocyte proliferation. The results of this study indicated that dietary CLA enhanced immune function in chicken, particularly those related to the IBDV-immunosupppressive status. Furthermore, at the molecular level, the immunoregulatory functions of CLA on chickens are attributable mainly to the anti-inflammatory properties of CLA and are mediated, at least in part, through suppressing IBDV-specific proinflammatory cytokines mRNA abundance. Another study of Zhang *et al.* (2005b) also suggested that dietary CLA could enhance innate and cellular immune response of broiler chicks.[115] Chicks fed up to 1% CLA produced more lysozyme activity in serum and spleen, showed an enhanced peripheral blood lymphocyte and mononuclear cell proliferation and an affected phagocytic ability. Systemic and peripheral blood lymphocytic synthesis of prostaglandin E_2 in chicks fed 10% CLA was significantly decreased, compared to non-supplemented chicken. Antibody production to sheep red blood cell and bovine serum albumin were elevated, but antibody titers to Newcastle disease virus were not influenced by CLA supplementation.

2.4.2   Layers

Efficient accumulation of CLA in yolk as 'functional' food, has not been feasible because excessive CLA intake resulted in a variety of adverse effects due to the change in physiological membrane constituents, especially on the reproduction processes, as well as changes in the egg quality in birds.

2.4.2.1   Zootechnical Performance

Literature on the zootechnical parameters when feeding laying hens a CLA diet is not straightforward. Shang *et al.* (2004) described that with increased dietary CLA, feed intake, body weight gain, rate of egg production, egg weight and feed efficiency all decreased linearly.[158] Suksombat *et al.* (2006), however, found that average daily feed intakes were similar when feeding 27-week-old layers 0, 1, 2, 3 or 4% CLA for 140 days, although hens fed with 4% CLA tended to consume less feed than the other hens.[159] Yin *et al.* (2008) also found that feed intake and egg weight of both Brown Dwarf and White Leghorn hens were decreased by 5% CLA supplementation, without affecting egg production and feed efficiency.[160] The reduced feed intake with CLA supplementation is in agreement with Ahn *et al.* (1999) and Szymczyk and Pisulewski (2003).[161,162] As feed consumption is positively related to egg weight, body weight gain and yolk weight, its reduction may account for that decrease. Body weight gain was decreased by feeding even 2.5% dietary CLA.[160] Shang *et al.* (2004) added up to 7% CLA for 4 weeks to a laying hen diet and recorded significant decreases in feed efficiency, egg weight and egg production.[158] The latter was in agreement with Jones *et al.* (2000), who reported that egg production was significantly decreased even with levels as low as 0.5 to 1% CLA in the diet.[163]

2.4.2.2   Reproduction

Egg production and fertility of laying hens were not significantly altered when feeding 0.5% CLA, however hatchability was reduced from 93 to 9.9% when the diet contained 0.5% CLA.[164] Embryonic mortality before 13 days of incubation was 70.77% for CLA eggs compared to 2.62% for control eggs.[164,165] Dietary CLA was shown to cause complete mortality in chick embryos from laying hens fed 0.5% CLA.[166–168] According to Aydin and Cook (2009) maternal CLA is not directly toxic for the developing chick embryo, but causes embryonic mortality by altering fatty acid composition of the egg yolk in the chicken.[169] When laying hens were fed a diet containing 0.5% CLA, yolks hardened and fertile eggs failed to hatch, possibly due to increased $16:0$ and $18:0$ and decreased $18:1$ n-9 concentrations in the yolk. The adverse effects of CLA may be due to the increased level of SFA.[167] If the high ratio of SFA to unsaturated fatty acids (UFA) significantly interfered with the absorption of yolk fatty acids by the developing embryo, embryonic mortality would ensue, because 90% of the embryonic energy requirement is derived from yolk fat oxidation. It is not known whether the changes in fatty acid profiles of yolk are the result of a mixture of CLA isomers or specific isomers. Lee *et al.* (1998) suggested that the *trans*-10, *cis*-12 CLA isomer reduced the ratio of MUFA to SFA by inhibiting liver stearoyl-coenzyme A desaturase enzyme activity, an enzyme that catalyses the insertion of a double bond between the C9 an C10 atoms of either $16:0$ or $18:0$ in the formation of $16:1$ n-7 and $18:1$ n-9, respectively.[142,170]

In addition, the adverse effects of dietary CLA on the hatchability of eggs were reversible within 6 days when the diet containing CLA supplied to the laying hens was replaced by a control diet.[167] Hatch of eggs obtained from CLA fed hens could be rescued if the hen's basal diet contains at least 3% oils exhibiting a FA composition of either MUFA or PUFA.[165,168] Maternal dietary CLA led to reduced body fat and heavier residual yolk sacs in newly hatched chicks in comparison to controls.[169,171,172] Lipid is primarily taken up from the yolk during the last week of incubation (days 14 to 21) before hatch.[173–176] Several of the enzymes, including lipoprotein lipase (lipid hydrolysis), stearoyl-coensyme A desatruase-1 (conversion of $18:0$ to $18:1$ n-9) and acyl coenzyme A:cholesterol acyltransferase (formation of cholesteryl esters) shown to be inhibited by CLA in other animal models, are heavily involved in the uptake, remodeling and repackaging of yolk lipid in the yolk sac membrane in the process of lipid transport to the embryo.[176–179] It is suggested that maternal dietary CLA could reduce the capacity of an avian embryo to transport yolk lipid during development, resulting in embryonic mortality due to lack of energy needed to maintain normal embryo growth and development. Indeed, Leone *et al.* (2010) showed that maternal feeding of CLA interferes with the normal transfer of lipid from the yolk sac to the developing embryo.[164]

Maternal feeding of CLA has been shown to improve the growth of the progeny in several mammalian model species.[120,180,181] However, Leone *et al.* (2009) reported no effect on progeny growth, feed intake or feed efficiency at 3 weeks in offspring of 0.5% CLA fed laying hens.[182]

2.4.2.3 Egg Quality

Opposing results on egg quality parameters were found in literature. Suksombat *et al.* (2006) reported a reduction in egg, yolk and albumen weights by 4% CLA supplementation.[159] Yolk weight and its ratio to egg weight were also decreased by 5% CLA addition in the trial of Yin *et al.* (2008), whereas the ratio of albumen to egg was increased.[160] Yolk color significantly decreased as dietary CLA increased. Shell thickness and Haugh units (albumen height) were not affected by CLA.[159] Whereas shell thickness, shell strength, yolk color, yolk index, egg diameter and Haugh units were negatively affected when CLA was fed alone, but the quality was improved when CLA was combined with some other fatty acids.[183] Fatty acid co-supplementation may have led to homeostasis of lipid metabolism in the liver and thereby maintained egg quality during CLA supplementation.

The increase in yolk FA saturation[141,162,167,184–186] affected firmness of the yolk of cooked eggs[161,167,186] and egg sensory properties.[186,187] CLA supplementation increased yolk moisture and firmness, impaired the sensory quality of eggs[167,187] and resulted in discoloration of yolk and albumen.[167,188] Lee (1996) suggested that the discoloration of eggs from hens fed a diet containing 0.5% CLA could result from an increased permeability of the vitelline membrane of eggs due to CLA-induced changes in fatty acid composition.[188] Abou-ashour and Edwards (1970) suggested that a higher 18:0 content of the egg yolk fatty acid would probably increase the permeability of the vitelline membrane.[189] The pink discoloration of albumen could be attributed to a combination of ovotransferin, egg albumen protein and yolk iron that diffuses into the egg albumen.[189]

Another prominent biological effect of dietary CLA was the effect on yolk and albumen pH of eggs stored at 4 °C for 2 or 10 weeks. Yolk pH was higher and albumen pH was lower in the eggs from CLA-fed hens. However, when eggs from CLA-fed hens were stored at room temperature (21 °C), the pH of the albumen and yolk was not different from control eggs.[167]

Eggs from 0.5% CLA-fed laying hens had greater concentration of Mg^{2+}, Na^+ and Cl^- and lower concentrations of Ca^{2+}, Zn^{2+} and Fe^{3+} in the yolk.[167] In contrast, eggs from laying hens fed the CLA diet had higher concentrations of Fe^{3+}, Ca^{2+} and Zn^{2+} and lower concentration of Mg^{2+}, Na^+ and Cl^- in the albumen. So, dietary CLA causes calcium and zinc to move from yolk into albumen and magnesium and sodium to move from albumen into yolk. Due to changes in permeability of vitelline membrane as a result of CLA supplementation, minerals can move down their concentration gradient. The migration of alkaline ions such as Na, K, and Mg, might permute with hydrogen ions in the yolk, and cause yolk pH to increase. This pH change could then induce denaturation of the protein in the yolk, increasing yolk firmness.

CLA co-supplemented with other FA was reported to prevent the apparent adverse effects on egg quality and egg production.[183]

2.4.2.4 Fatty Acid Composition

As a result of the benefits of CLA to human health, food products rich in CLA have been extensively investigated in recent years. Egg yolk is a good carrier of fatty acids. Feeding laying hens a diet containing CLA was shown to increase the level of yolk CLA in a dose-dependent manner.[161,162,169,184,190,191] Chamruspollert and Sell (1999) found that with increasing dietary CLA, the CLA concentration of egg yolk lipids increased linearly.[184] Shang *et al.* (2004), even concluded on a quadratically increase of CLA concentration in the yolk lipids, with increasing dietary CLA.[158] Chamruspollert and Sell (1999) showed that chicken eggs could be enriched in CLA to as high as 11% by feeding hens 5% CLA.[184] The average deposition rate was 18%, but this decreased as the rate of inclusion of other fats in the diet increased.[141,167,185,186] Results of the study of Yin *et al.* (2008) showed that the concentration of individual CLA isomers in egg yolks did not completely reflect those of the diet.[160] The percentages of *cis*-9, *trans*-11 and *trans*-10, *cis*-12 in the CLA source were similar, whereas their percentages in the yolk lipids differed, which was consistent with the report of Chamruspollert and Sell (1999).[184] Szymczyk and Pisulewski (2003) also observed that compared with other isomers of CLA, the *cis*-9, *trans*-11 isomer of CLA was preferentially incorporated into yolks, whereas the incorporation of *trans*-10, *cis*-12 was less efficient.[162]

The ingestion of CLA by laying hens not only resulted in the incorporation of CLA isomers into egg yolk, but also increased the level of SFA and decreased the level of MUFA.[158,159,161,162,168,184,187,188,190–194] Adding 0.5% CLA to the diet, resulted in about a 32% decrease in MUFA and a 34% increase in SFA of egg yolk.[167,168] Although dietary CLA had no effect on the concentration of yolk 18:2 n-6, the level of 20:4 n-6 in the yolk was significantly lower in eggs from CLA-fed hens. In the research performed by Yin *et al.* (2008) all fatty acids (except myristoleic acid) in yolk lipids were significantly altered by dietary CLA supplementation.[160] Feeding 2.5% or more of dietary CLA increased the contents of myristic, palmitic and stearic acids in the yolk lipids, but decreased the contents of palmitoleic, oleic, linoleic, linolenic, arachidonic, docosahexaenoic, and nervonic acids. Yang *et al.* (2002) reported that dietary CLA improved the concentration of linolenic acid in yolk lipids.[194] However, Raes *et al.* (2002) reported that the concentration of linolenic acid in yolk lipids was not affected by feeding 1% CLA.[141] In this study, concentrations of linolenic, arachidonic, and docosahexaenoic acids in yolk lipids decreased with increasing dietary CLA, which was consistent with other reports.[161,162,184] The reduction of these fatty acids might be related to inhibition of the stearoyl-coenzyme A desaturase enzyme system in the liver.[142,190]

2.5 Fish

As CLA has been shown to exert potential health benefits in mammals, studies have been conducted on fish to investigate the accumulation of CLA in their tissues when diets are supplemented with this particular fatty acid aiming production of 'functional food'. The discussion of the effect of CLA on growth performance and immunity, however, is rather difficult, due to research on many different fish species.

2.5.1   Zootechnical Performance

Zuo *et al.* (2013), Jiang *et al.* (2009) and Kennedy *et al.* (2007) failed to observe any effect of CLA supplementation on feed intake of large yellow croaker, Jian carp and rainbow trout, respectively.[195–197] In contrast, several studies reported that dietary CLA addition reduced feed intake of several fish species.[198–201]

Zuo *et al.* (2013) and Jiang *et al.* (2009) found no effect of CLA supplementation on feed efficiency of large yellow croaker and Jian carp, respectively.[195,196] However, Tan *et al.* (2013) noticed that CLA addition improved feed efficiency.[202] An ameliorated feed efficiency in Atlantic Salmon was also reported by Berge *et al.* (2004).[203] In contrast, feed efficiency was not affected by dietary CLA inclusion in other fish species.[197,200,205–210]

With respect to growth performance, results vary significantly among different fish species. It has been shown that growth performance was not affected by the inclusion of CLA.[196,197,200,203–211] However, growth performance was significantly decreased by CLA at the level of 1% in hybrid striped bass[198] and yellow catfish,[199] and by 2% CLA in gilthead sea bream[201] and tilapia.[215] Growth-promoting effects of dietary CLA have also been reported, with weight gain of common carp and channel catfish being increased by the level of 1% CLA.[202,213,214] The growth of fish was significantly enhanced by moderate inclusion of CLA, but decreased when the level of this particular fatty acid increased.[195,213,214] On the contrary, some studies have reported the deleterious effects of CLA on the growth performance of fish.[198,199,212] As dietary CLA increased from 0 to 0.42%, weight gain rate increased and then decreased with dietary CLA increasing from 0.42 to 1.7%.

2.5.2   Body Composition

According to Belury *et al.* (2003) *trans*-10, *cis*-12 CLA rather than *cis*-9, *trans*-11 CLA is the metabolite responsible for the lipid-lowering effect of CLA.[215] Decreased lipid concentration in the whole body or liver was observed in several studies.[197–199,202,204,216] Makol *et al.* (2009) reported also a reduced perivisceral fat concentration.[200] However, other studies report an increased whole body and muscle lipid content by feeding CLA,[195,209] or no effect on liver lipid content.[209,216,217] The differential physiological effect of CLA may be attributed to differences in metabolic rate, administered dose of CLA, relative proportion of each isomer in CLA mixtures, or duration of the feeding period, as pointed out by Terpstra *et al.* (2001).[218]

2.5.3   Fatty Acid Composition

Among all production animals, fish show the highest accumulation of CLA in their tissues when fed with diets supplemented with this particular fatty acid.[198,204,219] The increase of CLA content in fish is particularly interesting for producing 'functional' fish fillets enriched with CLA to enhance their quality for human consumption. Adding 0.5 or 1% CLA to the diet, resulted in 70 and 105 mg CLA per 100g flesh wet weight in trout,[197] which was lower compared to levels obtained in striped bass,[198] salmon,[203,209] trout[211] and sea bass.[210] However the fatty acids kinetics are more rapid in smaller fish with lower flesh lipid contents. Therefore it is possible that higher concentrations of CLA could be achieved in larger fish. Deposition of CLA isomers was increased in liver and muscle of the large yellow croaker with the increase of dietary CLA. CLA was incorporated into tissue lipids, with higher percentages found in flesh compared to liver.[197] In agreement, Leaver *et al.* (2006) measured a 2- to 3-fold higher accumulation of CLA in muscle than in liver of Atlantic salmon.[207] The sum of both isomer levels in flesh ranged from 1.8% of fatty acids in European sea bass fed 0.5%[213] to 8.4% of fatty acids in Atlantic salmon fed 4% CLA.[207] The minimal administration period of dietary CLA to obtain the desirable deposition levels of CLA in the muscle is 8 weeks.[220] The maximum level of 2.7% was reached after 12 weeks of CLA supplementation, however, deposition level already reached 2.2% after 8 weeks. Interestingly, after 2 weeks, CLA level in muscle of rainbow trout was already 1.3%, being higher than observed in most of the natural CLA sources.[12,221]

Total percentage of SFA increased and MUFA and PUFA percentage in the liver and muscle decreased as dietary CLA increased.[195,197,199,203,204,210,211,216,222] This shift in fatty acid composition could be attributed to a low conversion of SFA to MUFA, as a reduced activity of $\Delta 9$-desaturase, a rate-limiting enzyme converting SFA to MUFA, has been found.[210,211,213] Addition of CLA increased the proportion of n-3 fatty acids in liver, mainly due to elevated 20 : 5 n-3 and 22 : 6 n-3 concentrations.[197]

2.5.4 Immunomodulatory Effects

No differences in serum lysozyme activity were found by Zuo *et al.* (2013).[195] However, Jiang *et al.* (2010) and Makol *et al.* (2009), found increased lysozyme activity with increasing CLA.[196,200] They also found increasing hemagglutination titer, total iron-binding capacity, and blood cell counts. The transcriptional levels of COX-2 and IL-1β in the liver and kidney decreased with increasing dietary CLA. The transcription of TNF-α, however, was not affected.[195] These results suggest CLA modulates immuneresponse in fish.

2.6 Conclusion

Based on the available published data, it is clear that dietary supplementation of CLA results in a significant enrichment of CLA isomers in body tissues, organs and products (milk and eggs) of the major livestock species (ruminants, pigs, poultry and fish). In addition, and irrespective of the animal species considered, the tissue concentrations of saturated fatty acids and of mono/poly-unsaturated fatty acids are respectively increased and decreased due to CLA supplementation. This effect can be explained by an inhibitory effect of CLA isomers on Δ9-desaturase activity. Furthermore, CLA supplementation might also have a repartitioning effect towards a higher lean : fat ratio.

In contrast, the effects of CLA supplementation on zootechnical and re-production characteristics, and surely on immune competence were in-consistent between the published studies and this variability was observed in all livestock species under investigation. Indeed, no effects, as well as positive or negative influences of CLA supplementation on these parameters were reported. When looking to the experimental protocols with more scrutiny, it can be concluded that a whole array of factors, including animal-related (*e.g.* species, breed, gender, physiological status) and dietary-related (*e.g.* dosage and composition of supplemented CLA preparation, chemical form (free acid *versus* esterified in triacylglycerols) in which the CLA isomers are administered, composition of basal diet with special reference to fat level) factors are responsible for the apparent discrepancies between studies. It is also argued that it is of utmost importance that the exact isomer composition of the supplemented CLA supplement is known, not only to explain the variability between studies but, above all, to discriminate between the specific health effects of the major *cis*-9, *trans*-11 and *trans*-10, *cis*-12 CLA isomers.

References

1. J. C. Barlett and D. G. Chapman, *J. Agric. Food Chem.*, 1961, **9**, 50.
2. C. R. Kepler, H. P. Hirons, J. J. McNeill and S. B. Tove, *J. Biol. Chem.*, 1966, **242**, 3612.
3. T. C. Jenkins and M. A. McGuire, *J. Dairy Sci.*, 2006, **89**, 1302.
4. J. E. Santora, D. L. Palmquist and K. L. Roehrig, *J. Nutr.*, 2000, **130**, 208.
5. B. A. Corl, L. H. Baumgard, D. A. Dwyer, M. Griinari, B. S. Phillips and D. E. Bauman, *J. Nutr. Biochem.*, 2001, **12**, 622.
6. M. J. Azain, *Proc. Nutr. Soc.*, 2003, **62**, 319.
7. C. Ip, J. A. Scimeca and H. J. Thompson, *Cancer*, 1994, **74**(Suppl.), 1050.
8. C. M. Reynolds and H. M. Roche, *Prostaglandins Leukot. Essent. Fatty Acids*, 2010, **82**, 199.
9. K. Martinez, A. Kennedy, T. West, D. Milatovic, M. Aschner and M. McIntosh, *J. Biol. Chem.*, 2010, **285**, 17701.
10. Y. L. Ha, N. K. Grimm and M. W. Pariza, *J. Agric. Food Chem.*, 1989, **37**, 75.
11. S. F. Chin, W. Lui, J. M. Storkson, Y. L. Ha and M. W. Pariza, *J. Food Comp. Anal.*, 1992, **5**, 185.
12. H. Lin, T. D. Boylston, M. J. Chang, L. O. Luedecke and T. D. Shultz, *J. Dairy Sci.*, 1995, **78**, 2358.
13. Y. Yu, P. H. Correll and J. P. Vanden Heuvel, *Biochim Biophys Acta*, 2002, **1581**, 89.

14. J. J. Loor and J. H. Herbein, *J. Nutr.*, 1998, **128**, 2411.
15. P. A. Chouinard, L. Corneau, D. M. Barbano, L. E. Metzger and D. E. Bauman, *J. Nutr.*, 1999, **129**, 1579.
16. T. R. Mackle, J. K. Kay, M. J. Auldist, A. K. H. Mcgibbon, B. A. Philpott, L. H. Baumgard and D. E. Bauman, *J. Dairy Sci.*, 2003, **86**, 644.
17. L. H. Baumgard, B. A. Corl, D. A. Dwyer, A. Sæbo and D. E. Bauman, *Am. J. Physiol.*, 2000, **278**, R179.
18. L. H. Baumgard, E. Matitashvili, B. A. Corl, D. A. Dwyer and D. E. Bauman, *J. Dairy Sci.*, 2002a, **85**, 2155.
19. J. G. Giesy, M. A. McGuire, B. Shafii and T. W. Hanson, *J. Dairy Sci.*, 2002, **85**, 2023.
20. J. W. Perfield II, G. Bernal-Santos, T. R. Overton and D. E. Bauman, *J. Dairy Sci.*, 2002, **85**, 2609.
21. J. K. Kay, J. R. Roche and L. H. Baumgard, *J. Dairy Sci.* (Suppl.), 2004b, **87**, 73(Abstr.).
22. C. E. Moore, H. C. Hafliger, O. B. Mendivil, S. R. Sanders, D. E. Bauman and L. H. Baumgard, *J. Dairy Sci.*, 2004, **87**, 1886.
23. M. A. Sippel, R. S. Spratt and J. P. Cant, *Can. J. Anim. Sci.*, 2009, **89**, 393.
24. M. J. De Veth, J. M. Griinari, A. M. Pfeiffer and D. E. Bauman, *Lipids*, 2004, **39**, 365.
25. D. E. Bauman, J. W. Perfield II, K. J. Harvatine and L. H. Baumgard, *J. Nutr.*, 2008, **138**, 403.
26. D. E. Bauman and J. M. Griinari, *Livest. Prod. Sci.*, 2001, **70**, 15.
27. A. Saebo, P. C. Saebo, J. M. Griinari and K. J. Shingfield, *Lipids*, 2005, **40**, 823.
28. J. W. Perfield II, A. L. Lock, J. M. Griinari, A. Sæbo, P. Delmonte, D. A. Dwyer and D. E. Bauman, *J. Dairy Sci.*, 2007, **90**, 2211.
29. J. Pappritz, U. Meyer, R. Kramer, E.-M. Weber, G. Jahreis, J. Rehage, G. Flachowsky and S. Dänicke, *Arch. Anim. Nutr.*, 2011, **65**, 89.
30. M. Petzold, U. Meyer, S. Kersten, J. Spilke, R. Kramer, G. Jahreis and S. Dänicke, *Arch. Anim. Nutr.*, 2013, **67**, 185.
31. G. Bernal-Santos, J. W. Perfield II, D. M. Barbano, D. E. Bauman and T. R. Overton, *J. Dairy Sci.*, 2003, **86**, 3218.
32. I. A. Hutchinson, A. A. Hennnessy, R. J. Dewhurst, A. C. O. Evans, P. Lonergan and S. T. Butler, *J. Dairy Sci.*, 2012, **95**, 2442.
33. E. Castañcda-Guttiérrcz, T. R. Overton, W. R. Butler and D. E. Bauman, *J. Dairy Sci.*, 2005, **88**, 1078.
34. I. Hutchinson, M. J. de Veth, C. Stanton, R. J. Dewhurst, P. Lonergan, A. C. O. Evans and S. T. Butler, *J. Dairy Res.*, 2011, **78**, 308.
35. J. K. Kay, J. R. Roche and L. H. Baumgard, *J. Dairy Sci.*, 2006, **73**, 367.
36. S. R. Medeiros, D. E. Olivia, L. J. Aroeira, M. A. McGuire, D. E. Bauman and D. P. Lanna, *J. Dairy Sci.*, 2010, **93**, 1126.
37. L. H. Baumgard, C. E. Moore and D. E. Bauman, *Proc. Southwest Nutr. Conf.*, Tucson, Arizona, 2002b, p. 127, online: http://animal.cals.arizona.edu/swnmc/2002/proceedings.

38. L. H. Baumgards, L. J. Odens, J. K. Kay, R. P. Rhoads, M. J. VanBaale and R. J. Collier, *Proc. Southwest Nutr. Conf.*, 2006, p. 181, online: http://animal.cals.arizona.cdu/swnmc/2002/proceedings.

39. L. J. Odens, R. Burgos, M. Innocenti, M. J. VanBaale and L. H. Baumgard, *J. Dairy Sci.*, 2007, **90**, 293.

40. J. K. Kay, T. R. Mackle, D. E. Bauman, N. A. Thomson and L. H. Baumgard, *J. Dairy Sci.*, 2007, **90**, 721.

41. C. E. Moore, J. K. Kay, R. J. Collier, M. J. VanBaale and L. H. Baumgard, *J. Dairy Sci.*, 2005, **88**, 1732.

42. S. R. Medeiros, D. E. Oliveira, L. J. M. Aroeira, M. A. McGuire, D. E. Bauman and D. P. D. Lanna, *J. Dairy Sci.*, 2009, **93**, 1126.

43. G. Maxin, F. Glasser and H. Rulquin, *J. Dairy Res.*, 2010, 77, 295.

44. K. J. Shingfield and J. M. Griinari, *Eur. J. Lipid. Sci. Technol.*, 2007, **109**, 799.

45. K. J. Harvatine, Y. R. Boisclair and D. E. Bauman, *Animal*, 2009, 3, 40.

46. D. von Soosten, U. Meyer, E. M. Weber, J. Rehage, G. Flachowsky and S. Dänicke, *J. Dairy Sci.*, 2011, **94**, 2859.

47. D. von Soosten, U. Meyer, M. Piechotta, G. Flachowsky and S. Dänicke, *J. Dairy Sci.*, 2012, **95**, 1222.

48. D. von Soosten, R. Kramer, G. Jahreis, U. Meyer, G. Flachowsky and S. Dänicke, *Arch. Anim. Nutr.*, 2013, **67**, 119.

49. R. W. Canfield and W. R. Butler, *J. Dairy Sci.*, 1990, **73**, 2342.

50. W. R. Butler, *Br. Soc. Anim. Sci. Occ. Publ.*, 2001, **26**, 133.

51. M. J. De Veth, D. E. Bauman, W. Koch, G. E. Mann, A. M. Pfeiffer and W. R. Butler, *J. Dairy Sci.*, 2009, **92**, 2662.

52. B. Saremi, A. Al-Dawood, S. Winand, U. Müller, K. Pappritz, D. von Soosten, J. Rehage, S. Dänicke, S. Häussler, M. Mielenz and H. Sauerwein, *Vet. Immunol. Immunopathol.*, 2012, **146**, 201.

53. L. Renner, D. von Soosten, A. Sipka, S. Döll, A. Beineke, H. J. Schuberth and S. Dänicke, *Arch. Anim. Nutr.*, 2012a, **66**, 73.

54. L. Renner, J. Pappritz, R. Kramer, S. Kersten, G. Jahreis and S. Dänicke, *Lipids Health Dis.*, 2012b, **11**, 63.

55. S. Dänicke, J. Kowalczyk, L. Renner, J. Pappritz, U. Meyer, R. Kramer, E.-M. Weber, S. Döll, J. Rehage and G. Jahreis, *J. Dairy Sci.*, 2012, **95**, 3938.

56. H. Zhang, Y. Guo and J. Yuan, *Arch. Anim. Nutr.*, 2005a, **59**, 293.

57. E. A. Nunes, S. J. Bonatto, H. H. P. de Oliveira, N. L. M. Rivera, A. Maiorka, E. L. Krabbe, R. A. Tanhoffer and L. C. Fernandes, *Res. Vet. Sci.*, 2008, **84**, 277.

58. Y. Park, K. J. Albright, W. Liu, J. M. Storkson, M. E. Cook and M. W. Pariza, *Lipids*, 1997, **32**, 853.

59. M. W. Pariza, Y. Park and M. E. Cook, *Prog. Lipid Res.*, 2001, **40**, 283.

60. G. Schlegel, R. Ringseis, M. Shibani, E. Most, M. Schuster, F. J. Schwarz and K. Eder, *J. Anim. Sci.*, 2012, **90**, 1532.

61. P. Alberti, I. Gómez, J. A. Mendizabal, G. Ripoll, M. Barahona, V. Sarriés, K. Insausti, M. J. Beriain, A. Purroy and C. Realini, *Meat Sci.*, 2013, **94**, 208.

62. M. J. Azain, D. B. Hausman, M. B. Sisk, W. P. Flatt and D. E. Jewell, *J. Nutr.*, 2000, **130**, 1548.
63. M. A. Sugano, K. Akahoshi, K. Koba, K. Tanaka, T. Okumura, H. Matsuyama, Y. Goto, T. Miyazaki, K. Murao, M. Yamasaki, M. Nonaka and K. Yamada, *Biosci. Biotechnol. Biochem.*, 2001, **65**, 2535.
64. M. H. Gillis, S. K. Duckett, J. R. Sackmann, C. E. Realini, D. H. Keisler and T. D. Pringle, *J. Anim. Sci.*, 2004b, **85**, 851.
65. S. Schiavon, F. Tagliapietra, M. Dal Maso, L. Bailoni and G. Bittante, *J. Anim. Sci.*, 2010, **88**, 3372.
66. M. H. Gillis, S. K. Duckett and J. R. Sackmann, *J. Anim. Sci.*, 2004a, **82**, 1419.
67. M. H. Gillis, S. K. Duckett and J. R. Sackmann, *J. Anim. Sci.*, 2007, **85**, 1504.
68. C. S. Poulson, T. R. Dhiman, A. L. Ure, D. Comforth and K. C. Olson, *Livest. Prod. Sci.*, 2004, **91**, 117.
69. J. M. Eggert, M. A. Belury, A. Kempa-Steczko, S. E. Mills and A. P. Schinckel, *J. Anim. Sci.*, 2001, **79**, 2866.
70. W. Suksombat, T. Boonmee and P. Lounglawan, *Poult. Sci.*, 2007, **86**, 318.
71. H. M. White, B. T. Richert, J. S. Radcliffe, A. P. Schinckel, J. R. Burgess, S. L. Koser, S. S. Donkin and M. A. Latour, *J. Anim. Sci.*, 2009, **87**, 157.
72. M. E. Dugan, J. L. Aalhus and J. K. Kramer, *Am. J. Clin. Nutr.*, 2004, **79**, 1212S.
73. C. Corino, A. Di Giancamillo, R. Rossi and C. Domenghini, *J. Nutr.*, 2005, **135**, 1444.
74. F. Tischendorf, F. Schone, U. Kircheim and G. Jahreis, *J. Anim. Physiol. Anim. Nutr.*, 2002, **86**, 117.
75. M. E. R. Dugan, J. L. Aalhus, K. A. Lien, A. L. Schaefer and J. K. G. Kramer, *Can. J. Anim. Sci.*, 2001, **81**, 505.
76. M. E. R. Dugan, J. L. Aalhus, A. L. Schaefer and J. K. G. Kramer, *Can. J. Anim. Sci.*, 1997, 77, 723.
77. E. Ostrowska, M. Muralitharan, R. F. Cross, D. E. Bauman and F. R. Dunshea, *J. Nutr.*, 1999, **129**, 2037.
78. R. L. Thiel-Cooper, F. C. Parrish, J. C. Sparks, B. R. Wiegan and R. C. Ewan, *J. Anim. Sci.*, 2001, **79**, 1821.
79. D. Sun, X. Zhu, S. Qiao, S. Fan and D. Li, *Arch. Anim. Nutr.*, 2004, **58**, 277.
80. T. E. Weber, B. T. Richert, M. A. Belury, Y. Gu, K. Enright and A. P. Schinckel, *J. Anim. Sci.*, 2006, **84**, 720.
81. G. Bee, *Anim. Res.*, 2001, **50**, 383.
82. N. Tous, R. Lizardo, B. Vilà, M. Gispert, M. Font-i-Furnols and E. Esteva-Garcia, *Meat Sci.*, 2013, **93**, 517.
83. E. Ostrowska, R. F. Cross, M. Muralitharan, D. E. Bauman and F. R. Dunshea, *Br. J. Nutr.*, 2002, **88**, 625.
84. D. N. D'Sousa and B. M. Mullan, *Meat Sci.*, 2002, **60**, 95.
85. L. A. Gatlin, M. Y. See, D. K. Larick, X. Lin and J. Odle, *J. Nutr.*, 2002, **132**, 3105.

86. D. N. D. D'Souza, D. W. Pethick, F. R. Dunshea, J. R. Pluske and B. P. Mullan, *Austr. J. Agric. Res.*, 2003, **54**, 745.

87. M. E. R. Dugan, J. L. Aalhus, L. E. Jeremiah, J. F. G. Kramer and A. L. Schaefer, *J. Anim. Sci.*, 1999, **79**, 45.

88. B. R. Wiegand, F. C. Parrish, Jr, J. E. Swan, S. T. Larsen and T. J. Baas, *J. Anim. Sci.*, 2001, **79**, 2187.

89. B. R. Wiegand, J. C. Sparks, F. C. Parrish, Jr and D. R. Zimmerman, *J. Anim. Sci.*, 2002, **80**, 637.

90. L. Averette Gatlin, M. T. See, J. A. Hansen, D. Sutton and J Odle, *J. Anim. Sci.*, 2002, **80**, 1606.

91. S. T. Joo, J. L. Lee, Y. L. Ha and G. B. Park, *J. Anim. Sci.*, 2002, **80**, 108.

92. D. Martin, T. Antequera, E. Muriel, T. Perez-Palacios and J. Ruiz, *Meat Sci.*, 2008, **80**, 1309.

93. P. C. H. Morel, J. A. M. Janz, M. Zou, R. W. Purchas, W. H. Hendriks and B. H. P. Wilkinson, *J. Anim. Sci.*, 2008, **86**, 1145.

94. J. Jiang, M. J. Zhao, F. Lin, L. Yang and X. Q. Zhou, *Lipids*, 2010a, **45**, 531.

95. G. Cordero, B. Isabel, D. Menoyo, A. Daza, J. Morales, C. Piñeiro and C. J. López-Bote, *Meat Sci.*, 2010, **85**, 235.

96. W. Migdal, P. Pasciak, D. Wojtysiak, T. Barowicz, M. Pieszka and M. Pietras, *Meat Sci.*, 2004, **66**, 863.

97. C. Lauridsen and P. Henckel, *Meat Sci.*, 2005, **89**, 393.

98. D. P. Lo Fiego, P. Macchioni, P. Santoro, G. Pastorelli and C. Corino, *Meat Sci.*, 2005, **70**, 285.

99. C. Corino, M. Musella, G. Pastorelli, R. Rossi, K. Paolone, L. Costanza, A. Manchisi and G. Maiorano, *Meat Sci.*, 2008, **79**, 307.

100. K. O. Intarapichet, B. Maikhunthod and N. M. Thungmanee, *Meat Sci.*, 2008, **80**, 788.

101. G. Bee, *J. Nutr.*, 2000a, **130**, 2292.

102. E. Ostrowska, R. F. Cross, M. Mualitharan, D. E. Bauman and F. R. Dunshea, *Br. J. Nutr.*, 2003, **90**, 915.

103. G. Bee, S. Jacot, G. Guex and C. Biolley, *Animal*, 2008, **2**, 800.

104. C. Corino, V. Bontempo and D. Sciannimanico, *Can. J. Anim. Sci.*, 2002, **82**, 115.

105. C. Lai, J. Yin, D. Li, L. Zhao and X. Chen, *Arch. Anim. Nutr.*, 2005a, **59**, 41.

106. M. L. Moraes, A. M. L. Ribeiro, A. M. Kessler, V. S. Ledur, M. M. Fischer, L. Bockor, S. P. Cibulski and D. Gava, *J. Anim. Sci.*, 2012, **90**, 2590.

107. L. Zhao, J. Yin, D. Li, C. Lai, X. Chen and D. Ma, *Arch. Anim. Nutr.*, 2005, **59**, 429.

108. C. Corino, G. Pastorelli, F. Rosi, V. Bontempo and R. Rossi, *J. Anim. Sci.*, 2009, **87**, 2299.

109. D. DeVoney, M. Pariza and H. W. Cook, *Res. Vet. Sci.*, 1997, **47**, 387.

110. J. Bassanganya-Riera, R. Hontecillas-Magarzo, K. Bregendahl, M. J. Wannemuehler and D. Zimmerman, *J. Anim. Sci.*, 2001, **79**, 714.

111. J. Bassanganya-Riera, R. M. Pogranichniy, S. C. Jobgen, P. G. Halbur, K. J. Yoon, M. O'Shea, I. Mohede and R. Hontecillas, *J. Nutr.*, 2003, **133**, 3204.

112. R. Hontecillas, M. J. Wannemeulher, D. R. Zimmerman, D. L. Hutto, J. H. Wilson, D. U. Ahn and J. Bassanganya-Riera, *J. Nutr.*, 2002, **132**, 2019.

113. C. Lai, J. Yin, D. Li, L. Zhao, S. Qiao and J. Xing, *J. Nutr.*, 2005b, **135**, 239.

114. B. R. Wiegand, D. Pompeu, R. L. Thiel-Cooper, J. E. Cunnick and F. C. Parrisj, Jr, *J. Anim. Sci.*, 2011, **89**, 1588.

115. H. J. Zhang, Y. M. Guo and J. M. Yuan, *Br. J. Nutr.*, 2005b, **94**, 746.

116. G. Cordero, B. Isabel, J. Morales, D. Menoyo, C. Pineiro, A. Daza and C. J. Lopez-Bote, *Anim. Feed Sci. Technol.*, 2011, **168**, 232.

117. Y. Peng, F. Ren, J. D. Yin, Q. Fang, F. N. Li and F. D. Li, *J. Anim. Sci.*, 2010, **88**, 1741.

118. G. Bee, *J. Nutr.*, 2000b, **130**, 2981.

119. S. P. Poulos, M. J. Azain and G. J. Hausman, *Anim. Res.*, 2004, **53**, 275.

120. R. Patterson, M. L. Connor and C. M. Nyachoti, *J. Anim. Sci. (Suppl.)*, 2007, **85**, 50(Abstr.).

121. A. Schmid, M. Collomb, G. Bee, U. Butikofer, D. Wechsler, P. Everhard and R. Sieber, *Br. J. Nutr.*, 2008, **100**, 54.

122. U. Krogh, C. Flummer, S. K. Jensen and P. K. Theil, *J. Anim. Sci.*, 2012, **90**, 366.

123. V. Bontempo, D. Sciannimanico, G. Pastorelli, R. Rossi, F. Rosi and C. Corino, *J. Nutr.*, 2004, **134**, 817.

124. I. Halle, G. Jahreis and M. Henning, *J. Vervr. Lebensm.*, 2012, 7, 3.

125. M. Javadi, M. J. H. Geelen, H. Everts, R. Hovenier, E. Javadi, H. Kappert and A. C. Beynen, *Br. J. Nutr.*, 2007, **98**, 1152.

126. S. Kawahara, S. I. Takenoyama, K. Takuma, M. Muguruma and K. Yamauchi, *Anim. Sci. J.*, 2009, **80**, 468.

127. M. W. Pariza, Y. Park, M. E. Cook, K. Albright and W. Liu, *FASEB J.*, 1996, **10**, A560.

128. J. L. Sell, S. Jin and M. Jeffrey, *Poult. Sci.*, 2001, **80**, 209.

129. B. Szymczyk, P. M. Pisulewski, W. Szczurek and P. Hanczakowski, *Br. J. Nutr.*, 2001, **85**, 465.

130. G. M. Zhang, J. Wen, J. L. Chen, G. P. Zhao, M. Q. Zheng and W. J. Li, *Br. Poult. Sci.*, 2007, **48**, 217.

131. S. C. Bölükbasi, *Br. Poult. Sci.*, 2006, **47**, 470.

132. M. Du and D. U. Ahn, *Poult. Sci.*, 2002, **81**, 428.

133. A. Tanai, J. Peredi, E. Zsedely, T. Toth and J. Schmidt, *Arch. Geflügelk.*, 2011, **75**, 91.

134. M. Du and D. U Ahn, *Lipids*, 2003, **38**, 505.

135. B. Szymczyk, P. M. Pisulewski, P. Hanczakowski and W. Szczurek, *J. Sci. Food. Agric.*, 2000, **80**, 1553.

136. L. Badinga, K. T. Selberg, A. C. Dinges, C. W. Comer and R. D. Miles, *Poult. Sci.*, 2003, **82**, 111.

137. R. Bou, R. Codony, M. D. Baucells and F. Guardiola, *J. Agri. Food Chem.*, 2005, **53**, 7792.

138. F. Sirri, N. Tallarico, A. Meluzzi and A. Franchini, *Poult. Sci.*, 2003, **82**, 1356.

139. V. A. Aletor, K. Eder, K. Becker, B. R. Poulicks, F. X. Roth and D. A. Roth-Maier, *Poult. Sci.*, 2003, **82**, 796.

140. Y. Park, J. Albright, J. M. Storkson, W. Lieu, M. E. Cook and M. W. Pariza, *Lipids*, 1999, **34**, 243.

141. K. Raes, G. Huyghebaert, S. De Smet, L. Nollet, S. Amouts and D. Demeyer, *J. Nutr.*, 2002, **132**, 182.

142. K. N. Lee, M. W. Pariza and J. M. Ntambi, *Bichem. Biophys. Res. Commun.*, 1998, **248**, 817.

143. Y. Park, J. M. Storkson, J. M. Ntambi, M. E. Cook and C. J. Sih, *Biochem. Biophys. Acta*, 2000, **1486**, 285.

144. K. Eder, N. Slomma and K. Becker, *J. Nutr.*, 2002, **132**, 1115.

145. C. Ip, Chin, J. A. Scimeca and M. W. Pariza, *Cancer Res.*, 1991, **51**, 6118.

146. T. D. Shultz, B. P. Chew and W. R. Seaman, *Anticancer Res.*, 1992, **14**, 2143.

147. K. N. Lee, D. Kritchevsky and M. W. Pariza, *Atherosclerosis*, 1994, **108**, 19.

148. M. E. Cook, C. C. Miller, Y. Park and M. Pariza, *Poult. Sci.*, 1993, **72**, 1301.

149. K. Takahashi, K. Kawamata, Y. Akiba, T. Iwata and M. Kasai, *Br. Poult. Sci.*, 2002, **43**, 47.

150. M. Sugano, A. Tsujita, M. Yamasaki, M. Noguchi and K. Yamada, *Lipids*, 1998, **33**, 521.

151. M. Yamasaki, K. Kishihara, K. Mansho, Y. Ogino, M. Kasai, M. Sugano, H. Tachibana and K. Yamada, *Biosci. Biotechnol. Biochem.*, 2000, **64**, 2159.

152. Y. Takahashi, M. Kushiro, K. Shinohara and T. Ide, *Biochim, Biophys. Acta*, 2003, **1631**, 265.

153. F. Y. Long, Z. Wang, Y. M. Guo, D. Liu, X. Yang and P. Jiao, *Food Agric. Immunol.*, 2010, **4**, 295.

154. J. E. Hill, G. N. Rowland, K. S. Latimer and J. Brown, *Avian Dis.*, 1989, **33**, 86.

155. N. H. Sigal and F. J. Dumont, *Annu. Rev. Immunol.*, 1992, **10**, 519.

156. J. J. Turek, Y. Li, I. A. Schoenlein, K. Allen and B. A. Watkins, *J. Nutr. Biochem.*, 1998, **9**, 258.

157. F. Y. Long, Y. M. Guo, Z. Wang, D. Liu, B. K. Zhang and X. Yang, *Poult. Sci.*, 2011, **90**, 1926.

158. X. G. Shang, F. L. Wang, D. F. Li, J. D. Yin and J. Y. Li, *Poult. Sci.*, 2004, **83**, 1688.

159. W. Suksombat, S. Samitayotin and P. Lounglawan, *Poult. Sci.*, 2006, **85**, 1603.

160. J. D. Yin, X. G. Shang, D. F. Li, F. L. Wang, Y. F. Guan and Z. Y. Wang, *Poult. Sci.*, 2008, **87**, 284.

161. D. U. Ahn, J. L. Sell, C. Jo, M. Chamruspollert and M. Jeffrey, *Poult. Sci.*, 1999, **78**, 922.

162. B. Szymczyk and P. M. Pisulewski, *Br. J. Nutr.*, 2003, **90**, 93.

163. S. Jones, D. W. Ma, E. F. Robinson, C. J. Field and M. F. Clandinin, *J. Nutr.*, 2000, **130**, 2002.

164. V. A. Leone, S. P. Worzalla and M. E. Cook, *Poult. Sci.*, 2010, **89**, 621.

165. E. Muma, M. Palander, M. Nasi, A. M. Pfeiffer, T. Keller and J. M. Grinnari, *Poult. Sci.*, 2006, **85**, 712.

166. R. Aydin, M. W. Pariza and M. E. Cook, *FASEB J.*, 1999, **13**, A451.

167. R. Aydin, M. W. Pariza and M. E. Cook, *J. Nutr.*, 2001, **131**, 800.

168. R. Aydin and M. E. Cook, *Poult. Sci.*, 2004, **83**, 2016.

169. R. Aydin and M. E. Cook, *Anim. Feed Sci. Technol.*, 2009, **149**, 125.

170. H. W. Cook, *Biochemistry of Lipids, Lipoproteins and Membranes*, ed. E.D. Vance and J. Vance, Elsevier, New York, 1991, p.141.

171. G. Cherian, W. Ai and M. P. Goeger, *Lipids*, 2005, **40**, 130.

172. G. Cherian, *Eur. J. Lipid Sci. Technol.*, 2009, **111**, 546.

173. A. L. Romanoff, *The Avian Embryo: Structural and Functional Development*, Macmillan, New York, NY, 1960.

174. R. C. Noble and J. H. Moore, *Can. J. Biochem. Physiol.*, 1964, **42**, 1729.

175. B. M. Freeman and M. A. Vince, *Development of the Avian Embryo*, John Wiley and Sons Inc., New York, NY, 1974.

176. R. C. Noble and M. Cocchi, *Prog. Lipid Res.*, 1990, **29**, 107.

177. R. C. Noble, *J. Exp. Zool. Suppl.*, 1987, **1**, 65.

178. B. K. Speake, A. M. B. Murray and R. C. Noble, *Prog. Lipid Res.*, 1998a, **37**, 1.

179. B. K. Speake, R. C. Noble and A. M. B. Murray, *World's Poult. Sci. J.*, 1998b, **54**, 319.

180. S. F. Chin, J. M. Storkson, K. J. Albright, M. E. Cook and M. W. Pariza, *J. Nutr.*, 1994, **124**, 2344.

181. S. P. Poulos, M. Sisk, D. B. Hausman, M. J. Azain and G. J. Hausman, *J. Nutr.*, 2001, **131**, 2722.

182. V. A. Leone, D. L. Stranski, R. Aydin and M. E. Cook, *Poult. Sci.*, 2009, **88**, 1858.

183. J. H. Kim, J. Hwangbo, N. J. Choi, H. G. Park, D. H. A. Yoon, E. W. Park, S. H. Lee, B. K. Park and Y. J. Kim, *Poult. Sci.*, 2007, **86**, 1180.

184. M. Chamruspollert and J. L. Sell, *Poult. Sci.*, 1999, **78**, 1138.

185. C. Alvarez, P. Cachaldora, J. Méndez and P. Garcia-Rebollar, *Span. J. Agric. Res.*, 2004a, **2**, 203.

186. C. Alvarez, P. Cachaldora, J. Méndez, P. Garcia-Rebollar and J. C. De Blas, *Br. Poult. Sci.*, 2004b, **45**, 524.

187. C. Alvarez, P. Garcia-Rebollar, P. Cachaldor, J. Méndez and J. C. De Blas, *Br. Poult. Sci.*, 2005, **46**, 80.

188. K. N. Lee, *Conjugated Linoleic Acid and Lipid Metabolism*, PhD thesis, University of Wisconsin, Madison, WI, 1996.
189. A. M. Abou-ashour and H. M. Edwards, *J. Nutr.*, 1970, **100**, 757.
190. X. G. Shang, F. L. Wang, D. F. Li, J. D. Yin, X. J. Li and G. F. Yi, *Poult. Sci.*, 2005, **84**, 1886.
191. G. Cherian, T. B. Holsonbake, M. P. Goeger and R. Bidfell, *Lipids*, 2002, **37**, 751.
192. G. Cherian, M. G. Traber, M. P. Goeger and S. W. Leonard, *Poult. Sci.*, 2007, **86**, 953.
193. K. Schafer, K. Manner, A. Sagredos, K. Eder and O. Simon, *Lipids*, 2001, **11**, 1217.
194. L. Yang, Y. Huang, A. E. James, L. W. Lam and Z. Y. Chen, *J. Agric. Food Chem.*, 2002, **17**, 4941.
195. R. Zuo, Q. Ai, K. Mai and W. Xu, *Br. J. Nutr.*, 2013, **110**, 1220.
196. W. D. Jiang, L. Feng, Y. Liu, J. Jiang and X. Q. Zhou, *Aquac. Res.*, 2009, **39**, 955.
197. S. R. Kennedy, R. Bickerdike, R. K. Berge, A. R. Porter and D. R. Tocher, *Aquaculture*, 2007, **264**, 372.
198. R. G. Twibell, B. A. Watkins, L. Rogers and P. B. Brown, *Lipids*, 2000, **35**, 155.
199. X. Y. Tan, Z. Luo, P. Xie, X. D. Li, X. J. Liu and W. Q. Xi, *Aquaculture*, 2010, **310**, 186.
200. A. Makol, S. Torrecillas, A. Fernandez-Vaquero, L. Robaina, D. Montero, M. J. Caballero, L. Tort and M. Izquierdo, *Comp. Biochem. Physiol.*, 2009, **154B**, 179.
201. A. Diez, D. Menoyo, S. Pérez-Benavente, J. A. Calduch-Giner, S. Vega-Rubin de Celis, A. Obach, L. Favre-Krey, E. Boukouvala, M. J. Leaver, D. R. Tocher, J. Perez-Sanchez, G. Krey and J. M. Bautista, *J. Nutr.*, 2007, **137**, 1363.
202. X. Y. Tan, Z. Luo and Q. Zeng, *Lipids*, 2013, **48**, 505.
203. G. M. Berge, B. Ruyter and T. Asgard, *Aquaculture*, 2004, **237**, 365.
204. R. G. Twibell, B. A. Watkins and P. B. Brown, *J. Nutr.*, 2001, **131**, 2322.
205. A. C. Figueiredo-Silva, P. Rema, N. M. Bandarra, M. L. Nunes and L. M. P. Valente, *Aquaculture*, 2005, **248**, 163.
206. S. R. Kennedy, M. J. Leaver, P. J. Campbell, X. Zheng, J. R. Dick and D. R. Tocher, *Lipids*, 2006, **41**, 423.
207. M. J. Leaver, D. R. Tocher, A. Obach, L. Jensen, R. J. Henderson, A. R. Porter and G. Krey, *Comp. Biochem. Physiol.*, 2006, **145A**, 258.
208. Z. Luo, X. Y. Tan, C. X. Liu, X. D. Li, X. J. Liu and W. Q. Xi, *Aquacult. Res.*, 2012, **43**, 1392.
209. S. R. Kennedy, P. J. Campbell, A. Porter and D. R. Tocher, *Comp. Biochem. Physiol.*, 2005, **141**, 168.
210. L. M. P. Valente, N. M. Bandarra, A. C. Figueiredo-Silva, A. R. Cordeiro, R. M. Simoes and M. L. Nunes, *Aquaculture*, 2007b, **267**, 225.

211. L. M. P. Valente, N. M. Bandarra, A. C. Figueiredo-Silva, P. Rema, P. Vaz-Pires, S. Martins, J. A. M. Prates and M. L. Nunes, *Br. J. Nutr.*, 2007a, **97**, 289.

212. A. Yasmin, T. Takeuchi, M. Hayashi, T. Hirota, W. Ishizuka and S. Ishida, *Fish Sci.*, 2004, **70**, 473.

213. B. D. Choi, S. J. Kang and Y. L. Ha, *Quality Attributes of Muscle Foods*, ed. Y. L. Xiong, C. T. Ho and F. Shahidi, Kluwer Academic, Plenum Publisher, New York, 1999, pp. 61–71.

214. B. C. Peterson, B. Manning and M. Li, *Glob. Aquac. Advocate*, 2003, **6**, 48.

215. M. A. Belury, A. Mahon and S. Banni, *J. Nutr.*, 2003, **133**, 257S.

216. M. Bandarra, M. L. Nunes, A. M. Andrade, J. A. M. Prates, S. Pereira, M. Monteiro, P. Rema and L. M. P. Valente, *Aquaculture*, 2006, **254**, 496.

217. R. G. Twibell and R. P. Wilson, *Aquaculture*, 2003, **221**, 621.

218. A. H. M. Terpstra, *J. Nutr.*, 2001, **131**, 2067.

219. M. E. Evans, J. M. Brown and M. K. McIntosh, *J. Nutr. Biochem.*, 2002, **13**, 508.

220. A. Ramos, N. M. Bandarra, P. Rema, P. Vaz-Pires, P. Nunes, A. M. Andrade, A. R. Cordeiro and L. M. P. Valente, *Aquaculture*, 2008, **274**, 366.

221. A. Schmid, M. Collomb, R. Sieber and G. Bee, *Meat Sci.*, 2006, **73**, 29.

222. L. D. Dos Santos, W. M. Furuya and L. C. R. da Silva, *Aquacult. Nutr.*, 2011, **17**, 70.

CHAPTER 3

Health Benefits of Conjugated Fatty Acids

YEONHWA PARK* AND YAN WU

Department of Food Science, University of Massachusetts, Amherst,
102 Holdsworth Way, Amherst, MA 01003, USA
*Email: ypark@foodsci.umass.edu

3.1 Introduction

The first report of the bioactivity of conjugated linoleic acids (CLAs) was in the 1980s by Dr Pariza's group at the University of Wisconsin-Madison and dealt with its anti-cancer effects.[1,2] It was an unexpected discovery as CLA was isolated from ground beef extract during the time when the discovery of the relationship between eating overcooked beef and mutagens in meat was made. Since the initial discovery, a number of other biological functions of CLA have been reported, which include its effects on body composition, immune and inflammatory responses, bone health, and cardiovascular diseases.[3,4] This chapter will review CLA's bioactivities along with current health concerns and also review bioactivities of other naturally occurring conjugated fatty acids.

3.2 Discovery of CLA and its Natural Origins

The earliest report of CLA in scientific papers can be tracked to 1932, when increased absorption at 230 nm in summer milk over winter milk was observed.[5] Definitive structural determination was not possible then due to limited analytical techniques, but it can be inferred due to the presence of

RSC Catalysis Series No. 19
Conjugated Linoleic Acids and Conjugated Vegetable Oils
Edited by Bert Sels and An Philippaerts
© The Royal Society of Chemistry 2014
Published by the Royal Society of Chemistry, www.rsc.org

conjugated double bonds in CLA. Now it is known that the concentration of CLA in summer milk when cows were fed pastures would be higher when compared to winter milk, supporting earlier observation of increased absorption at 230 nm in summer milk.[6] CLA is derived from biohydrogenation by rumen bacteria when linoleic acid is saturated to stearic acid.[7] The *cis*-9, *trans*-12 CLA is one of two main intermediates of this process, also called *rumenic acid,* and it is believed that this isomer of CLA can escape part of the biohydrogenation process and can be found in beef and dairy products.[7,8] Others reported that the alternative significant intermediate of biohydrogenation, *trans*-11 vaccenic acid, can be converted to the *cis*-9, *trans*-11 CLA by Δ9-desaturation in mammalian tissue.[9,10]

In addition to the *cis*-9,*trans*-11 CLA, other positional and geometric CLA isomers have been reported from natural sources.[11] CLA isomers with double bonds starting at 7, 8, 9, 10, or 11 with either *cis* or *trans* configuration have been reported previously.[11] Among them the most important CLA isomer is the *trans*-10,*cis*-12 isomer[3]. This isomer is present at low levels in food, but when CLA is prepared from linoleic acid by chemical and heat processes, it is present in significant quantity, similar to that of the *cis*-9,*trans*-11 CLA.[3,12] Often the CLA preparation that contains primarily these two isomers in similar amounts is referred to as a 'CLA mixture' or '50 : 50 mix'. Most studies have used this preparation of CLA rather than the individual isomers.

Foods from ruminant origins are relatively good sources of naturally occurring CLA since the origin of CLA is primarily from rumen bacteria.[12] The levels of CLA found in food are less than 10 mg per g fat, mostly in beef, milk, cheese, or other dairy products.[12] Thus, intakes of CLA from diet vary depending on one's dietary pattern, ranging from 49 to 659 mg per day.[13] In the US, the average CLA intake from foods is estimated to be 176–212 mg per day for men and 104–151 mg per day for women.[14–17]

It is noteworthy to point out that the current rule of *trans* fat labeling by the US Food and Drug Administration, specifically defines 'All unsaturated fatty acids that contain one or more isolated double bonds (*i.e.* nonconjugated) in a *trans* configuration.'[18] Thus, the 'conjugated' *trans* fatty acids, mainly CLA, have been excluded from the *trans* fat labeling. In addition, CLA as a 50 : 50 mixture has been approved as generally recognized as safe (GRAS) for certain types of food in the US since 2008.

3.3 CLA's Health Impact

Currently almost 100 human studies have been published regarding CLA's effects including reduction of body fat, prevention of cancer, prevention of cardiovascular diseases, modulation of immune and inflammatory responses, and improvement of bone health. Most animal studies used 0.5 w/w% of CLA in the diets, with ranges of 0.01–2% CLA. In human trials, the majority of studies used 3–4 g CLA per day, ranging from 1.7 to 6.8 g per day.[3,4] It was previously reported that when rats were fed 0.5% CLA containing diet for 4 weeks, serum levels of CLA were 23–150 μM.[19] Mele *et al.*[20] reported recently that the plasma concentrations of CLA after taking 0.8–3.2 g CLA per day for 2 months ranged from 50 to 200 μM in humans, suggesting that this dose of CLA may have relevance in its biological activities in humans.

Most of CLA studies used either free fatty acid or triglyceride (TG) form. These forms of CLA were equally effective in animal models and in human studies.[21–24] Thus, this chapter will not necessarily distinguish the form of CLA used.

3.3.1 Antiobesity Effects

Since the first report on CLA's anti-obesity effects in 1997, it is the most known and studied biological effect of CLA.[19] Based on animal studies showing dramatic body fat reduction after CLA supplementation, a number of studies tested its effects in humans.[3,17] Most of them reported changes in body weight, body mass index, body fat, lean body mass, waist, or sagittal abdominal diameter. Overall effects of CLA are much less dramatic in humans than mice, less than a 2 kg change in body weight or less than 8% reduction in body fat, while a number of reports cited no changes by CLA supplementation on any of these markers.[3,17] These inconsistent results may be explained by differences in doses, durations, subjects, and/or study designs.

Three meta-analysis publications reported CLA supplementation resulted in significant, although small, improvement in body weight, body fat, or lean body mass compared to control groups.[25–27] This was particularly evident when CLA was supplemented at 3.4 g per day for at least 6 months.[25–27] In particular, Schoeller *et al.*[27] reported that CLA had small but significant effects on enhancing lean body mass, which can be perceived as weight gain, thus resulting in less net body weight changes. This improved lean body mass observation is particularly important during the weight gain/regain period. Kamphuis *et al.*[28] reported when CLA was supplemented during the weight regain period following weight loss, CLA improved lean mass gain while reducing fat mass gain without changing body weight. Consistently, CLA supplementation prevented holiday weight and fat mass gain compared to control groups.[29]

It is important to note that a number of animal studies were completed while animals were in rapid growth period (thus considered as positive energy balance), while most human studies were completed with adults along with hypocaloric diet (thus negative energy balance).[3] Previously, the effects of CLA during negative energy balance were determined using a pair-feeding method in an animal model, where no benefits of CLA on body fat reduction was reported.[30] This suggests that CLA can be useful to prevent fat mass gain during weight gain period, while CLA may have minimum to no effects during weight loss period.[28,29,31–34]

The improved lean mass due to CLA supplementation may serve as a very important tool to help prevent or alleviate age associated muscle loss, such as sarcopenia. Sarcopenia is described as the age-dependent gradual loss of skeletal muscle mass and is prevalent in more than 50% of the elderly over 80 years of age.[35-37] As reported by Tarnipolsky *et al.*,[38] subjects older than 65 years had improved exercise outcomes after supplemental CLA along with creative monohydrate. This implies that CLA may have potential to be applied to improve or preserve muscle mass for those at risk of significant muscle loss.

Among the two major CLA isomers, it was determined that the *trans*-10, *cis*-12 CLA isomer is solely responsible for CLA's effects on body fat reduction, regardless of the presence of the *cis*-9,*trans*-11 isomer. Ten human studies that used mainly the *cis*-9, *trans*-11 CLA isomer, either from synthetic preparations or enriched foods, consistently reported no effects on the markers listed above.[39-48] This confirms that the *trans*-10, *cis*-12 CLA is critical for its effects on weight and body fat control.[49]

CLA may exert its effects on body composition *via* multiple mechanisms. Firstly, it has been reported that CLA increased energy expenditure in animal models.[50-55] Similar results have been reported in human studies,[28,29,56,57] however, the studies are rather limited and others reported no differences in energy expenditure even when changes in body fat or body weights were observed.[58-61] Secondly, CLA increased fatty acid oxidation in skeletal muscle, which was further supported by Close *et al.*[57] who reported increased fat oxidation during sleep after supplementing CLA in humans.[19,49,62-66] This was independent of fatty acid oxidation of CLA itself as it has been shown that the oxidation rate of the CLA isomer was similar to that of oleic acid.[67] Thirdly, CLA targets lipid and adipocyte metabolisms, such as inhibiting adipocyte lipoprotein lipase activity, inhibiting stearoyl-CoA desaturase activity, potentiating adipocyte apoptosis, promoting adipocyte lipolysis, limiting preadipocyte differentiation, and modulating a number of cytokines and/or adipokines.[21,68] Lastly, it has been suggested that CLA may influence overall food consumption. It has been proven that the body fat reduction due to CLA in animals was independent from its effect on temporary reduction of food intake.[30] Three human studies reported positive influence of CLA supplementation on appetite and reduced adverse effects associated with hypocaloric diets.[32,34,69] This may be helpful for those on weight loss regimes, although the studies were rather limited and warrant further investigation.

3.3.2 Anticancer Effects

As CLA was originally discovered as an anticancer component, there have been plentiful reports on CLA and its anticancer effects in animals and cell culture experiments.[70-72] Not only in a number of different cancer types, but CLA has been shown to be involved in all stages of cancer to reduce cancer development and incidences, as well as metastasis.[70-72] Moreover, both major isomers of CLA have been reported to have anticancer activities.[73-75] In contrast to these, there are only a limited number of human studies involving CLA and cancer, all of which focus on the *cis*-9, *trans*-11 isomer, the primary isomer of naturally occurring CLA. There are seven human studies on CLA where breast cancer incidences and CLA intake or tissue levels were analysed. Among them two reported an inverse correlation between CLA intake or CLA tissue levels and breast cancer incidence,[76,77] while others found no correlation between them.[78-82]

An additional study reported an inverse correlation between CLA intake and colorectal cancer incidences.[83] Recently, it was reported in rectal cancer patients that CLA supplementation along with chemo-radiotherapy significantly reduced matrix metalloproteinase types 2 and 9, both of which are biomarkers of angiogenesis and tumor metastasis.[84] Moreover, earlier work with CLA in immune challenged animals showed reduced cachexia and prevented muscle wasting in tumor bearing mice,[85-87] which would be beneficial to cancer patients. These findings suggest that supplementation with CLA may have great significance to be used along with current cancer therapies to improve efficacy of cancer treatment as well as patients' wellbeing.

3.3.3 Cardiovascular Disease Prevention

Cardiovascular diseases are the leading cause of death in the US and are considered to be preventable diseases through dietary modulation.[88] Thus, a number of nutritional studies targeted food bioactives for regulation of risk factors of cardiovascular diseases. CLA was also tested for its effects on those markers starting in the 1990s.[89] Reports of CLA preventing atherosclerosis development were very promising; where CLA treated animals had reduced scores of atherosclerotic plaque formations along with positive changes in blood markers.[89-92]

There are plentiful clinical studies with CLA supplementation reporting on blood markers for cardiovascular diseases.[3] In contrast to animal studies, most human studies with CLA reported no to minimal effects within normal ranges of serum total, LDL or HDL cholesterol, or triacylglyceride.[3] However, there were six human studies reporting CLA's positive effect on blood pressure.[93-98] In particular, Zhao *et al.*[97] co-supplemented CLA along with an antihypertensive drug, ramipril, where greater hypotensive effects were observed compared to those of ramipril alone. In addition, two studies reported hypotensive effects of CLA on patients with pregnancy-induced hypertension.[94,95] This hypotensive effect of CLA itself can be considered to be beneficial, reducing risk for cardiovascular diseases.

There are several reports involving CLA and increased C-reactive protein (CRP),[29,60,99-103] which is an inflammatory marker associated with cardiovascular disease risk.[104] Other studies reported no changes of CRP after CLA supplementation.[38,42,46,105-111] Increased CRP along with elevated other risk factors for cardiovascular diseases is considered to enhanced the risk for cardiovascular diseases. However, since increased CRP after CLA supplementation was not accompanied by changes of other risk factors, it is likely that increased CRP due to CLA may not be indicative of increased risk for cardiovascular diseases.

3.3.4 Modulation of Immune and Inflammatory Responses

Earlier CLA studies reported that CLA reduced adverse effects associated with immune stimulation, reduced inflammatory responses, and reduced hypersensitivity responses in animal models.[4] While most human studies reported no significant differences in markers of immune and inflammatory responses after CLA supplementation,[3,17,110-112] there are limited studies reporting potential benefits of CLA, such as reduced allergic responses, improved antibody production following hepatitis B vaccination, reduced symptoms of atopic dermatitis, and reduced symptoms of rhinovirus infection.[109,113-115]

Particularly, there are several reports of a preventive role of CLA in inflammatory bowel diseases, which have been suggested to be linked to its effect on peroxisome proliferator-activated receptor-γ (PPARγ).[116-118] Alternatively, changes of microbiomes by CLA supplementation may have contributed to its beneficial effects on colitis, a form of inflammatory bowel disease.[119] Similarly, CLA supplementation reduced disease activity compared to the beginning of the study in patients with Crohn's diseases.[120] Since there are limited treatment options for patients with inflammatory bowel diseases, CLA may give alternative or supplemental options with reduced adverse effects to those suffering with inflammatory bowel diseases. Additionally, this further supports potential beneficial effects of CLA on certain types of colorectal cancer as inflammatory bowel diseases are associated with colorectal cancer.[121]

3.3.5 Improving Bone Health

With regards to CLA's effects on bone mass, results were not consistent in animal studies.[3,72,122] It was suggested that co-supplementation of CLA and calcium may improve its effects on bone as well as increased calcium absorption.[123-128] Currently there are nine human studies reporting on CLA's effect on bone markers.[22,38,94,99,103,129-132] Among them, Brownbill *et al.*[129] reported that consuming CLA rich food (primarily the *cis*-9,*trans*-11 isomer) was positively associated with improving bone mineral density in postmenopausal women along with calcium supplements. Kreider *et al.*[99] reported a non-significant increase (p $=$ 0.08) in bone mass when CLA supplementation was provided along with resistance training. However, others reported no effects of CLA on either bone mineral mass or density in humans.[38,103,131,132] Thus, it is not conclusive whether CLA improves bone health. However, along with observations that co-supplementation of CLA and calcium improves bone mass and CLA improves osteoblastogenesis (bone formation) while reducing osteoclastogenesis (bone resorption) in animal and cell culture models,[133-138] it is possible that CLA has great potential to improve bone health in specifically targeted populations, such as those at risk of developing osteoporosis associated with ageing and/or obesity.

3.3.6 Potential Health Concerns

Since CLA has been approved for use as a food additive, the potential health concerns are very important issues to address.[3,26] Current regulation in US FDA is that consumption of CLA, as a mixture of about 60–90% of the *cis*-9,*trans*-11 and *trans*-10, *cis*-12 isomers in approximately 50 : 50 ratio, up to 6 g CLA per day for 1 year or 3.4 g CLA per day for up to 2 years is considered to be safe.[22,33,34,131] Currently four aspects of CLA supplementation are of concern; effects on oxidative stress, insulin sensitivity, fatty liver, and milk fat depression.[4,100,139–143]

3.3.6.1 *CLA and Oxidative Stress*

CLA supplementation has been consistently associated with increased oxidative markers in human clinical trials; increased serum or urine isoprostanes; 8-iso-prostaglandin $F_{2\alpha}$ and/or 15-ketodihydro-PGF$_{2\alpha}$, non-enzymatic and enzymatic oxidative markers, respectively.[41,98,100,109,111,144–148] Since increased oxidative stress is associated with development of certain chronic diseases, this can be of concern. However, no other biomarkers of oxidative stress were linked to CLA and CLA has been reported to be metabolized to structural analogues of isoprostanes that cannot be distinguished with methods used to measure the above mentioned isoprostanes.[149,150] Furthermore, CLA supplementation did not alter oxidized LDL, serum lipid peroxidation, nor lipid antioxidant concentrations, which are alternative markers of oxidative stress.[98,151]

3.3.6.2 *CLA and Insulin Sensitivity*

There are no consistent reports of CLA with regard to glucose metabolism in animal studies, not only in normal but also in obese and diabetic models.[21,72] Similarly, most human studies reported no effects of CLA on plasma levels of glucose or insulin with few exceptions.[96,105,152] Among long-term studies with CLA (6 months to 2 years), Gaullier *et al.*[131] reported elevated insulin levels: however, insulin C-peptide levels were not altered by CLA supplementation. Among two major isomers, the *trans*-10,*cis*-12 CLA isomer is particularly associated with altered glucose metabolism, but not the *cis*-9,*trans*-11 isomer. Moreover, when CLA mixture was used in the same study, no adverse effects of CLA on glucose metabolisms were observed.[100,153] This suggests that use of CLA mixture is beneficial over use of individual isomer preparation, particularly the *trans*-10,*cis*-12 CLA isomer. Overall, it is not conclusive whether CLA supplementation would adversely influence glucose metabolism.

3.3.6.3 CLA and Fatty Liver

There are reports that CLA supplementation caused fatty liver, in severe cases lipodystrophy, in animal studies.[140,154–156] Human studies found that CLA supplementation either caused no changes or changes within normal ranges for serum markers of liver functions or no changes in liver morphology with one exception of a single case of acute hepatitis.[3,26,40,157,158] When high doses of CLA were supplemented in healthy subjects for 3 weeks (19.3 g per day – 14.6 g cis-9, trans-11 and 4.7g trans-10,cis-12 CLA), no adverse effects of liver functions were observed.[157] This suggests that effects of CLA on the liver observed in animal studies are likely specific to rodents, particularly mice, but may not be a major concern for humans with the doses and durations tested currently. However, it will be necessary to closely monitor any potential adverse effects of CLA with regard to liver functions, particularly for long-term use.

3.3.6.4 CLA and Milk Fat Depression

Along with dramatic reduction in body fat, CLA reduced milk fat content, called milk fat depression, in animal studies, particularly cows.[159,160] This may be of concern for lactating women. Currently there are four studies published with regard to CLA and human milk fat content.[161–164] Among them, Masters et al.[163] reported that CLA supplementation caused reduction of milk fat content within normal range, while two other studies reported no effects of CLA on milk fat content.[161,162] Another study used the cis-9,trans-11 CLA, which is known to be inactive in for fat reduction, thus no changes were reported.[164] The different observations between lactating cows and humans with regard to milk fat after CLA supplementation may be due to the different origin of milk fat. It is known that ruminant milk fat is mainly derived from *de novo* fatty acid synthesis, while human milk fat is mainly derived from diet and stored fat.[159,160,165] Thus, the milk fat reduction by CLA in cows is significant since CLA inhibits *de novo* fatty acid synthesis. It is also important to note that all human studies described above were relatively short term (about 5 days of supplementation), thus it is not clear if longer-term CLA supplementation would cause significant effects on milk fat content in humans.

3.4 Other Conjugated Fatty Acids

In addition to CLA, other conjugated fatty acids are currently known, either metabolites of CLA or naturally occurring conjugated fatty acids. Studies of biological activities as well as biochemical mechanisms of these fatty acids are rather limited, however, they may possess great potential for future applications.

3.4.1 CLA Metabolites

The levels of CLA in blood and tissue correlate well with dietary levels of CLA and decline without supplementation of CLA.[20,166] Decline of CLA levels in the biological system suggests that CLA may be metabolized as is the case with any other fatty acids. In fact, CLA metabolites have been reported, such as products of elongation and/or desaturation (Scheme 3.1).[20,167,168] There is only limited knowledge about the biological activities of these CLA metabolites. The elongated form of CLA, conjugated eicosadienoic acid as *cis*-11,*trans*-13 and *trans*-12,*cis*-14 isomers, exhibited equal or less effects on inhibition of adipocyte lipoprotein lipase activity and reduction of body fat in mice compared to effects of CLA.[168] It has been determined that conjugated eicosadienoic acid is converted to CLA in the biological system, thus any effects of this fatty acid are due to effects of CLA itself. Conjugated eicosatrienoic acid (*cis*-8,*trans*-12,*cis*-14 20:3), the desaturated and elongated form of the *trans*-10,*cis*-12 CLA isomer, had no or minimal effect on inhibition of lipoprotein, suggesting that unmetabolized CLA is important for its activity.[168]

It has also been reported that CLA itself can undergo fatty acid β-oxidation with the oxidation rate of CLA reported to be similar to those of other fatty acids.[20,67,168] Oxidative metabolites of CLA were conjugated hexadecadienoic (*cis*-7,*trans*-9 & *trans*-8,*cis*-10 16:2), conjugated tetradecadienoic (*cis*-5,*trans*-7 & *trans*-6,*cis*-8 14:2), and conjugated dodecadienoic acids (*cis*-3,*trans*-5 &

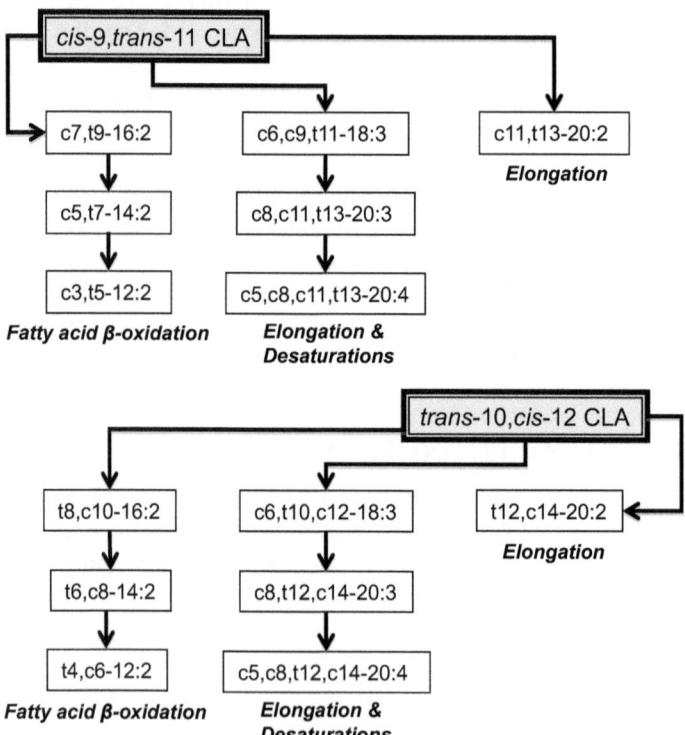

Scheme 3.1   Summary of metabolisms of CLA isomers.

trans-4,*cis*-6 12 : 2) (Scheme 3.1).[20,168] Nonetheless, the levels of these metabolites were very low and their bioactivities are yet to be determined.

3.4.2 Odd-Carbon Conjugated Fatty Acids

As CLA cognates, two odd-carbon conjugated fatty acids were previously
tested compared to CLA; conjugated heneicosadienoic acid (*cis*-12,*trans*-14 &
trans-13,*cis*-15 21 : 2) and conjugated nonadecadienoic acid (*cis*-10,*trans*-12 &
trans-11,*cis*-13 19 : 2).[168,169] Conjugated heneicosadienoic acid did not show
any effects, while conjugated nonadecadienoic acid (CNA) effectively inhibited lipoprotein lipase activity in 3T3-L1 adipocytes.[168,169] CNA was further tested and it effectively reduced body fat in mice model. CNA was five
times more active than CLA, and had similar biochemical mechanisms.[170]
Due to a one-carbon difference between CLA and CNA, it is unlikely that
these two fatty acids share similar metabolites, suggesting there may be
common biochemical pathways these two conjugated fatty acids share to
exhibit their biological functions. It is also important to note that unlike
CLA, which is found in natural products, CNA is not. Thus, application of
CNA is rather limited to pharmacological applications, however since it is
more effective than CLA itself, it may be used to provide additional benefits
to control body fat along with currently used antiobesity approaches.

3.4.3 Other Naturally Occurring Conjugated Fatty Acids

There are other naturally occurring conjugated fatty acids and their bioactivities are summarized in Table 3.1. Most of these oils have been tested for
their antitumor activities, with limited studies for other bioactivities, such as
antidiabetes, antiobesity, anti-inflammation, hypolipidemic, antioxidant,
and hypocholesterolemic effects (Table 3.1). Among them, α-eleostearic acid
(*cis*-9,*trans*-11,*trans*-13 18 : 3) and punicic acid (*cis*-9,*trans*-11,*cis*-13 18 : 3)
have been reported to be converted to the *cis*-9,*trans*-11 CLA isomer by Δ13-
saturation in animal tissues.[171,172] Similarly, jacaric acid (*cis*-8,*trans*-10,*trans*-
13 18 : 3) has been reported to converted to the *cis*-8,*trans*-10 CLA isomer by
Δ12-saturation in the small intestine of rats.[173] These naturally occurring
conjugated fats may serve as independent sources of CLA in the diet.

In addition to studies on the effects of these conjugated fatty acids on a
variety of bioactivities, one non-conjugated unsaturated fatty acid, pinolenic
acid (*cis*-5,*cis*-9,*cis*-12 18 : 3), has been tested for its biological activities.
Pinolenic acid primarily originates from pine nut oil (either from *Pinus
pinaster* or *Pinus koraiensis*) and has been reported to have antitumor,
hypocholesterolemic, and antiobesity effects.[174–182] Moreover, it has been
reported that consumption of pine nut oil reduces appetite in post-meno-
pausal overweight women.[183] Overall, additional research is needed to
confirm effects as well as determine mechanisms of action for these conjugated and non-conjugated fatty acids in the future.

Table 3.1 Summary of bioactivities from naturally occurring conjugated fatty acids

Oil	Fatty Acid	Bioactivities		Ref.
Pomegranate seed oil or **Snake gourd seed oil**	Punicic Acid (*c*9,*t*11,*c*13-18:3)	• Antitumor	– *In vitro*: Mouse fibroblast (A31, SV-T2), Human monocytic leukemia (U-937), Colon (DLD-1), Breast (MDA-MB-231, MDA-ERα7), Prostate (PC3, LNCaP); *In vivo* (CD1, TPA-induced skin tumor mice; Azoxymethane-induced colon tumor rats)	184–191
		• Anti-inflammation	– *In vitro* (3T3-L1, human neutrophils); *In vivo* (db/db, CD1, C57Bl/J6, IL-10⁻/⁻, PPARγ-null mice; Streptozotocin-induced diabetic rats, NaAsO₂-induced oxidative stress rats, chemically-induced inflammatory bowel rodent models)	192–197
		• Antidiabetes	– *In vitro* (3T3-L1); *In vivo* (db/db, CD1, C57Bl/J6, PPARγ-null mice; Streptozotocin-induced diabetic rats)	192,194,197,198
		• Antiobesity	– *In vivo* (CD1, C57Bl/J6 mice; OLETF rats)	197–199
		• Hypolipidemic	– *In vitro* (HepG2); *In vivo* (NaAsO₂-induced oxidative stress rats, OLETF rats)	199–201
		• Antioxidant	– *In vivo* (Streptozotocin-induced diabetic rats, NaAsO₂-induced oxidative stress rats)	194,195,202–205
Bitter gourd seed oil or **Tung oil**	α-Eleostearic Acid (*c*9,*t*11,*t*13-18:3)	• Antitumor	– *In vitro*: Mouse fibroblast (A31, SV-T2), Human leukemia (HL60, U-937), colon (DLD-1, Caco-2, HT-29), Breast (MDA-MB-231, MDA-ERα7, MCF-7), Liver (HepG2), Cervix (HeLa), Stomach (MKN-7), Lung (A549); *In vivo* (Transplant model, Azoxymethane-induced colon tumor rats)	184,189,206–217

Source	Fatty acid	Health benefit	Model	References
		• Anti-inflammation	– *In vitro*: Mouse macrophage (RAW 264.7); *In vivo* (chemical induced colon inflammation mice; Streptozotocin-induced diabetic rats, NaAsO$_2$-induced oxidative stress rats)	194,195,218,219
		• Antidiabetes	– *In vivo* (Streptozotocin-induced diabetic rats)	194
		• Antiobesity	– *In vitro* (3T3-L1)	220
		• Hypolipidemic	– *In vivo* (NaAsO$_2$-induced oxidative stress rats)	201
		• Antioxidant	– *In vitro*; *In vivo* (Streptozotocin-induced diabetic rats, NaAsO$_2$-induced oxidative stress rats)	194,195,201–204, 207,221,222
Tung oil	β-Eleostearic Acid (t9,t11,t13-18:3)	• Antitumor	– *In vitro* Human bladder (T24), Colon (DLD-1, Caco-2)	189,209,223
Jacaranda seed oil	Jacaric Acid (c8,t10,c12-18:3)	• Antitumor	– *In vitro*: Human colon (DLD-1), prostate (PC-3, LNCaP); *In vivo* (Transplant model)	188,189
Catalpa seed Oil	Catalpic Acid (t9,t11,c13-18:3)	• Antidiabetes	– *In vivo* (ICR mice)	224
		• Antiobesity	– *In vivo* (ICR mice)	224
		• Antitumor	– *In vitro*: Mouse fibroblast (A31, SV-T2), Human monocytic leukemia (U-937), prostate (PC-3, LNCaP), Colon (DLD-1)	184,188,189
Pot marigold seed oils	α- & β-Calendic Acid (t8,t10,c12 & t8,t10,t12-18:3)	• Antidiabetes	– *In vivo* (C57BL/6J, db/wt mice)	225
		• Antiobesity	– *In vivo* (C57BL/6J, db/wt mice)	225
		• Antitumor	– *In vitro*: Human prostate (PC-3, LNCaP), Colon (DLD-1, Caco-2)	188,189,209
Garden balsam seed oil	Parinaric Acid (c9,t11,t13,c15-18:4)	• Antitumor	– *In vitro*: Human leukemia (THP-1, HL60), retinoblastoma (Y79), Monocytic leukemia (U-937)	226

3.5 Conclusion

With the current obesity epidemic, CLA supplementation may provide additional benefit to control body fat and improve lean mass along with current treatment options for obesity. In addition, CLA may be used to help reducing other chronic diseases. The application of CLA in foods is expect to increase considerably in the next few years. Thus, close monitoring for potential health concerns including unforeseen adverse effects are necessary.

Acknowledgements

The author thanks Ms Jayne M. Storkson for help preparing this manuscript. This project is supported in part by the Department of Food Science and F. J. Francis Endowment at the University of Massachusetts, Amherst. Yeonhwa Park is one of the inventors of CLA use patents that are assigned to the Wisconsin Alumni Research Foundation.

References

1. M. W. Pariza and W. A. Hargraves, *Carcinogenesis*, 1985, **6**, 591.
2. Y. L. Ha, N. K. Grimm and M. W. Pariza, *Carcinogenesis*, 1987, **8**, 1881.
3. A. Dilzer and Y. Park, *Crit. Rev. Food Sci. Nutr.*, 2012, **52**, 488.
4. Y. Park, *J. Food Comp. Anal.*, 2009, **22S**, S4.
5. P. W. Parodi, in *Advances in Conjugated Linoleic Acid Resear, Volume 1*, ed. M. P. Yurawecz, M. M. Mossoba, J. K. G. Kramer, M. W. Pariza and G. J. Nelson, AOCS Press, Champaign, IL, 1999, p. 1.
6. T. R. Dhiman, G. R. Anand, L. D. Satter and M. W. Pariza, *J. Dairy Sci.*, 1999, **82**, 2146.
7. C. R. Kepler, K. P. Hirons, J. J. McNeill and S. B. Tove, *J. Biol. Chem.*, 1966, **241**, 1350.
8. J. K. Kramer, P. W. Parodi, R. G. Jensen, M. M. Mossoba, M. P. Yurawecz and R. O. Adlof, *Lipids*, 1998, **33**, 835.
9. J. K. Kay, T. R. Mackle, M. J. Auldist, N. A. Thomson and D. E. Bauman, *J. Dairy Sci.*, 2004, **87**, 369.
10. B. A. Corl, D. M. Barbano, D. E. Bauman and C. Ip, *J. Nutr.*, 2003, **133**, 2893.
11. J. K. Kramer, N. Sehat, M. E. Dugan, M. M. Mossoba, M. P. Yurawecz, J. A. Roach, K. Eulitz, J. L. Aalhus, A. L. Schaefer and Y. Ku, *Lipids*, 1998, **33**, 549.
12. S. F. Chin, J. M. Storkson, Y. L. Ha and M. W. Pariza, *J. Food Comp. Anal.*, 1992, **5**, 185.
13. Y. Park, M. K. McGuire, R. Behr, M. A. McGuire, M. A. Evans and T. D. Shultz, *Lipids*, 1999, **34**, 543.
14. S. Mushtaq, E. Heather Mangiapane and K. A. Hunter, *Br. J. Nutr.*, 2010, **103**, 1366.

15. K. L. Ritzenthaler, M. K. McGuire, R. Falen, T. D. Shultz, N. Dasgupta and M. A. McGuire, *J. Nutr.*, 2001, **131**, 1548.
16. B. K. Herbel, M. K. McGuire, M. A. McGuire and T. D. Shultz, *Am. J. Clin. Nutr.*, 1998, **67**, 332.
17. T. A. McCrorie, E. M. Keaveney, J. M. Wallace, N. Binns and M. B. Livingstone, *Nutr. Res. Rev.*, 2011, **24**, 206.
18. US Food and Drug Administration, *Nutrition Labeling: Trans fat labeling.*, http://www.fda.gov/Food/GuidanceRegulation/GuidanceDocuments RegulatoryInformation/LabelingNutrition/ucm064904.htm#transfat.
19. Y. Park, K. J. Albright, W. Liu, J. M. Storkson, M. E. Cook and M. W. Pariza, *Lipids*, 1997, **32**, 853.
20. M. C. Mele, G. Cannelli, G. Carta, L. Cordeddu, M. P. Melis, E. Murru, C. Stanton and S. Banni, *Prostaglandins Leukot. Essent. Fatty Acids*, 2013, **89**, 115.
21. Y. Park and M. W. Pariza, *Food Res. Int.*, 2007, **40**, 311.
22. J. M. Gaullier, J. Halse, K. Hoye, K. Kristiansen, H. Fagertun, H. Vik and O. Gudmundsen, *Am. J. Clin. Nutr.*, 2004, **79**, 1118.
23. A. H. Terpstra, M. Javadi, A. C. Beynen, S. Kocsis, A. E. Lankhorst, A. G. Lemmens and I. C. Mohede, *J. Nutr.*, 2003, **133**, 3181.
24. T. Porsgaard, X. Xu and H. Mu, *J. Nutr.*, 2006, **136**, 2201.
25. L. D. Whigham, A. C. Watras and D. A. Schoeller, *Am. J. Clin. Nutr.*, 2007, **85**, 1203.
26. I. J. Onakpoya, P. P. Posadzki, L. K. Watson, L. A. Davies and E. Ernst, *Eur. J. Nutr.*, 2012, **51**, 127.
27. D. A. Schoeller, A. C. Watras and L. D. Whigham, *Appl. Physiol. Nutr. Metab.*, 2009, **34**, 975.
28. M. M. Kamphuis, M. P. Lejeune, W. H. Saris and M. S. Westerterp-Plantenga, *Int. J. Obes. Relat. Metab. Disord.*, 2003, **27**, 840.
29. A. C. Watras, A. C. Buchholz, R. N. Close, Z. Zhang and D. A. Schoeller, *Int. J. Obes. (Lond.)*, 2007, **31**, 481.
30. Y. Park, K. J. Albright, J. M. Storkson, W. Liu and M. W. Pariza, *J. Food Sci.*, 2007, **72**, S612.
31. R. L. Atkinson, in *Advances in Conjugated Linoleic Acid Research*, ed. M. P. Yurawecz, M. M. Mossoba, J. K. G. Kramer, M. W. Pariza and G. J. Nelson, AOCS Press, Champaign, IL, 1999, vol. 1, p. 348.
32. M. M. Kamphuis, M. P. Lejeune, W. H. Saris and M. S. Westerterp-Plantenga, *Eur. J. Clin. Nutr.*, 2003, **57**, 1268.
33. T. M. Larsen, S. Toubro, O. Gudmundsen and A. Astrup, *Am. J. Clin. Nutr.*, 2006, **83**, 606.
34. L. D. Whigham, M. O'Shea, I. C. Mohede, H. P. Walaski and R. L. Atkinson, *Food Chem. Toxicol.*, 2004, **42**, 1701.
35. T. J. Doherty, *J. Appl. Physiol.*, 2003, **95**, 1717.
36. D. Glass and R. Roubenoff, *Ann. N. Y. Acad. Sci.*, 2010, **1211**, 25.
37. M. V. Narici and N. Maffulli, *Br. Med. Bull.*, 2010, **95**, 139.
38. M. Tarnopolsky, A. Zimmer, J. Paikin, A. Safdar, A. Aboud, E. Pearce, B. Roy and T. Doherty, *PLoS ONE*, 2007, **2**, e991.

39. E. J. Noone, H. M. Roche, A. P. Nugent and M. J. Gibney, *Br. J. Nutr.*, 2002, **88**, 243.
40. C. Malpucch-Brugere, W. P. Verboeket-van de Venne, R. P. Mensink, M. A. Arnal, B. Morio, M. Brandolini, A. Saebo, T. S. Lassel, J. M. Chardigny, J. L. Sebedio and B. Beaufrere, *Obes. Res.*, 2004, **12**, 591.
41. U. Riserus, B. Vessby, J. Arnlov and S. Basu, *Am. J. Clin. Nutr.*, 2004, **80**, 279.
42. S. Desroches, P. Y. Chouinard, I. Galibois, L. Corneau, J. Delisle, B. Lamarche, P. Couture and N. Bergeron, *Am. J. Clin. Nutr.*, 2005, **82**, 309.
43. S. Tricon, G. C. Burdge, E. L. Jones, J. J. Russell, S. El-Khazen, E. Moretti, W. L. Hall, A. B. Gerry, D. S. Leake, R. F. Grimble, C. M. Williams, P. C. Calder and P. Yaqoob, *Am. J. Clin. Nutr.*, 2006, **83**, 744.
44. J. Herrmann, D. Rubin, R. Hasler, U. Helwig, M. Pfeuffer, A. Auinger, C. Laue, P. Winkler, S. Schreiber, D. Bell and J. Schrezenmeir, *Lipids Health. Dis.*, 2009, **8**, 35.
45. M. Raff, T. Tholstrup, S. Toubro, J. M. Bruun, P. Lund, E. M. Straarup, R. Christensen, M. B. Sandberg and S. Mandrup, *J. Nutr.*, 2009, **139**, 1347.
46. I. Sluijs, Y. Plantinga, B. de Roos, L. I. Mennen and M. L. Bots, *Am. J. Clin. Nutr.*, 2010, **91**, 175.
47. S. V. Joseph, H. Jacques, M. Plourde, P. L. Mitchell, R. S. McLeod and P. J. Jones, *J. Nutr.*, 2011, **141**, 1286.
48. S. Venkatramanan, S. V. Joseph, P. Y. Chouinard, H. Jacques, E. R. Farnworth and P. J. Jones, *J. Am. Coll. Nutr.*, 2010, **29**, 152.
49. Y. Park, J. M. Storkson, K. J. Albright, W. Liu and M. W. Pariza, *Lipids*, 1999, **34**, 235.
50. D. B. West, J. P. Delany, P. M. Camet, F. Blohm, A. A. Truett and J. Scimeca, *Am. J. Physiol.*, 1998, **275**, R667.
51. Y. Park and Y. Park, *Food Chem.*, 2012, **133**, 400.
52. K. Nagao, N. Inoue, Y. M. Wang, J. Hirata, Y. Shimada, T. Nagao, T. Matsui and T. Yanagita, *Biochem. Biophys. Res. Commun.*, 2003, **306**, 134.
53. K. Ohnuki, S. Haramizu, K. Oki, K. Ishihara and T. Fushiki, *Lipids*, 2001, **36**, 583.
54. K. Ohnuki, S. Haramizu, K. Ishihara and T. Fushiki, *Biosci. Biotechnol. Biochem.*, 2001, **65**, 2200.
55. A. H. Terpstra, A. C. Beynen, H. Everts, S. Kocsis, M. B. Katan and P. L. Zock, *J. Nutr.*, 2002, **132**, 940.
56. J. A. Nazare, A. B. de la Perriere, F. Bonnet, M. Desage, J. Peyrat, C. Maitrepierre, C. Louche-Pelissier, J. Bruzeau, J. Goudable, T. Lassel, H. Vidal and M. Laville, *Br. J. Nutr.*, 2007, **97**, 273.
57. R. N. Close, D. A. Schoeller, A. C. Watras and E. H. Nora, *Am. J. Clin. Nutr.*, 2007, **86**, 797.

58. C. Pinkoski, P. D. Chilibeck, D. G. Candow, D. Esliger, J. B. Ewaschuk, M. Facci, J. P. Farthing and G. A. Zello, *Med. Sci. Sports Exerc.*, 2006, **38**, 339.
59. E. V. Lambert, J. H. Goedecke, K. Bluett, K. Heggie, A. Claassen, D. E. Rae, S. West, J. Dugas, L. Dugas, S. Meltzeri, K. Charlton and I. Mohede, *Br. J. Nutr.*, 2007, **97**, 1001.
60. S. E. Steck, A. M. Chalecki, P. Miller, J. Conway, G. L. Austin, J. W. Hardin, C. D. Albright and P. Thuillier, *J. Nutr.*, 2007, **137**, 1188.
61. K. L. Zambell, N. L. Keim, M. D. Van Loan, B. Gale, P. Benito, D. S. Kelley and G. J. Nelson, *Lipids*, 2000, **35**, 777.
62. J. C. Bouthegourd, P. C. Even, D. Gripois, B. Tiffon, M. F. Blouquit, S. Roseau, C. Lutton, D. Tome and J. C. Martin, *J. Nutr.*, 2002, **132**, 2682.
63. P. Degrace, L. Demizieux, J. Gresti, J. M. Chardigny, J. L. Sebedio and P. Clouet, *J. Nutr.*, 2004, **134**, 861.
64. K. Nagao, N. Inoue, Y. M. Wang, B. Shirouchi and T. Yanagita, *J. Nutr.*, 2005, **135**, 9.
65. J. M. Peters, Y. Park, F. J. Gonzalez and M. W. Pariza, *Biochim. Biophys. Acta*, 2001, **1533**, 233.
66. J. Ribot, M. P. Portillo, C. Pico, M. T. Macarulla and A. Palou, *Br. J. Nutr.*, 2007, **97**, 1074–1082, DOI: 10.1017/S0007114507682932.
67. C. Malpuech-Brugere, R. P. Mensink, O. Loreau, A. Maret, C. E. Fernie, T. S. Lassel, J. M. Chardigny, C. M. Scrimgeour, J. L. Sebedio and B. Beaufrere, *Lipids*, 2010, **45**, 1047.
68. A. Kennedy, K. Martinez, S. Schmidt, S. Mandrup, K. LaPoint and M. McIntosh, *J. Nutr. Biochem.*, 2010, **21**, 171.
69. H. Blankson, J. A. Stakkestad, H. Fagertun, E. Thom, J. Wadstein and O. Gudmundsen, *J. Nutr.*, 2000, **130**, 2943.
70. K. W. Lee, H. J. Lee, H. Y. Cho and Y. J. Kim, *Crit. Rev. Food Sci. Nutr.*, 2005, **45**, 135.
71. N. S. Kelley, N. E. Hubbard and K. L. Erickson, *J. Nutr.*, 2007, **137**, 2599.
72. A. Bhattacharya, J. Banu, M. Rahman, J. Causey and G. Fernandes, *J. Nutr. Biochem.*, 2006, **17**, 789.
73. P. A. Masso-Welch, D. Zangani, C. Ip, M. M. Vaughan, S. F. Shoemaker, S. O. McGee and M. M. Ip, *J. Nutr.*, 2004, **134**, 299.
74. S. H. Lee, K. Yamaguchi, J. S. Kim, T. E. Eling, S. Safe, Y. Park and S. J. Baek, *Carcinogenesis*, 2006, **27**, 972.
75. C. Ip, Y. Dong, M. M. Ip, S. Banni, G. Carta, E. Angioni, E. Murru, S. Spada, M. P. Melis and A. Saebo, *Nutr. Cancer*, 2002, **43**, 52.
76. P. Knekt, R. Jarvinen, R. Seppanen, E. Pukkala and A. Aromaa, *Br. J. Cancer*, 1996, **73**, 687.
77. A. Aro, S. Mannisto, I. Salminen, M. L. Ovaskainen, V. Kataja and M. Uusitupa, *Nutr. Cancer*, 2000, **38**, 151.
78. V. Chajes, F. Lavillonniere, P. Ferrari, M. L. Jourdan, M. Pinault, V. Maillard, J. L. Sebedio and P. Bougnoux, *Cancer Epidemiol. Biomarkers Prev.*, 2002, **11**, 672.

79. L. E. Voorrips, H. A. Brants, A. F. Kardinaal, G. J. Hiddink, P. A. van den Brandt and R. A. Goldbohm, *Am. J. Clin. Nutr.*, 2002, **76**, 873.
80. V. Chajes, F. Lavillonniere, V. Maillard, B. Giraudeau, M. L. Jourdan, J. L. Sebedio and P. Bougnoux, *Nutr. Cancer*, 2003, **45**, 17–23.
81. H. Rissanen, P. Knekt, R. Jarvinen, I. Salminen and T. Hakulinen, *Nutr. Cancer*, 2003, **45**, 168.
82. S. E. McCann, C. Ip, M. M. Ip, M. K. McGuire, P. Muti, S. B. Edge, M. Trevisan and J. L. Freudenheim, *Cancer Epidemiol. Biomarkers Prev.*, 2004, **13**, 1480.
83. S. C. Larsson, L. Bergkvist and A. Wolk, *Am. J. Clin. Nutr.*, 2005, **82**, 894.
84. M. Mohammadzadeh, E. Faramarzi, R. Mahdavi, B. Nasirimotlagh and M. Asghari Jafarabadi, *Integr. Cancer. Ther.*, 2013, **12**, 496.
85. M. E. Cook, D. L. Jerome, T. D. Crenshaw, D. R. Buege, M. W. Pariza, K. J. Albright, S. P. Schmidt, J. A. Scimeca, P. A. Lofgren and E. J. Hentges, *FASEB J*, 1998, **12**, A836.
86. C. C. Miller, Y. Park, M. W. Pariza and M. E. Cook, *Biochem. Biophys. Res. Commun.*, 1994, **198**, 1107.
87. E. Graves, A. Hitt, M. W. Pariza, M. E. Cook and D. O. McCarthy, *Res. Nurs. Health*, 2005, **28**, 48.
88. K. Rees, M. Dyakova, K. Ward, M. Thorogood and E. Brunner, *Cochrane Database Syst. Rev.*, 2013, **3**, CD002128.
89. K. N. Lee, D. Kritchevsky and M. W. Pariza, *Atherosclerosis*, 1994, **108**, 19.
90. R. J. Nicolosi, E. J. Rogers, D. Kritchevsky, J. A. Scimeca and P. J. Huth, *Artery*, 1997, **22**, 266.
91. D. Kritchevsky, S. A. Tepper, S. Wright, S. K. Czarnecki, T. A. Wilson and R. J. Nicolosi, *Lipids*, 2004, **39**, 611.
92. T. A. Wilson, R. J. Nicolosi, A. Saati, T. Kotyla and D. Kritchevsky, *Lipids*, 2006, **41**, 41.
93. G. Berven, A. Bye, O. Hals, H. Blankson, H. Fagertun, E. Thom, J. Wadstein and O. Gudmundsen, *Eur. J. Lipid Sci. Tech.*, 2000, **102**, 455.
94. J. A. Herrera, M. Arevalo-Herrera, A. K. Shahabuddin, G. Ersheng, S. Herrera, R. G. Garcia and P. Lopez-Jaramillo, *Am. J. Hypertens.*, 2006, **19**, 381.
95. J. A. Herrera, A. K. Shahabuddin, G. Ersheng, Y. Wei, R. G. Garcia and P. Lopez-Jaramillo, *Int. J. Gynaecol. Obstet.*, 2005, **91**, 221.
96. T. Iwata, T. Kamegai, Y. Yamauchi-Sato, A. Ogawa, M. Kasai, T. Aoyama and K. Kondo, *J. Oleo Sci.*, 2007, **56**, 517.
97. W. S. Zhao, J. J. Zhai, Y. H. Wang, P. S. Xie, X. J. Yin, L. X. Li and K. L. Cheng, *Am. J. Hypertens.*, 2009, **22**, 680.
98. M. Pfeuffer, K. Fielitz, C. Laue, P. Winkler, D. Rubin, U. Helwig, K. Giller, J. Kammann, E. Schwedhelm, R. H. Boger, A. Bub, D. Bell and J. Schrezenmeir, *J. Am. Coll. Nutr.*, 2011, **30**, 19.
99. R. B. Kreider, M. P. Ferreira, M. Greenwood, M. Wilson and A. L. Almada, *J. Strength Cond. Res.*, 2002, **16**, 325.

100. U. Riserus, S. Basu, S. Jovinge, G. N. Fredrikson, J. Arnlov and B. Vessby, *Circulation*, 2002, **106**, 1925.
101. U. Riserus, B. Vessby, P. Arner and B. Zethelius, *Diabetologia*, 2004, **47**, 1016.
102. A. Smedman, S. Basu, S. Jovinge, G. N. Fredrikson and B. Vessby, *Br. J. Nutr.*, 2005, **94**, 791.
103. J. M. Gaullier, J. Halse, H. O. Hoivik, K. Hoye, C. Syvertsen, M. Nurminiemi, C. Hassfeld, A. Einerhand, M. O'Shea and O. Gudmundsen, *Br. J. Nutr.*, 2007, **97**, 550.
104. M. Madjid and J. T. Willerson, *Br. Med. Bull.*, 2011, **100**, 23.
105. F. Moloney, T. P. Yeow, A. Mullen, J. J. Nolan and H. M. Roche, *Am. J. Clin. Nutr.*, 2004, **80**, 887.
106. S. Tricon, G. C. Burdge, S. Kew, T. Banerjee, J. J. Russell, R. F. Grimble, C. M. Williams, P. C. Calder and P. Yaqoob, *Am. J. Clin. Nutr.*, 2004, **80**, 1626.
107. J. D. Ramakers, J. Plat, J. L. Sebedio and R. P. Mensink, *Lipids*, 2005, **40**, 909.
108. M. Raff, T. Tholstrup, S. Basu, P. Nonboe, M. T. Sorensen and E. M. Straarup, *J. Nutr.*, 2008, **138**, 509.
109. A. M. Turpeinen, N. Ylonen, E. von Willebrand, S. Basu and A. Aro, *Br. J. Nutr.*, 2008, **100**, 112.
110. M. L. Asp, A. L. Collene, L. E. Norris, R. M. Cole, M. B. Stout, S. Y. Tang, J. C. Hsu and M. A. Belury, *Clin. Nutr.*, 2011, **30**, 443.
111. L. A. Smit, M. B. Katan, A. J. Wanders, S. Basu and I. A. Brouwer, *J. Nutr.*, 2011, **141**, 1673.
112. F. Sofi, A. Buccioni, F. Cesari, A. M. Gori, S. Minieri, L. Mannini, A. Casini, G. F. Gensini, R. Abbate and M. Antongiovanni, *Nutr. Metab. Cardiovasc. Dis.*, 2010, **20**, 117.
113. R. Albers, R. P. van der Wielen, E. J. Brink, H. F. Hendriks, V. N. Dorovska-Taran and I. C. Mohede, *Eur. J. Clin. Nutr.*, 2003, **57**, 595.
114. K. M. Peterson, M. O'Shea, W. Stam, I. C. Mohede, J. T. Patrie and F. G. Hayden, *Antivir. Ther.*, 2009, **14**, 33.
115. C. Thijs, A. Muller, L. Rist, I. Kummeling, B. E. Snijders, M. Huber, R. van Ree, A. P. Simoes-Wust, P. C. Dagnelie and P. A. van den Brandt, *Allergy*, 2011, **66**, 58.
116. R. Hontecillas, M. J. Wannemeulher, D. R. Zimmerman, D. L. Hutto, J. H. Wilson, D. U. Ahn and J. Bassaganya Ricra, *J. Nutr.*, 2002, **132**, 2019.
117. J. Bassaganya-Riera, K. Reynolds, S. Martino-Catt, Y. Cui, L. Hennighausen, F. Gonzalez, J. Rohrer, A. U. Benninghoff and R. Hontecillas, *Gastroenterology*, 2004, **127**, 777.
118. J. Bassaganya-Riera and R. Hontecillas, *Clin. Nutr.*, 2006, **25**, 454.
119. J. Bassaganya-Riera, M. Viladomiu, M. Pedragosa, C. De Simone, A. Carbo, R. Shaykhutdinov, C. Jobin, J. C. Arthur, B. A. Corl, H. Vogel, M. Storr and R. Hontecillas, *PLoS One*, 2012, 7, e31238.
120. J. Bassaganya-Riera, R. Hontecillas, W. T. Horne, M. Sandridge, H. H. Herfarth, R. Bloomfeld and K. L. Isaacs, *Clin. Nutr.*, 2012, **31**, 721.

121. N. P. Evans, S. A. Misyak, E. M. Schmelz, A. J. Guri, R. Hontecillas and J. Bassaganya-Riera, *J. Nutr.*, 2010, **140**, 515.
122. B. A. Watkins, Y. Li, H. E. Lippman, S. Reinwald and M. F. Seifert, *Am. J. Clin. Nutr.*, 2004, **79**, 1175S.
123. Y. Park, M. W. Pariza and Y. Park, *J. Food Sci.*, 2008, **73**, C556.
124. C. Jewell, S. Cusack and K. D. Cashman, *Prostaglandins Leukot. Essent. Fatty Acids*, 2005, **72**, 163.
125. C. Jewell and K. D. Cashman, *Br. J. Nutr.*, 2003, **89**, 639.
126. E. F. Murphy, C. Jewell, G. J. Hooiveld, M. Muller and K. D. Cashman, *Prostaglandins Leukot. Essent. Fatty Acids*, 2006, **74**, 295.
127. Y. Park, M. Terk and Y. Park, *J. Bone Miner. Metab.*, 2011, **29**, 268.
128. E. F. Murphy, G. J. Hooiveld, M. Muller, R. A. Calogero and K. D. Cashman, *Genes Nutr.*, 2009, **4**, 103.
129. R. A. Brownbill, M. Petrosian and J. Z. Ilich, *J. Am. Coll. Nutr.*, 2005, **24**, 177.
130. L. Doyle, C. Jewell, A. Mullen, A. P. Nugent, H. M. Roche and K. D. Cashman, *Eur. J. Clin. Nutr.*, 2005, **59**, 432.
131. J. M. Gaullier, J. Halse, K. Hoye, K. Kristiansen, H. Fagertun, H. Vik and O. Gudmundsen, *J. Nutr.*, 2005, **135**, 778.
132. N. M. Racine, A. C. Watras, A. L. Carrel, D. B. Allen, J. J. McVean, R. R. Clark, A. R. O'Brien, M. O'Shea, C. E. Scott and D. A. Schoeller, *Am. J. Clin. Nutr.*, 2010, **91**, 1157.
133. M. M. Rahman, G. V. Halade, P. J. Williams and G. Fernandes, *J. Cell. Physiol.*, 2011, **226**, 2406.
134. M. M. Rahman, A. Bhattacharya, J. Banu and G. Fernandes, *J. Nutr. Biochem.*, 2007, **18**, 467.
135. M. M. Rahman, A. Bhattacharya and G. Fernandes, *J. Lipid Res.*, 2006, **47**, 1739.
136. S. W. Ing and M. A. Belury, *Nutr. Rev.*, 2011, **69**, 123.
137. G. V. Halade, M. M. Rahman, P. J. Williams and G. Fernandes, *J. Nutr. Biochem.*, 2011, **22**, 459.
138. J. Kim, Y. Park, S. H. Lee and Y. Park, *J. Nutr. Biochem.*, 2013, **24**, 672.
139. D. S. Kelley and K. L. Erickson, *Lipids*, 2003, **38**, 377.
140. T. M. Larsen, S. Toubro and A. Astrup, *J. Lipid Res.*, 2003, **44**, 2234.
141. S. Tricon and P. Yaqoob, *Curr. Opin. Clin. Nutr. Metab. Care*, 2006, **9**, 105.
142. M. W. Pariza, *Am. J. Clin. Nutr.*, 2004, **79**, 1132S.
143. L. Clement, H. Poirier, I. Niot, V. Bocher, M. Guerre-Millo, S. Krief, B. Staels and P. Besnard, *J. Lipid Res.*, 2002, **43**, 1400.
144. S. Basu, U. Riserus, A. Turpeinen and B. Vessby, *Clin. Sci. (Lond)*, 2000, **99**, 511.
145. S. Basu, A. Smedman and B. Vessby, *FEBS Lett.*, 2000, **468**, 33.
146. A. Smedman, B. Vessby and S. Basu, *Clin. Sci. (Lond)*, 2004, **106**, 67.
147. J. S. Taylor, S. R. Williams, R. Rhys, P. James and M. P. Frenneaux, *Arterioscler. Thromb. Vasc. Biol.*, 2006, **26**, 307.

148. S. M. Cornish, D. G. Candow, N. T. Jantz, P. D. Chilibeck, J. P. Little, S. Forbes, S. Abeysekara and G. A. Zello, *Int. J. Sport Nutr. Exerc. Metab.*, 2009, **19**, 79.

149. S. Banni, A. Petroni, M. Blasevich, G. Carta, L. Cordeddu, E. Murru, M. P. Melis, A. Mahon and M. A. Belury, *Lipids*, 2004, **39**, 1143.

150. J. L. Sebedio, E. Angioni, J. M. Chardigny, S. Gregoire, P. Juaneda and O. Berdeaux, *Lipids*, 2001, **36**, 575.

151. J. Kim, H. D. Paik, M. J. Shin and E. Park, *Eur. J. Nutr.*, 2012, **51**, 135.

152. U. Riserus, L. Berglund and B. Vessby, *Int. J. Obes. Relat. Metab. Disord.*, 2001, **25**, 1129.

153. U. Riserus, P. Arner, K. Brismar and B. Vessby, *Diabetes Care*, 2002, **25**, 1516.

154. M. A. Belury and A. Kempa-Steczko, *Lipids*, 1997, **32**, 199.

155. A. Jaudszus, P. Moeckel, E. Hamelmann and G. Jahreis, *Ann. Nutr. Metab.*, 2010, **57**, 103.

156. N. Tsuboyama-Kasaoka, M. Takahashi, K. Tanemura, H. J. Kim, T. Tange, H. Okuyama, M. Kasai, S. Ikemoto and O. Ezaki, *Diabetes*, 2000, **49**, 1534.

157. A. J. Wanders, L. Leder, J. D. Banga, M. B. Katan and I. A. Brouwer, *Food Chem. Toxicol.*, 2010, **48**, 587.

158. R. Ramos, J. Mascarenhas, P. Duarte, C. Vicente and C. Casteleiro, *Dig. Dis. Sci.*, 2009, **54**, 1141.

159. L. Bernard, C. Leroux and Y. Chilliard, *Adv. Exp. Med. Biol.*, 2008, **606**, 67.

160. D. E. Bauman and J. M. Griinari, *Annu. Rev. Nutr.*, 2003, **23**, 203.

161. A. Hasin, J. M. Griinari, J. E. Williams, A. M. Shahin, M. A. McGuire and M. K. McGuire, *Lipids*, 2007, **42**, 835.

162. S. A. Mosley, A. M. Shahin, J. Williams, M. A. McGuire and M. K. McGuire, *Lipids*, 2007, **42**, 723.

163. N. Masters, M. A. McGuire, K. A. Beerman, N. Dasgupta and M. K. McGuire, *Lipids*, 2002, **37**, 133–138.

164. K. L. Ritzenthaler, M. K. McGuire, M. A. McGuire, T. D. Shultz, A. E. Koepp, L. O. Luedecke, T. W. Hanson, N. Dasgupta and B. P. Chew, *J. Nutr.*, 2005, **135**, 422.

165. D. L. Hachey, G. H. Silber, W. W. Wong and C. Garza, *Pediatr. Res.*, 1989, **25**, 63.

166. Y. Park, *Regulation of Energy Metabolism and the Catabolic Effects of Immune Stimulation by Conjugated Linoleic Acid*, PhD thesis, University of Wisconsin-Madison, Madison, WI, 1996.

167. J. P. Sergiel, J. M. Chardigny, J. L. Sebedio, O. Berdeaux, P. Juaneda, O. Loreau, B. Pasquis and J. P. Noel, *Lipids*, 2001, **36**, 1327.

168. Y. Park, J. M. Storkson, K. J. Albright, W. Liu and M. W. Pariza, *Biochim. Biophys. Acta*, 2005, **1687**, 120.

169. Y. Park and M. W. Pariza, *Biochim. Biophys. Acta*, 2001, **1533**, 171.

170. Y. Park and Y. Park, *J. Nutr. Biochem.*, 2010, **21**, 764.

171. T. Tsuzuki, Y. Tokuyama, M. Igarashi, K. Nakagawa, Y. Ohsaki, M. Komai and T. Miyazawa, *J. Nutr.*, 2004, **134**, 2634.
172. T. Tsuzuki, M. Igarashi, M. Komai and T. Miyazawa, *J. Nutr. Sci. Vitaminol. (Tokyo)*, 2003, **49**, 195.
173. R. Kijima, T. Honma, J. Ito, M. Yamasaki, A. Ikezaki, C. Motonaga, K. Nishiyama and T. Tsuduki, *J. Oleo Sci.*, 2013, **62**, 305.
174. S. Chen, C. Hsu, C. Li, J. Lu and L. Chuang, *Food Chem.*, 2011, **126**, 1708.
175. J. Lee, K. Lee, S. Lee, I. Kim and C. Rhee, *Lipids*, 2004, **39**, 383.
176. N. H. Le, S. Shin, T. H. Tu, C. S. Kim, J. H. Kang, G. Tsuyoshi, K. Teruo, S. N. Han and R. Yu, *J. Agric. Food Chem.*, 2012, **60**, 11935.
177. S. Michihiro and T. Oka, *Br. J. Nutr.*, 1994, **72**, 175.
178. G. Asset, E. Baugé, R. Wolff, J. Fruchart and J. Dallongeville, *Prostaglandins Leukot. Essent. Fatty Acids*, 2000, **62**, 307.
179. G. Asset, A. Leroy, E. Bauge, R. L. Wolff, J. Fruchart and J. Dallongeville, *Br. J. Nutr.*, 2000, **84**, 353.
180. G. Asset, E. Baugé, R. Wolff, J. Fruchart and J. Dallongeville, *Eur. J. Nutr.*, 2001, **40**, 268.
181. G. M. Hughes, E. J. Boyland, N. J. Williams, L. Mennen, C. Scott, T. C. Kirkham, J. A. Harrold, H. G. Keizer and J. Halford, *Lipids Health Dis*, 2008, 7, 1.
182. G. Asset, B. Staels, R. L. Wolff, E. Bauge, Z. Madj, J. C. Fruchart and J. Dallongeville, *Lipids*, 1999, **34**, 39.
183. W. J. Pasman, J. Heimerikx, C. M. Rubingh, R. van den Berg, M. O'Shea, L. Gambelli, H. F. Hendriks, A. W. Einerhand, C. Scott, H. G. Keizer and L. I. Mennen, *Lipids Health. Dis.*, 2008, 7, 10.
184. R. Suzuki, R. Noguchi, T. Ota, M. Abe, K. Miyashita and T. Kawada, *Lipids*, 2001, **36**, 477.
185. M. E. Grossmann, N. K. Mizuno, T. Schuster and M. P. Cleary, *Int. J. Oncol.*, 2010, **36**, 421.
186. E. P. Lansky, G. Harrison, P. Froom and W. G. Jiang, *Invest. New Drugs*, 2005, **23**, 121.
187. J. Gasmi and J. T. Sanderson, *J. Agric. Food Chem.*, 2010, **58**, 12149.
188. J. Gasmi and J. Thomas Sanderson, *Phytomedicine*, 2013, **20**, 734.
189. N. Shinohara, T. Tsuduki, J. Ito, T. Honma, R. Kijima, S. Sugawara, T. Arai, M. Yamasaki, A. Ikezaki and M. Yokoyama, *Biochim. Biophys. Acta*, 2012, **1821**, 980.
190. J. J. Hora, E. R. Maydew, E. P. Lansky and C. Dwivedi, *J. Med. Food*, 2003, **6**, 157.
191. H. Kohno, R. Suzuki, Y. Yasui, M. Hosokawa, K. Miyashita and T. Tanaka, *Cancer. Sci.*, 2004, **95**, 481.
192. R. Hontecillas, M. O'Shea, A. Einerhand, M. Diguardo and J. Bassaganya-Riera, *J. Am. Coll. Nutr.*, 2009, **28**, 184.
193. T. Boussetta, H. Raad, P. Lettéron, M. Gougerot-Pocidalo, J. Marie, F. Driss and J. El-Benna, *PLoS One*, 2009, **4**, e6458.
194. S. S. Saha and M. Ghosh, *Br. J. Nutr.*, 2012, **108**, 974.

195. S. S. Saha, P. Dasgupta, S. Sengupta Bandyopadhyay and M. Ghosh, *Biochim. Biophys. Acta*, 2012, **1820**, 1951.
196. J. Bassaganya-Riera, M. DiGuardo, M. Climent, C. Vives, A. Carbo, Z. E. Jouni, A. W. Einerhand, M. O'Shea and R. Hontecillas, *Br. J. Nutr.*, 2011, **106**, 878.
197. B. K. McFarlin, K. A. Strohacker and M. L. Kueht, *Br. J. Nutr.*, 2009, **102**, 54.
198. I. O. Vroegrijk, J. A. van Diepen, S. van den Berg, I. Westbroek, H. Keizer, L. Gambelli, R. Hontecillas, J. Bassaganya-Riera, G. Zondag and J. A. Romijn, *Food Chem. Toxicol.*, 2011, **49**, 1426.
199. K. Arao, Y. Wang, N. Inoue, J. Hirata, J. Cha, K. Nagao and T. Yanagita, *Lipids Health Dis*, 2004, **3**, 1.
200. K. Arao, H. Yotsumoto, S. Han, K. Nagao and T. Yanagita, *Biosci. Biotechnol. Biochem.*, 2004, **68**, 2643.
201. S. S. Saha, A. Chakraborty, S. Ghosh and M. Ghosh, *Eur. J. Nutr.*, 2012, **51**, 483.
202. S. S. Saha and M. Ghosh, *Food Chem. Toxicol.*, 2009, **47**, 2551.
203. S. Saha and M. Ghosh, *Food Chem. Toxicol.*, 2010, **48**, 3398.
204. S. S. Saha and M. Ghosh, *Chem. Biol. Interact.*, 2011, **190**, 109.
205. C. Mukherjee, S. Bhattacharyya, S. Ghosh and D. K. Bhattacharyya, *J. Oleo Sci.*, 2002, **51**, 513.
206. M. Kobori, M. Ohnishi-Kameyama, Y. Akimoto, C. Yukizaki and M. Yoshida, *J. Agric. Food Chem.*, 2008, **56**, 10515.
207. T. Tsuzuki, Y. Tokuyama, M. Igarashi and T. Miyazawa, *Carcinogenesis*, 2004, **25**, 1417.
208. Y. Yasui, M. Hosokawa, H. Kohno, T. Tanaka and K. Miyashita, *Chemotherapy*, 2006, **52**, 220.
209. Y. Yasui, M. Hosokawa, H. Khono, T. Tanaka and K. Miyashita, *Anticancer Res.*, 2006, **26**, 1855.
210. Y. Yasui, M. Hosokawa, T. Sahara, R. Suzuki, S. Ohgiya, H. Kohno, T. Tanaka and K. Miyashita, *Prostaglandins Leukot. Essent. Fatty Acids*, 2005, **73**, 113.
211. M. E. Grossmann, N. K. Mizuno, M. L. Dammen, T. Schuster, A. Ray and M. P. Cleary, *Cancer Prevention Research*, 2009, **2**, 879.
212. M. Igarashi and T. Miyazawa, *Cancer Lett.*, 2000, **148**, 173.
213. J. Eom, M. Seo, J. Baek, H. Chu, S. H. Han, T. S. Min, C. Cho and C. Yun, *Biochem. Biophys. Res. Commun.*, 2010, **391**, 903.
214. H. Moon, D. Guo, H. Lee, Y. Choi, J. Kang, K. Jo, J. Eom, C. Yun and C. Cho, *Cancer science*, 2010, **101**, 396.
215. H. Kohno, Y. Yasui, R. Suzuki, M. Hosokawa, K. Miyashita and T. Tanaka, *Int. J. Cancer*, 2004, **110**, 896.
216. T. Tsuzuki and Y. Kawakami, *Carcinogenesis*, 2008, **29**, 797.
217. T. Zhang, Y. Gao, Y. Mao, Q. Zhang, C. Lin, P. Lin, J. Zhang and X. Wang, *J. Nat. Med.*, 2012, **66**, 77.
218. S. N. Lewis, L. Brannan, A. J. Guri, P. Lu, R. Hontecillas, J. Bassaganya-Riera and D. R. Bevan, *PloS one*, 2011, **6**, e24031.

219. M. Kobori, H. Nakayama, K. Fukushima, M. Ohnishi-Kameyama, H. Ono, T. Fukushima, Y. Akimoto, S. Masumoto, C. Yukizaki and Y. Hoshi, *J. Agric. Food Chem.*, 2008, **56**, 4004.
220. Y. Chou, H. Su, T. Lai, J. Chyuan and P. Chao, *Nutrition*, 2012, **28**, 803.
221. P. Dhar, K. Chattopadhyay, D. Bhattacharyya, A. Roychoudhury, A. Biswas and S. Ghosh, *J. Oleo Sci.*, 2006, **56**, 19.
222. P. Dhar, S. Ghosh and D. Bhattacharyya, *Lipids*, 1999, **34**, 109.
223. Z. Sun, H. Wang, S. Ye, S. Xiao, J. Liu, W. Wang, D. Jiang, X. Liu and J. Wang, *Prostaglandins Other Lipid Mediat.*, 2012, **99**, 1.
224. N. Shinohara, J. Ito, T. Tsuduki, T. Honma, R. Kijima, S. Sugawara, T. Arai, M. Yamasaki, A. Ikezaki, M. Yokoyama, K. Nishiyama, K. Nakagawa, T. Miyazawa and I. Ikeda, *J. Oleo Sci.*, 2012, **61**, 433.
225. R. Hontecillas, M. Diguardo, E. Duran, M. Orpi and J. Bassaganya-Riera, *Clin. Nutr.*, 2008, **27**, 764.
226. A. S. Cornelius, N. R. Yerram, D. A. Kratz and A. A. Spector, *Cancer Res.*, 1991, **51**, 6025.

CHAPTER 4

Commercial CLA and its Chemical Use

RAFAEL LOPES QUIRINO

Chemistry Department, Georgia Southern University, Statesboro, GA – 30460, USA
Email: rquirino@georgiasouthern.edu

4.1   Introduction

Vegetable oils are triglycerides of various fatty acid compositions, naturally produced in the seeds of a variety of plants. The main difference between distinct natural oils is the length of the fatty acid chains, the number and the position of the carbon–carbon double bonds along those fatty acid chains and, in some cases, the presence of specific functional groups. The fatty acid composition of vegetable oils varies according to the corresponding growing conditions experienced by the parent plant.[1] The generic chemical structure of triglycerides is given in Figure 4.1, while Table 4.1 lists a series of fatty acids of commercial relevance. Naturally occurring fatty acids contain an even number of carbon atoms, and the vast majority of the carbon–carbon double bonds in the unsaturated fatty acids have a *cis* configuration. In polyunsaturated fatty acids, the carbon–carbon double bonds are usually non-conjugated. Conjugated vegetable oils from the isomerization of carbon–carbon double bonds have been initially reported as by-products of reactions involving vegetable oils.[2] Since then, the catalytic production of conjugated fatty acids and triglycerides has been studied and optimized.[3,4] For the remainder of this chapter, the term "conjugated" refers to carbon–carbon double bonds that are conjugated (as in a 1,3-diene).

RSC Catalysis Series No. 19
Conjugated Linoleic Acids and Conjugated Vegetable Oils
Edited by Bert Sels and An Philippaerts
© The Royal Society of Chemistry 2014
Published by the Royal Society of Chemistry, www.rsc.org

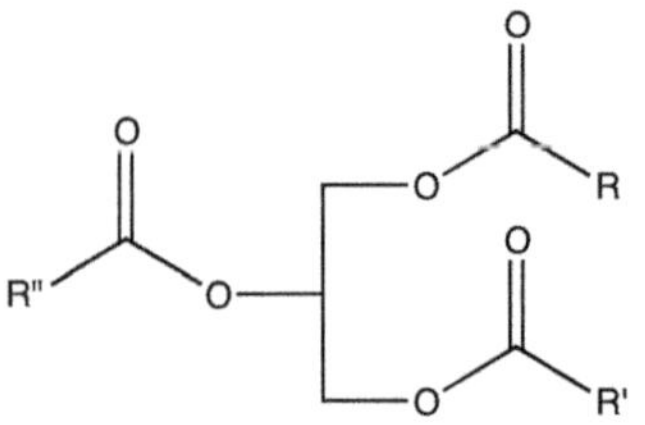

Figure 4.1 Generic chemical structure of a triglyceride. R, R′, and R″ are distinct hydrocarbon chains.

Table 4.1 Most common fatty acids and their natural source.

Fatty acid	Natural source
Linolenic acid	Linseed oil
Linoleic acid	Walnut and Safflower oils
Conjugated linoleic acid (CLA)	Ruminant animals
Oleic acid	Olive oil
Stearic acid	Cocoa and shea butters
Palmitic acid	Palm oil
Palmitoleic acid	Human adipose tissue
Vaccenic acid	Ruminant animals
Arachidonic acid	Phospholipids
Arachidic acid	Peanut oil
α-eleostearic acid	Tung oil
Ricinoleic acid	Castor oil
Eicosa pentaenoic acid (EPA)	Algae oil
Docosa hexaenoic acid (DHA)	Algae oil

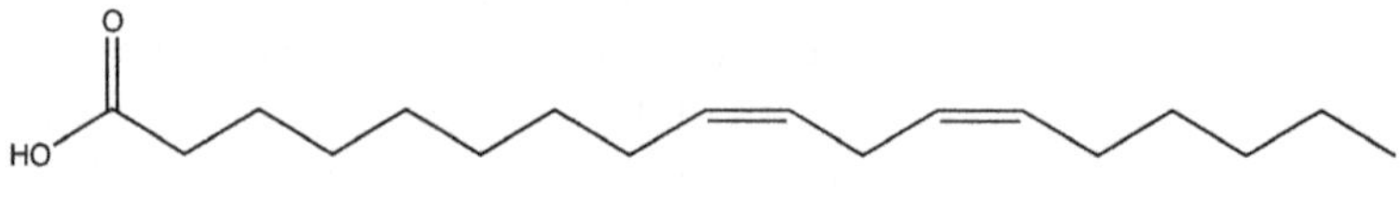

Figure 4.2 Chemical structure of linoleic acid.

Linoleic acid is one of the most abundant fatty acids naturally found in vegetable oils. The chemical structure of linoleic acid is presented in Figure 4.2. Table 4.2 lists the percentage of linoleic acid present in the most common natural oils. As a result of a system of non-conjugated carbon–carbon double bonds, regular linoleic acid is fairly unreactive. Upon isomerization and conjugation of these bonds, many interesting and useful properties arise. All of the possible positional isomers of linoleic acid with conjugated carbon–carbon double bonds are denoted by the term conjugated linoleic acid (CLA). Over the years, the production of conjugated fatty acids and triglycerides has been studied and optimized.[4,5] Vegetable oils can also be used as a source of CLA.[6] Indeed, vegetable oils with a high content of linoleic acid have great potential for the production of CLA upon conjugation of the carbon–carbon double bonds.

Table 4.2 Percentage of linoleic acid in the most commonly used natural oils.

Oil	Linoleic acid content (%)
Safflower	78
Walnut	73
Grapeseed	63
Corn	60
Low saturation soybean	57
Sunflower	54
Soybean	53
Sesame	43
Peanut	32
Canola	21
Linseed	15
Tung	9
Olive	6
Castor	4
Fish	–

In this chapter, the many chemical applications of conjugated vegetable oils and CLA will be covered, as well as their industrial production.[6] There are two main sections: Section 4.2 covers the industrial production of CLA and conjugated vegetable oils, while Section 4.3 is dedicated to their various chemical applications. Each section will include a thorough literature review of the most relevant published research and technologies in the corresponding area, followed by a more detailed overview of the advances made within the last decade.

4.2 Industrial Production of CLA and Conjugated Vegetable Oils

4.2.1 Industrial Production of CLA

CLA production has been mainly targeted towards the food industry, and it is predominantly used as a nutritional supplement for humans and cattle.[7] Additionally, there are reports on the use of CLA as a drying oil in replacement of tung oil,[7] as will be discussed in more detail later in the text. The synthesis of CLA can be achieved through various routes. One of the earliest industrial methods for CLA production consists of the dehydration of ricinoleic acid, the major fatty acid present in castor oil.[7] The procedure yields 83% of 9-*cis*,11-*trans* CLA.[8] The CLA products obtained from this process are not interesting for nutritional uses. In fact, the product of the direct dehydration of castor oil has been commercialized in the US since the early 1930s as a regular drying oil.[9] Around the same time, it was found that the carbon–carbon double bonds of unsaturated triglycerides can be isomerized in the presence of an alkaline solution.[10] The process results in conjugated carboxylates that can be protonated followed by separation of the glycerol to afford free fatty acids with conjugated carbon–carbon double bonds, at 65% yield, under optimized conditions.[11] This process can be employed with oils that are rich in linoleic acid for the production of CLA.[11]

More recently, a direct process for the production of CLA, as free fatty acids, from castor oil triglycerides utilizing *Lactobacillus plantarum* JCM 1551 has been developed.[11] The process yields a mixture of *cis*-9, *trans*-11, and *trans*-9, *trans*-11 CLA isomers with a maximum productivity of 0.044 mg per mL per hour.[11]

Metal catalysts have also been employed for the conjugation of carbon–carbon double bonds in methyl and ethyl linoleates.[7,11,12] Under certain conditions, some CLA isomers may undergo an intramolecular sigmatropic rearrangement to yield the desired, and more interesting, from a nutritional point of view, 9-*cis*,11-*trans* CLA isomer.[7]

4.2.2 Industrial Production of Conjugated Vegetable Oils

Conjugated vegetable oils are triglycerides in which the polyunsaturated fatty acid chains exhibit carbon–carbon double bonds that are conjugated, like in a 1,3-diene. Despite the occurrence of some rare, naturally conjugated oils, like tung oil, in which 84% of the fatty acid chains consist of α-eleostearic acid (Figure 4.3), the majority of conjugated vegetable oils have to be prepared through synthetic routes that involve the isomerization of the carbon–carbon double bonds.[3,11] Many of these processes involve successive addition of metal hydrides across carbon–carbon double bonds, and their subsequent elimination until a thermodynamically stable product is formed. As expected, conjugated π systems are more stable than isolated carbon–carbon double bonds. The interest in conjugated vegetable oils relates to their overall lower activation energy for conventional chemical transformations when compared to their non-conjugated counterparts. It has been shown, for example, that the free radical polymerization of conjugated vegetable oils proceeds through the formation of more stable intermediates than in the case of non-conjugated systems.[11] In these cases, the intermediate free radicals can be stabilized by the electrons on the adjacent carbon–carbon double bonds through resonance.

The preparation of conjugated triacylglycerides is nowadays industrially achieved by the enzymatic esterification of CLA with glycerol.[13] As explained previously the industrial process, using homogeneous bases, is unable to directly produce conjugated vegetable oils due to the competitive hydrolysis of triglycerides.

Alternatively, metal catalysts can be used. The conjugation of a variety of vegetable oils was successfully achieved in the presence of a rhodium

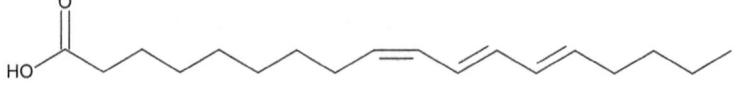

Figure 4.3 Chemical structure of α-eleostearic acid.

catalyst in a process that is well described in the literature.[3,6] The use of as little as 0.1 mol% of $[RhCl(C_8H_{14})_2]_2$, 0.25 mol% of $PtCl_2(PPh_3)_2$, or 0.5 mol% of $RuHCl(CO)(PPh_3)_3$, where Ph = phenyl, under mild reaction conditions, efficiently isomerizes the carbon–carbon double bonds of soybean oil, resulting in a fully conjugated π system with no hydrogenation by-products.[6] Conjugated soybean oil prepared by this method has exceptional drying properties, and coatings made from it exhibit good solvent resistance.[6] The $[RhCl(C_8H_{14})_2]_2$ catalyst provides similar high yields of CLA, other conjugated vegetable oils, and conjugated ethyl linoleate.[6] Other well-documented rhodium catalysts, such as Wilkinson's catalyst, $RhCl(PPh_3)_3$, have also been found to be effective for the conjugation of ethyl linoleate.[6]

The structural isomers formed by the homogeneous rhodium-catalysed isomerization of vegetable oils have been carefully identified and studied.[8] It has been shown that $[RhCl(C_8H_8)_2]_2$ is a selective isomerization catalyst for the production of highly conjugated vegetable oils with a high CLA content. The combined fractions of the two major CLA isomers [(9Z,11E)-CLA and (10E,12Z)-CLA] in the overall CLA mixture corresponds to a minimum of 76% of the total. The high efficiency and selectivity of this catalyst along with the straightforward purification process render this approach highly promising for the preparation of conjugated oils and CLA.

Recent improvements in the aforementioned catalyst recovery and reusability helped making this conjugation method more appealing.[11] The biphasic conjugation of soybean and other natural oils, as well as the isomerization of various alkenes, has been examined using a modified version of the aforementioned Rhodium catalyst.[11] A maximum yield of 96% of conjugated soybean oil has been obtained when the reaction is run at 80 °C, under argon, with ethanol as the polar solvent, using triphenylphosphine monosulfonate sodium salt (tppms) as the ligand, and the surfactant sodium dodecyl sulfate (SDS).[11] The improved process allows for easy separation of catalyst from products for subsequent re-use.

Historically, it has been observed that the partial hydrogenation of polyunsaturated vegetable oils carried out in the presence of heterogeneous catalysts, yielded by-products in which the carbon–carbon double bonds are conjugated.[14] This observation led to the development of processes and reaction conditions that favor isomerization of carbon–carbon double bonds over hydrogenation.[12] A recent review covers the latest advancements in active metals and different supports evaluated for the conjugation of vegetable oils.[12]

Other methods for the conjugation of triglycerides have been investigated to date. The quantitative conjugation of soybean oil has been achieved in diluted hexanes solution after irradiation with visible light at 68 °C.[15] The process uses I_2 as a catalyst and yields predominantly the E9, E11 isomer, as opposed to the Z9, E11 obtained as the major product in processes that involve organometallic catalysts.[15] A related process, using UV radiation and I_2 catalyst, and targeting conjugated oils for nutritional applications, yielded 20% conjugation after twelve hours at 48 °C.[16] Despite the low conjugation yield, the process was designed for a continuous production using a laminar flow reactor and represented a significant reduction in reaction time in comparison to previous technologies.[16] Further studies showed that minor components present in soybean oil, such as tocopherols, phospholipids, magnesol, and peroxides have an effect on the CLA yields and on the oxidative stability of the products.[17]

4.3   Chemical Use of CLA and Conjugated Vegetable Oils

4.3.1   The Use of CLA and Conjugated Vegetable Oils as Drying Oil

One of the major industrial uses of CLA, besides applications related to nutrition and the biomedical field, is in coatings and inks as a substitute for tung oil, a relatively expensive natural oil mainly produced in Asia. CLA behaves as a drying oil when exposed to air at ambient conditions. It is known that drying oils interact with atmospheric oxygen and initiate a slow polymerization process under mild conditions, referred to as auto-oxidation. When a drying oil is applied as a thin layer on a substrate, its auto-oxidation product constitutes a polymeric film suitable for applications as coatings.

Chemically, the drying properties of CLA rely on its increased reactivity towards free radicals. In fact, it is speculated that conjugated dienes can readily react with singlet oxygen or other reactive species.[18] In that sense, either plain CLA, CLA methyl esters, or CLA-rich triglycerides can be used as drying oils. One plausible mechanism for the auto-oxidation of regular oils has been recently proposed, and is adapted here to conjugated systems (Figure 4.4).[19] Figure 4.4 clearly shows that the auto-oxidation process involves a free radical chain reaction at the conjugated carbon–carbon double bond system of the starting material. Therefore, the remainder of the chemical structure of the starting material may have important effects on other physical–chemical properties, but is not directly involved in the auto-oxidation

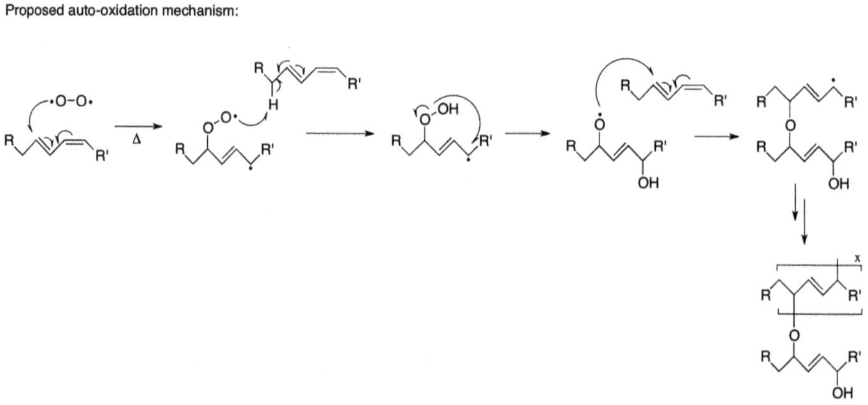

Proposed auto-oxidation mechanism:

Figure 4.4   Plausible proposed mechanism for the auto-oxidation of conjugated oils.

process. In conclusion, it is possible to use any polyunsaturated system as a drying oil, provided that the carbon–carbon double bonds are conjugated. For systems that are non-conjugated, free radicals preferentially abstract bis-allylic hydrogens, forming radicals that are greatly stabilized by resonance with the carbon–carbon double bonds. As a consequence, for practical considerations, non-conjugated systems undergo auto-oxidation at a very slow rate.

The effect of surface catalysis by glass vials on the auto-oxidation of CLA has been studied and revealed to be significant.[18] Furthermore, the oxidative stability of the different CLA isomers has been evaluated, and the results indicate that the *cis*-9, *cis*-11 isomer is the least stable.[18] The preparation of triglycerides containing 42.5 mol% of the fatty acid chains being CLA resulted in oils with a lower oxidative stability than the parent extra-virgin olive oil.[20] Similar results were obtained with soybean oil.[21]

4.3.2 Use of CLA and Conjugated Vegetable Oils in Diels–Alder Reactions

Next to their high reactivity in auto-oxidation processes, CLA is also highly reactive in Diels–Alder reactions. The use of CLA-based Diels–Alder adducts with maleic anhydride as a plasticizer for PVC was already reported in 1958.[22] Another application of CLA in that field consists of a grinding aid and dispersant for pigments in ink formulations.[23] The dispersant was obtained by pericyclic and radical reactions of CLA with maleic anhydride, acrylic, methacrylic, or fumaric acids.[23] In bio-based ink formulations, up to 9 wt% CLA was unintentionally found when thermosetting inks were prepared from tall oil fatty acid,[24] or vegetable oil-based rosin ester.[25]

Maleinized triglycerides for use as plasticizers for PVC have been obtained by the Diels–Alder and ene reactions of conjugated fatty acid esters from CLA, tung, and calendula oils, and maleic anhydride.[26] As discussed earlier for the auto-oxidation process, it is crucial that the carbon–carbon double bonds in oils be conjugated in order to successfully obtain maleinized triglycerides. As a matter of fact, the Diels–Alder reaction requires a 1,3 diene that is able to adopt a *syn* conformation (aliphatic, conjugated carbon–carbon double bonds satisfy that) and a dienophile (maleic anhydride, for example). The reaction is easily carried out under heat, resulting in overall good yields. From a fundamentals perspective, when non-conjugated oils, fatty acids, or their methyl esters are considered, the 1,3 diene is missing, and therefore, the Diels–Alder product cannot be formed.

4.3.3 Application of Conjugated Vegetable Oils in Bio-based Polymer and Composites

4.3.3.1 Cationic Polymerization

A strong electrophile is required to promote the cationic polymerization of triglycerides and fatty acids. Among the several initiators tested, $BF_3 \cdot OEt_2$ is the most commonly used in the cationic copolymerization of natural oils with vinyl comonomers. With naturally conjugated carbon–carbon double bonds, tung oil is fairly reactive towards cationic polymerization. Very strong materials, with storage moduli at room temperature of approximately 2 GPa, are obtained from the cationic copolymerization of 50–55 wt% of tung oil with divinylbenzene (DVB).[27] The gel time of these resins can be controlled by the substitution of up to 25 wt% of tung oil with less reactive vegetable oils, such as soybean (SOY), low saturation soybean (LSS), or conjugated low saturation soybean (CLS) oils.[29] Similarly, mixtures containing 35–55 wt% of corn or conjugated corn oil with various amounts of DVB and styrene (ST) afford polymeric materials whose mechanical properties range from soft and rubbery to rigid plastics, depending on the stoichiometry of the comonomers used.[28] The gel times are exponentially longer with increasing amounts of oil, especially beyond 40 wt%.[28] The gel times for resins containing conjugated corn oil are significantly shorter than for resins made from regular corn oil, indicating the higher reactivity of conjugated triglycerides.[28] The cure temperatures also affect the gel time of the cationic resins. For example, when a resin containing 45 wt% of corn oil, 32 wt% of styrene (ST), 15 wt% of divinylbenzene (DVB), and 8 wt% of $BF_3 \cdot OEt_2$ is cured at room temperature, the gel time observed is 116 minutes. At 15 °C, the gel time for the same resin is longer than 24 hours.[28]

Extensive research has been conducted on the cationic copolymerization of SOY, LSS, and CLS with various crosslinking agents.[29] Whenever 50–60 wt% of SOY, LSS, or CLS is copolymerized with DVB, a densely crosslinked polymer network, interpenetrated by 12–31 wt% of unreacted oil or oligomers, is produced.[29] The amount of unreacted free oil left after complete cure of the resin is directly dependent on the amount and reactivity of the oil initially employed in the preparation of the resin. Poor miscibility between the oil and the $BF_3 \cdot OEt_2$ results in a micro-phase separation in the SOY and LSS-based copolymers, with distinctly different crosslink densities in different parts of the bulk copolymer.[29] It has been also observed that an increase in the room temperature storage modulus of the final vegetable oil-based copolymers is seen when employing higher amounts of DVB in the original composition.[29] For example, the storage modulus at room temperature increases from 0.78 GPa to 1.20 GPa when the DVB content increases from 25 wt% to 35 wt% in CLS-based copolymers.[29] The unreacted free oil or oligomers present in the final copolymers largely affects the thermal stability of the thermosets. Between SOY, LSS, and CLS polymers, the latter have the highest storage moduli and thermal stabilities because they contain the least unreacted oil.[29]

In order to increase structural uniformity of crosslinked copolymers, the monofunctional monomer ST has been added to the original composition using SOY, LSS, or CLS/DVB.[30] With the substitution of 25–50 wt% of DVB by ST, the overall properties of the resulting plastics are significantly improved.[30] The thermophysical properties of the thermosets are considerably affected by the concentration and reactivity of monomers that have the ability to crosslink.[30] Among the crosslinking agents investigated, DVB exhibits the highest reactivity and thus gives the most promising, crosslinked materials with good damping and shape memory properties.[30] The overall properties of soybean oil-based thermosets are maximized when the concentration of the oil is equivalent to that of the other comonomers.[31] An isothermal cure study of these systems established the ideal cure temperature to be in the range 12–66 °C, depending on the actual composition of the resin[32] and a DSC study of the cure of resins initiated by different concentrations of $BF_3 \cdot OEt_2$ revealed that 2 wt% is the optimum concentration of initiator.[33] The cationic copolymerization of SOY and conjugated soybean oil (CSO) with dicyclopentadiene (DCPD) results in thermosetting materials with elastomeric properties when the DCPD content is lower than 42 wt%.[34] The conjugation of a variety of other vegetable oils, with varying levels of polyunsaturates, including olive, peanut, sesame, canola, grapeseed, sunflower, safflower, walnut, and linseed oils have been monitored through ^{1}H NMR and their conjugated products have been cationically copolymerized with DVB and/or ST to form a range of thermosets with properties that can be tailored for specific applications.[35] Overall, the properties of these polymers, with the exception of the gelation time, gradually increase with the degree of unsaturation of the oil.[35]

4.3.3.2   Radical Polymerization

Conjugated vegetable oils have also been employed in the preparation of translucent bio-based polymers through their free radical copolymerization with vinyl monomers.[36] The bulk free radical copolymerization of mixtures containing 30–75 wt% of conjugated linseed oil with various amounts of acrylonitrile (AN) and DVB affords a range of materials from flexible to rigid thermosets.[36] The free radical reaction, in this case, was initiated by 1 wt% of azobisisobutyronitrile (AIBN).[36] Although the materials exhibit promising thermophysical and mechanical properties, Soxhlet extraction with methylene chloride for 24 hours revealed that only 61–96 wt% of the oil initially added was incorporated into the final polymer network.[36] Alternatively, benzoyl peroxide has been used as a free radical initiator for the preparation of rigid thermosets containing 30–65 wt% of regular linseed oil, ST, and DVB.[37] Similar to the cationic thermosets described earlier, it has been observed that there is a tendency for the properties to decrease when the content of vegetable oil increases, regardless of the free radical initiator used.[36,37]

Soybean oil-based thermosets with 100% incorporation of CLS have been prepared by the copolymerization of mixtures containing 40–85 wt% of CLS with various amounts of AN, DVB, and/or DCPD in the presence of AIBN.[38] This higher oil incorporation in DCPD-containing thermosets is related to the similar reactivity of both components. When the oil content exceeds 70 wt%, a large amount of unreacted oil is recovered from the final thermoset.[38]

Drying oils undergo auto-oxidation in the presence of oxygen, and form peroxides, which then undergo crosslinking through a free radical process.

This characteristic of drying oils makes them very useful biorenewable starting materials for coatings applications. It has been shown that grafted copolymers of methyl methacrylate (MMA) and *n*-butyl methacrylate (BMA) can be made from conjugated linseed and soybean oils.[39,40]

In 1940, a copolymer of tung oil and styrene, having 0.1–2.0 wt% of the former, was produced by simply heating these two compounds at 125 °C for 3 days.[41] Later on, thermosets containing 30–70 wt% of tung oil (TUN) have been prepared through the thermal copolymerization of TUN with various amounts of ST and DVB at 85–160 °C.[42] These fully cured thermosets, prepared by simply heating the co-monomer mixture, exhibit higher crosslink densities and overall better properties when metallic salts of Co, Ca, and Zr are added to the resin during the cure.[42] Similar materials have been prepared using 30–70 wt% of a partially conjugated linseed oil.[43] The use of a less reactive oil, in comparison to TUN, resulted in the appearance of two distinct glass transition temperatures (T_g's), as determined by dynamic mechanical analysis (DMA), which indicates a micro-phase separation of the resin.[43] A scanning electron microscopy (SEM) analysis of solvent extracted samples of such thermosets revealed that the polymer matrix consists of a material with evenly distributed nanopores.[43]

4.3.3.3   Polyamides

There is precedence for the use of conjugated vegetable oil dimers in the preparation of polyamides. In this case, vegetable oil-based polyamides have also been prepared from linseed oil with ethylenediaminetetraacetic acid (EDTA).[44] The system was further cured at ambient conditions in the presence of poly(styrene-*co*-maleic anhydride) to give a material with good corrosion resistance and physico-mechanical properties.[44] Co-polyamides of soybean oil and amino acids have been prepared.[45] It has been observed that the structure and content of the amino acids affected the crystallinity of the final materials.[45]

4.3.3.4   Production of Composites

Particularly stiff composites have been obtained by reinforcing the previously described cationic, conjugated corn,[46] and soybean[47] oil-based copolymers with continuous glass fibers. Another class of bio-based composites has been prepared by reinforcing cationic thermosets from conjugated oils with organomodified montmorillonite clay.[48] The filler has been prepared by the cationic exchange of sodium montmorillonite with (4-vinylbenzyl)triethylammonium chloride in aqueous solution.[48] Wide angle X-ray (WAX) and transmission electron microscopy (TEM) experiments have revealed the nanocomposite character and the intercalated and exfoliated morphology of the materials prepared.[48] Clay loadings of 1–2 wt% result in the best overall properties.[48]

Composites with a high bio-based content have been prepared, by compression molding, using a variety of ligno-cellulosic fillers. Spent germ has been used to reinforce a tung oil-based free radical resin, initiated by t-butyl peroxide.[49] When compared to the pure resin, the reinforced thermosets exhibit higher storage moduli, and thermal stability.[49] The use of shorter particles results in better properties.[49] When the filler load is increased from 40 wt% to 60 wt%, agglomeration and formation of micro-voids in the composite cause a decrease in the properties.[49] This phenomenon can be compensated for by an increase in the molding pressure.[49] It was also found that residual corn oil from the spent germ acts as a plasticizer in the composite and if it is removed prior to preparation of the composite, better mechanical properties are obtained.[49]

Along the same lines, a conjugated soybean oil (CSO)-based free radical resin has been reinforced with soybean hulls.[50] An optimum cure sequence of 5 hours at 130 °C, followed by a post-cure of 2 hours at 150 °C, has been established by differential scanning calorimetry (DSC).[50] In this particular system, a decrease in the properties was observed when an excessive molding pressure was applied.[50] Free radical resins containing 50 wt% of either CSO or conjugated linseed oil (CLO), and various amounts of DVB and BMA have been reinforced with 20–80 wt% of corn stover.[51] Increasing the amount of corn stover and decreasing the length of the fiber results in an overall improvement of the mechanical properties and a decrease in the thermal stability of the biocomposites.[53] As expected, water uptake experiments have confirmed that water absorption increases with the fiber content of the composite.[51]

Wheat straw has also been used as a natural filler in the preparation of CLO-based biocomposites.[52] In this system, maleic anhydride (MA) was employed as a compatibilizer between the hydrophobic matrix and the hydrophilic filler, resulting in significant improvements in the mechanical properties.[52] Similarly, a CLO-based free radical resin has been used to determine the optimal conditions for the preparation of rice hull biocomposites.[53] The same CLO-based free radical resin has been reinforced with switch grass.[54] It has been observed that beyond loadings of 70 wt%, excessive agglomeration of the filler compromises the composites' properties.[54]

Finally, a thorough study has been carried out of the effect of different conjugated vegetable oils, and different natural fillers on the properties of cationic composites.[55] Cationic resins made from CSO, CLO, and conjugated corn oils have been reinforced with corn stover, wheat straw, and switch grass.[55] Among the different fillers studied, wheat straw afforded composites with the most promising properties.[55]

4.4 Conclusions

Linoleic acid is one of the most abundant fatty acids naturally found in vegetable oils. Its most remarkable feature is a fairly unreactive, non-conjugated diene. Upon isomerization and conjugation of the carbon–carbon double bonds in linoleic acid, many interesting and useful properties arise.

The collection of various positional and geometric isomers of linoleic acid with conjugated carbon–carbon double bonds is denoted by the term conjugated linoleic acid (CLA). Despite being found as a free fatty acid, linoleic acid is most commonly a natural component of triglycerides in plant oils and animal fat. Just like for CLA, conjugated oils are obtained when the carbon–carbon double bonds in triglycerides containing linoleic acid are brought into conjugation. There exist many processes for the preparation of CLA and conjugated oils. Conjugated vegetable oils have been initially reported as byproducts of bleaching of soybean oil with NaOH. Since then, the catalytic production of conjugated fatty acids and triglycerides has been studied and optimized.

CLA and conjugated vegetable oils are largely used in the chemical industry as drying oils for coating and ink applications. In addition, CLA Diels–Alder adducts are employed as plasticizers or dispersants in ink formulations. CLA and conjugated vegetable oils are also used as building blocks for the synthesis of various bio-based thermosetting materials for applications in the automobile, and construction industries. Thermosets from CLA and conjugated vegetable oils can be obtained through their cationic or free radical direct co-polymerization with other reactive species, or from their polycondensation with amines and amino-acids to form polyamides. The reinforcement of such polymers with solid particles results in bio-based composites with promising mechanical and thermal properties. Despite the great accomplishments in the development of bio-based materials from conjugated vegetable oils and CLA, future research efforts in this area should include improvements in the current conjugation processes, exploration of other polymerization methods, examination of these new materials in various industrial applications, and the development of functional nano-composites for commercialization. As new incentives arise for the production and use of bio-based materials and biorenewable chemicals, it is expected, and needed, to carry out studies on the biodegradability/recyclability and life cycle analysis of the most successful CLA- and conjugated vegetable oil-based systems.

References

1. M. A. R. Meier, J. O. Metzger and U. S. Schubert, *Chem. Soc. Rev.*, 2007, **36**, 1788.
2. M. O. Jung, S. H. Yoon and M. Y. Jung, *J. Agric. Food Chem.*, 2001, **49**, 3010.
3. D. D. Andjelkovic, B. Min, D. Ahn and R. C. Larock, *J. Agric. Food Chem.*, 2006, **54**, 9535.
4. O. A. Simakova, A. Leino, B. Campo, P. Maki-Arvela, K. Kordas, J. Mikkola and D. Y. Murzin, *Catalysis Today*, 2010, **150**, 32.
5. P. Pakdeechanuan, K. Intarapichet, L. N. Fernando and I. U. Grun, *J. Agric. Food Chem.*, 2005, **53**, 923.

6. R. C. Larock, X. Dong, S. Chung, C. K. Reddy and L. E. Ehlers, *J. Am. Oil Chem. Soc.*, 2001, **78**, 447.
7. A. Saebo, in *Advances in Conjugated Linoleic Acid Research*, ed. J. Sebedio, W. W. Christie and R. Adlof, 2003, vol. 1, AOCS Press, Champaign, IL, p. 71.
8. O. Berdeaux, W. W. Christie, F. D. Gunstone and J. Sebedio, *J. Am. Oil Chem. Soc.*, 1997, **74**, 1011.
9. J. Scheiber, *U.S. Patent* **1942778**, 1934.
10. H. G. Kirschenbauer and N. J. Allendale, *U.S. Patent* **23893260**, 1945.
11. R. L. Quirino and R. C. Larock, *J. Am. Oil Chem. Soc.*, 2012, **89**, 1113.
12. A. Philippaerts, S. Goossens, W. Vermandel, M. Tromp, S. Turner, J. Geboers, G. V. Tendeloo, P. A. Jacobs and B. F. Sels, *ChemSusChem.*, 2011, **4**, 757.
13. A. Saebo, C. Skarie, D. Jerome and G. Haroldsson, *U.S. Patent* **6410761B1**, 2002.
14. A. Philippaerts, S. Goossens, P. A. Jacobs and B. F. Sels, *ChemSusChem.*, 2011, **4**, 684.
15. V. R. Chintareddy, R. E. Oshel, K. M. Doll and J. G. Verkade, *Am. Chem. Soc. Nat. Meeting*, 2010, 142.
16. V. P. Jain, A. Proctor and R. Lall, *J. Food Sci.*, 2008, **73**, 183.
17. T. Tokle, V. P. Jain and A. Proctor, *J. Agric. Food Chem.*, 2009, **57**, 8989.
18. M. P. Yurawecz, P. Delmonte, T. Vogel and J. K. G. Kramer, in *Advances in Conjugated Linoleic Acid Research*, ed. J. Sebedio, W. W. Christie and R. Adlof, AOCS Press, Champaign, IL, 2003, Vol. 2, p. 56.
19. R. L. Quirino and R. C. Larock, in ACS books, Symposium Series, Vol. 1063, *Renewable and Sustainable Polymers*, ed. G. F. Payne and P. B. Smith, ACS Publications, Washington, DC, 2011, p. 37.
20. J. H. Lee, K. Lee, C. C. Akoh, S. K. Chung and M. R. Kim, *J. Agric. Food Chem.*, 2006, **54**, 5416.
21. B. Yang, W. Wang, F. Zeng, T. Li, Y. Wang and L. Li, *J. Food Biochem.*, 2011, **35**, 1612.
22. H. M. Teeter, E. W. Bell, J. L. O'Donnell, M. J. Danzig and J. C. Cowan, *J. Am. Oil Chem. Soc.*, 1958, **35**, 238.
23. P. J. LeBlanc and P. Schilling, *U.S. Patent* **5182326**, 1993.
24. J. T. Moynihan, *U.S. Patent* **4519841**, 1985.
25. R. H. Reiter and N. M. Patel, *U.S. Patent* **5552467**, 1996.
26. U. Biermann, A. Jungbauer and J. O. Metzger, *Eur. J. Lipid Sci. Technol.*, 2012, **114**, 49.
27. F. Li and R. C. Larock, *J. Appl. Polym. Sci.*, 2000, **78**, 1044.
28. F. Li, J. Hasjim and R. C. Larock, *J. Appl. Polym. Sci.*, 2003, **90**, 1830.
29. F. Li, M. V. Hanson and R. C. Larock, *Polymer*, 2001, **42**, 1567.
30. F. Li and R. C. Larock, *J. Appl. Polym. Sci.*, 2001, **80**, 658.
31. F. Li and R. C. Larock, *J. Polym. Sci.: Part B: Polym. Phys.*, 2001, **39**, 60.
32. F. Li and R. C. Larock, *Polym. Int.*, 2003, **52**, 126.
33. P. Badrinarayanan, Y. Lu, R. C. Larock and M. R. Kessler, *J. Appl. Polym. Sci.*, 2009, **113**, 1042.

34. D. D. Andjelkovic, Y. Lu, M. R. Kessler and R. C. Larock, *Macromol. Mater. Eng.*, 2009, **294**, 472.
35. D. D. Andjelkovic, M. Valverde, P. Henna, F. Li and R. C. Larock, *Polymer*, 2005, **46**, 9674.
36. P. H. Henna, D. D. Andjelkovic, P. P. Kundu and R. C. Larock, *J. Appl. Polym. Sci.*, 2007, **104**, 979.
37. V. Sharma, J. S. Banait, R. C. Larock and P. P. Kundu, *eXpress Polym. Letters*, 2008, **2**, 265.
38. M. Valverde, D. Andjelkovic, P. P. Kundu and R. C. Larock, *J. Appl. Polym. Sci.*, 2008, **107**, 423.
39. B. Cakmakli, B. Hazer, I. O. Tekin and F. B. Comert, *Biomacromolecules*, 2005, **6**, 1750.
40. B. Cakmakli, B. Hazer, I. O. Tekin, S. Kizgut, M. Koksal and Y. Menceloglu, *Macromol. Biosci.*, 2004, **4**, 649.
41. S. M. Stoesser and A. R. Gabel, *US Pat.* **2 190 906**, 1940.
42. F. Li and R. C. Larock, *Biomacromolecules*, 2003, **4**, 1018.
43. P. P. Kundu and R. C. Larock, *Biomacromolecules*, 2005, **6**, 797.
44. N. Alam and M. Alandis, *J. Polym. Environ.*, 2011, **19**, 391.
45. Y. Deng, X. D. Fan and J. Waterhouse, *J. Appl. Polym. Sci.*, 1999, 73, 1081.
46. Y. Lu and R. C. Larock, *J. Appl. Polym. Sci.*, 2006, **102**, 3345.
47. Y. Lu and R. C. Larock, *Macromol. Mater. Eng.*, 2007, **292**, 1085.
48. Y. Lu and R. C. Larock, *Biomacromolecules*, 2006, 7, 2692.
49. D. P. Pfister, J. R. Baker, P. H. Henna, Y. Lu and R. C. Larock, *J. Appl. Polym. Sci.*, 2008, **108**, 3618.
50. R. L. Quirino and R. C. Larock, *J. Appl. Polym. Sci.*, 2009, **112**, 2033.
51. D. P. Pfister and R. C. Larock, *Bioresour. Technol.*, 2010, **101**, 6200.
52. D. P. Pfister and R. C. Larock, *Composites: Part A*, 2010, **41**, 1279.
53. R. L. Quirino and R. C. Larock, *J. Appl. Polym. Sci.*, 2011, **121**, 2039.
54. D. P. Pfister and R. C. Larock, *J. Appl. Polym. Sci.*, 2013, **127**, 1921.
55. D. P. Pfister and R. C. Larock, *J. Appl. Polym. Sci.*, 2012, **123**, 1392.

CHAPTER 5

Recent Advances in the Production of CLA and Conjugated Vegetable Oils: Microbial and Enzymatic Production of Conjugated Fatty Acids and Related Fatty Acids in Biohydrogenation Metabolism

JUN OGAWA,*[a,c,d] MICHIKI TAKEUCHI[a] AND SHIGENOBU KISHINO[a,b]

[a] Division of Applied Life Sciences, Kyoto University, Kyoto, 606-8502, Japan; [b] Laboratory of Industrial Microbiology, Graduate School of Agriculture, Kyoto University, Kyoto, 606-8502, Japan; [c] Research Unit for Physiological Chemistry, the Center for the Promotion of Interdisciplinary Education and Research, Kyoto University, Kyoto, 606-8502, Japan; [d] Research Division of Microbial Sciences, Kyoto University, Kyoto, 606-8502, Japan
*Email: ogawa@kais.kyoto-u.ac.jp

5.1 Introduction

Conjugated fatty acid' is a collective term for positional and geometric iso-mers of fatty acids with conjugated double bonds. In particular, conjugated

RSC Catalysis Series No. 19
Conjugated Linoleic Acids and Conjugated Vegetable Oils
Edited by Bert Sels and An Philippaerts
© The Royal Society of Chemistry 2014
Published by the Royal Society of Chemistry, www.rsc.org

linoleic acids (CLAs) such as *cis*-9,*trans*-11-CLA and *trans*-10,*cis*-12-CLA reduce carcinogenesis,[1,2] atherosclerosis,[3] and body fat.[4] In regard to lipid metabolism, CLA is a potent peroxisome proliferator-activated receptor α (PPARα) agonist,[5] and treatment with CLA increases the catabolism of lipids in the liver of rodents.[6] Based on these findings, CLA is now commercialized as a functional food for control of body weight, especially in the USA and European countries.

CLA is contained in edible fats derived from ruminant animals, such as milk fat and beef tallow, and mainly consists of *cis*-9,*trans*-11- and *trans*-10,*cis*-12-octadecadienoic acid (18 : 2).[7–9] Today, CLA, as a dietary supplement, is produced through chemical isomerization of linoleic acid, which results in the by-production of unexpected isomers. However, recent studies have revealed that each isomer can have different effects on metabolism and cell functions, and acts through different cell signaling pathways.[10] To date, *cis*-9,*trans*-11 and *trans*-10,*cis*-12 isomers have been paid particular attention because of their remarkable biological activities.[11] Considering the use of CLA for medicinal and nutraceutical purposes, a safe isomer-selective process is required.[12] A bioprocess, *i.e.*, fermentation, microbial, or enzymatic processes, is of potential use for this purpose.

5.2 Biosynthesis of CLA

Dairy products are major natural sources of CLA, of which *cis*-9,*trans*-11–18 : 2 is the main isomer.[7] CLA in dairy products is produced by certain rumen microorganisms such as *Butyrivibrio* species from polyunsaturated fatty acids.[13] *cis*-9,*trans*-11–18 : 2 May be an intermediate in the biohydrogenation (saturation of unsaturated fatty acids) of linoleic acid to stearic acid by the anaerobic rumen bacterium *Butyrivibrio fibrisolvens*.[14] The reaction sequence of linoleic acid biohydrogenation to stearic acid involves at least two steps (Figure 5.1). The sequence begins with the isomerization of linoleic acid to *cis*-9,*trans*-11–18 : 2, followed by the hydrogenation of the *cis*-double bond of the conjugated diene to yield a *trans*-11-octadecenoic acid (*trans*-vaccenic acid). Successive reduction of the *trans*-monoenoic acid appears to be rate limiting for complete biohydrogenation. Therefore, *trans*-vaccenic acid is accumulated together with CLA through linoleic acid biohydrogenation. The *trans*-vaccenic acid is converted to *cis*-9,*trans*-11–18 : 1 within mammalian cells through Δ9 desaturation.[15]

In addition to rumen bacteria, several strains of foodgrade and probiotic microorganisms have been identified as potential producers of CLA, including strains of *Bifidobacterium*, *Enterococcus*, *Lactobacillus*, *Lactococcus*, *Propionibacterium* and *Streptococcus*.[12,16–18] These cultures may be very important for the food industry, in particular to the dairy industry, as they can be used for the production of traditional and or novel CLA-enriched foods through *in situ* fermentation. Despite the high isomer specificity for *cis*-9,*trans*-11–18 : 1 in fermentative production, only limited CLA yields were obtained, because free linoleic acid, that is the substrate for CLA during

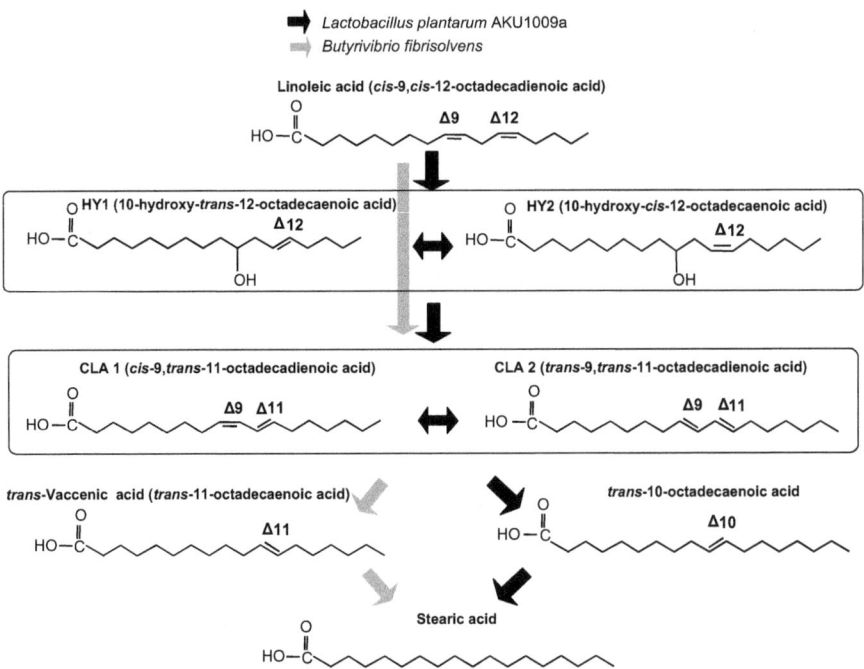

Figure 5.1   Proposed pathway of linoleic acid hydrogenation to stearic acid by *Butyrivibrio fibrisolvens* and *Lactobacillus plantarum*.

fermentation, inhibits the growth of bacteria. In order to avoid the inhibitory effects of linoleic acid, we examined washed cell (resting cell) reactions catalysed by some lactic acid bacteria.[19]

5.3   Useful Reactions for CLA Production by Washed Cells of Lactic Acid Bacteria

5.3.1   Isomerization of Linoleic Acid to CLA

To establish efficient processes for CLA production, we screened lactic acid bacteria for the ability to produce CLA from linoleic acid.[20] Firstly, we evaluated CLA productivity during cultivation in linoleic acid-containing medium. However, the productivity was low (less than 0.5 mg ml^{-1}), as reported previously, because of the inhibitory effect of linoleic acid on the growth of lactic acid bacteria. Thus, we investigated CLA productivity of the washed cells of lactic acid bacteria to avoid the inhibitory effect of linoleic acid on cell growth. The washed cells were prepared just by washing the cultured cell by suitable buffer. We used the washed cells as resting cells, which are the bags of enzymes, so that we could neglect the growth inhibition by linoleic acid and could increase the substrate, linoleic acid, concentration during the reactions. Our preliminary experiments revealed three interesting points: (i) the washed cells of lactic acid bacteria exhibiting high levels of CLA production were obtained by cultivation in medium containing a small amount of linoleic acid; (ii) the production of CLA by the washed cells of lactic acid bacteria was clearly observed under microaerobic conditions (O_2 concentration of less than 1%); and (iii) linoleic acid should be pretreated with a detergent or albumin so that it is well dispersed in the reaction mixture and available for the washed cells of lactic acid bacteria. Through extensive screening that considered the above three points, many strains were found to produce CLA from linoleic acid.[19,20] All strains produced two specific isomers of CLA, *i.e.*, *cis*-9,*trans*-11–18:2 (CLA1) and *trans*-9,*trans*-11–18:2 (CLA2), together with two hydroxy fatty acid, *i.e.*, 10-hydroxy-*trans*-12-octadecenoic acid (18:1) (HY1) and 10-hydroxy-*cis*-12–18:1 (HY2).[21]

The mechanism of CLA production from linoleic acid in relation to hydroxy fatty acid production was investigated using *Lactobacillus acidophilus* AKU 1137 as a representative strain.[22] Preceding the production of CLA isomers, hydroxy fatty acids were accumulated. As the reaction proceeded, the amount of accumulated hydroxy fatty acids rapidly decreased, followed by an increase in the level of CLA production. The isolated hydroxy fatty acids were transformed to CLA isomers upon incubation with the washed cells of *L. acidophilus*. These results suggest that the hydroxy fatty acids are intermediates of CLA production from linoleic acid (Figure 5.1). Therefore, linoleic acid isomerization to CLA by lactic acid bacteria was found to consist of at least two successive reactions, *i.e.*, the hydration of linoleic acid to 10-hydroxy-18:1 and the dehydrating isomerization of the hydroxy fatty acid to CLA. Further investigations revealed that the produced CLA was saturated to stearic acid *via trans*-10–18:1 (Figure 5.1). These results indicated that fatty acid isomerization is a part reaction of fatty acid saturation metabolism. Only the free form of linoleic acid acted as a substrate for CLA production by lactic acid bacteria, *i.e.*, the ester and triacylglycerol of linoleic acid did not.

5.3.2   Dehydration of Ricinoleic Acid to CLA

On the basis of the results described above, the transformation of hydroxy fatty acids by lactic acid bacteria was investigated using *Lactobacillus plantarum* AKU 1009a as a representative strain. Among the various hydroxy fatty acids examined, this strain transformed ricinoleic acid (12-hydroxy-*cis*-9–18:1) into CLA (a mixture of CLA1 and CLA2).[23] The ability to produce CLA from ricinoleic acid was found to be widely distributed in lactic acid bacteria.[19,24] The CLA-producing activity of washed cells increased when they were cultivated in medium containing linoleic acid, α-linolenic acid or both. There are two possible pathways for CLA synthesis from ricinoleic acid by lactic acid bacteria: (i) direct transformation of ricinoleic acid into CLA through dehydration at the Δ11 position; and (ii) a two-step transformation *via* linoleic acid through dehydration at the Δ12 position and successive isomerization of linoleic acid (Figure 5.2). The existence of linoleic acid and

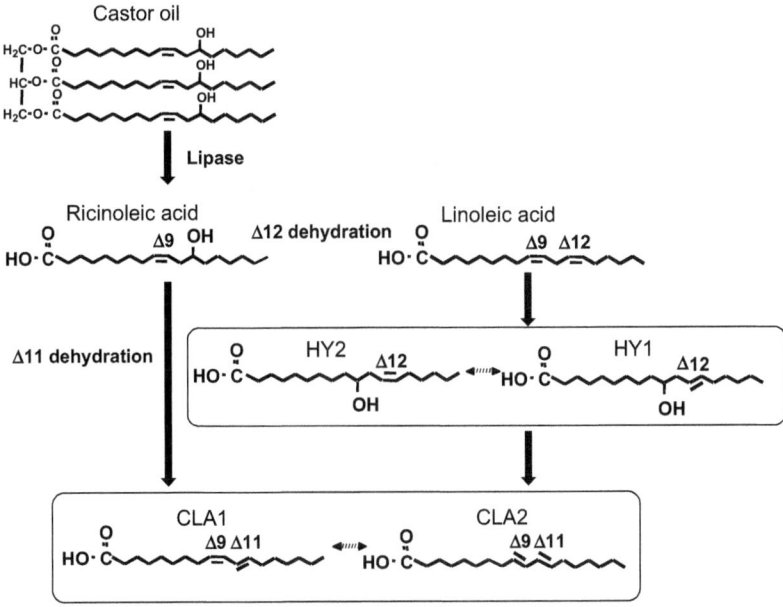

Figure 5.2   Proposed pathway of ricinoleic acid and castor oil transformation to CLA by *Lactobacillus plantarum*.

10-hydroxy-18:1 in the reaction mixture with ricinoleic acid as the substrate indicated the participation of the latter pathway. Only the free form of ricinoleic acid acted as a substrate for CLA production by lactic acid bacteria, *i.e.*, the ester and triacylglycerol of ricinoleic acid did not.

5.4 Preparative Production of CLA by Washed Cells of Lactic Acid Bacteria

5.4.1 CLA production from Linoleic Acid

More than 250 bacterial strains from 14 genera were examined, and strains belonging to the genera *Enterococcus, Pediococcus, Propionibacterium*, and *Lactobacillus* were found to produce considerable amounts of CLA from linoleic acid.[19,20] From these strains, *L. plantarum* AKU 1009a was selected for its potential to produce CLA from linoleic acid.[20] *L. plantarum* AKU 1009a was simple to cultivate and showed a high growth rate even under aerobic conditions. Washed cells of *L. plantarum* exhibiting a high level of CLA production were obtained by cultivation in nutrient medium containing free linoleic acid [0.06% (w/v)], but such cells were not obtained by cultivation in medium containing the methyl ester of linoleic acid. The cells at the late log phase showed significant CLA productivity, but further incubation resulted in a decrease in productivity. The CLA-producing reaction using the washed cells as a catalyst proceeded well even under aerobic conditions in 0.1 M

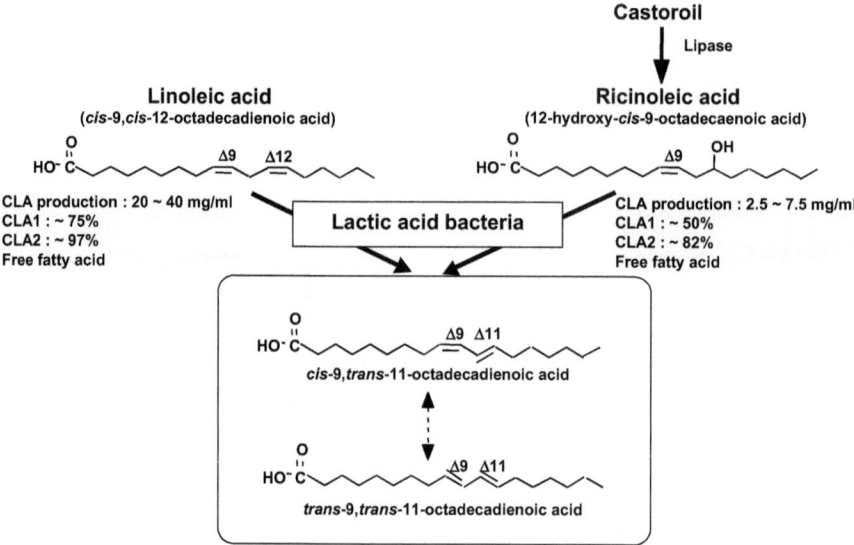

Figure 5.3   CLA production from linoleic acid, ricinoleic acid and castor oil by *Lactobacillus plantarum*.

potassium phosphate buffer (pH 6.5) at 37 °C with free linoleic acid mixed with bovine serum albumin in the ratio of 5 : 1 as the substrate. Under the optimum reaction conditions using the free form of albumin-mixed linoleic acid as the substrate, the washed cells of *L. plantarum* (33% [wet w/v]) produced 40 mg ml⁻¹ CLA from 120 mg ml⁻¹ of linoleic acid in 108 h; thus, the molar yield from linoleic acid was 33% (Figure 5.3). The resulting CLA comprised a mixture of CLA1 (38% of total CLA) and CLA2 (62% of total CLA), and accounted for 50% of the total fatty acids obtained. A higher yield (80% molar yield from linoleic acid) was attained using the washed cells of *L. plantarum* (23% [wet w/v]) and 2.6% (w/v) of the free form of albumin-mixed linoleic acid as the substrate in 96 h, resulting in CLA production of 20 mg ml⁻¹ [consisting of CLA1 (2%) and CLA2 (98%), and accounting for 80% of the total fatty acids obtained]. Most of the CLA produced was accumulated as intracellular or cell-associated lipids in the free form; thus, it was simple to recover CLA by centrifugation, and the cells themselves could be used as a source of CLA.[20]

5.4.2   CLA Production from Ricinoleic Acid

More than 250 bacterial strains from 14 genera were examined, and strains belonging to the genera *Streptococcus*, *Leuconostoc*, *Pediococcus*, *Propionibacterium*, and *Lactobacillus* were found to produce considerable amounts of CLA from ricinoleic acid.[19,24] From these strains, *L. plantarum* JCM 1551 was selected for its potential to produce CLA from ricinoleic acid.[24] This strain had the highest CLA-producing activity when it was cultivated in medium supplemented with 0.2% (w/v) of a mixture of α-linolenic acid and linoleic acid in the ratio of 1:5. The CLA-producing activity was highest in the middle logarithmic to early stationary phase, and decreased thereafter. The CLA-producing reaction using the washed cells as a catalyst proceeded well under microaerobic conditions in 0.5 M sodium citrate buffer (pH 6.0) at 37 °C with free ricinoleic acid mixed with bovine serum albumin in the ratio of 5:1 as the substrate. Under the optimum reaction conditions, using the free form of albumin-mixed ricinoleic acid as the substrate and washed cells of *L. plantarum* (12% [wet w/v]) as the catalyst, 2.4 mg ml^{-1} CLA was produced from 3.4 mg ml^{-1} ricinoleic acid in 90 h; thus, the molar yield from ricinoleic acid was 71% (Figure 5.3). The CLA produced, which was obtained in the free fatty acid form, consisted of CLA1 (21% of total CLA) and CLA2 (79% of total CLA), and accounted for 72% of the total fatty acids obtained.[24] Of the total amount of CLA produced 70% was accumulated as intracellular or cell-associated lipids, the remainder was found in the reaction supernatant. The unreacted ricinoleic acid was mainly found in the supernatant.

5.4.3 CLA Production from Castor Oil

Castor oil, a vegetable oil, is an economical source of ricinoleic acid. About 88% of the total fatty acids in castor oil is ricinoleic acid. Unfortunately, CLA cannot be directly produced from castor oil by lactic acid bacteria. Lactic acid bacteria only use the free form of ricinoleic acid for CLA production, *i.e.*, not the triacylglycerol form, which is the major form found in castor oil. However, the coexistence of lipase in the reaction mixture, which generates the free form of ricinoleic acid, made triacylglycerol available for lactic acid bacteria (Figure 5.2). In particular, Lipase M Amano 10 effectively supports CLA production. The origin of the lipase is *Mucor javanicus* and it is 1,3-regiospecific. Nonspecific lipase also increased the CLA productivity; however, the best results were obtained with Lipase M Amano 10, which exhibits 1,3-regiospecificity.[23,25] In the case of CLA production from linoleic acid or ricinoleic acid, the substrate was mixed with albumin. However, the dispersion of castor oil with albumin was not effective. We examined various kinds of detergents and found that nonionic detergents, especially poly-hydroxy-type detergents, such as Lubrol PX, well dispersed the castor oil. The CLA-producing reaction using the washed cells of *L. plantarum* JCM 1551 as the catalyst proceeded well under microaerobic conditions in 1.0 M sodium citrate buffer (pH 6.0) at 37 °C with castor oil mixed with Lubrol PX in the ratio of 5:1 as the substrate. Under the optimum conditions using detergent-mixed castor oil as the substrate and washed cells of *L. plantarum* JCM 1551 (12% [wet w/v]) as the catalyst, 2.7 mg ml^{-1} CLA was produced from 5.0 mg ml^{-1} castor oil in 99 h (Figure 5.3). The CLA produced accounted for 46% of the total fatty acids obtained, and consisted of CLA1 (26%) and CLA2 (74%).[25] Of the CLA produced 70% and 30% were accumulated intracellularly (or associated with cells) and extracellularly, respectively, mainly as the free form. The unreacted ricinoleic acid was chiefly found in the supernatant, mostly as the free from.

5.5 Production of Conjugated Trienoic Fatty Acids by Washed Cells of Lactic Acid Bacteria

5.5.1 Transformation of C18 Polyunsaturated Fatty Acids by Lactic Acid Bacteria

Several unique biological effects have been found not only for CLA but also for conjugated trienoic acids,. Conjugated trienoic acid produced from α-linolenic acid through alkali-isomerization exhibited cytotoxicity toward human tumor cells.[26] 9,11,13-octadecatrienoic acid (18:3) isomers in pomegranate, tung and catalpa oils have also been found to be cytotoxic toward mouse tumor and human monocytic leukemia cells.[27] These findings led us to develop a novel microbial method for conjugated trienoic acid production.

Firstly, we investigated the substrate spectrum of polyunsaturated fatty acid transformation by the washed cells of *L. plantarum* AKU 1009a. Among various polyunsaturated fatty acids examined, α-linolenic acid (*cis*-9,*cis*-12,*cis*-15–18:3), γ-linolenic acid (*cis*-6,*cis*-9,*cis*-12–18:3), pinolenic acid (*cis*-5,*cis*-9,*cis*-12–18:3), and stearidonic acid (*cis*-6,*cis*-9,*cis*-12,*cis*-15-octadecatetraenoic acid [18:4]) were found to be transformed.[19,28] The fatty acids transformed by the strain had the common structure of C18 fatty acids with a *cis*-9,*cis*-12 diene system. Three major fatty acids were produced from α-linolenic acids, which were identified as *cis*-9,*trans*-11,*cis*-15–18:3, *trans*-9,*trans*-11,*cis*-15–18:3, and *trans*-10,*cis*-15–18:2.[29] Four major fatty acids were produced from γ-linolenic acid, which were identified as *cis*-6,*cis*-9,*trans*-11–18:3, *cis*-6,*trans*-9,*trans*-11–18:3, *cis*-6,*trans*-10–18:2, and *trans*-10–18:1.[28] The time course of changes in fatty acid composition during α-linolenic acid transformation was studied. In the initial stage of the transformation, the conjugated trienoic acids of *cis*-9,*trans*-11,*cis*-15–18:3 and *trans*-9,*trans*-11,*cis*-15–18:3 were accumulated. As the reaction proceeded, the amount of accumulated conjugated trienoic acid gradually decreased, followed by an increase in the amount of the dienoic acid of *trans*-10,*cis*-15–18:2. Similar results were obtained for γ-linolenic acid transformation. In the initial stage of γ-linolenic acid transformation, the conjugated trienoic acids of *cis*-6,*cis*-9,*trans*-11–18:3 and *cis*-6,*trans*-9,*trans*-11–18:3 were accumulated. As the reaction proceeded, the amount of accumulated conjugated trienoic acid gradually decreased, followed by an increase in the amounts of the dienoic and monoenoic acids of *cis*-6,*trans*-10–18:2, and *trans*-10–18:1. On the basis of these results, we proposed the pathways of α- and γ-linolenic acid transformation shown in Figure 5.4.[29] *L. plantarum* AKU 1009a transformed the *cis*-9,*cis*-12 diene system of C18 fatty acids to the conjugated diene systems of *cis*-9,*trans*-11 and *trans*-9,*trans*-11.

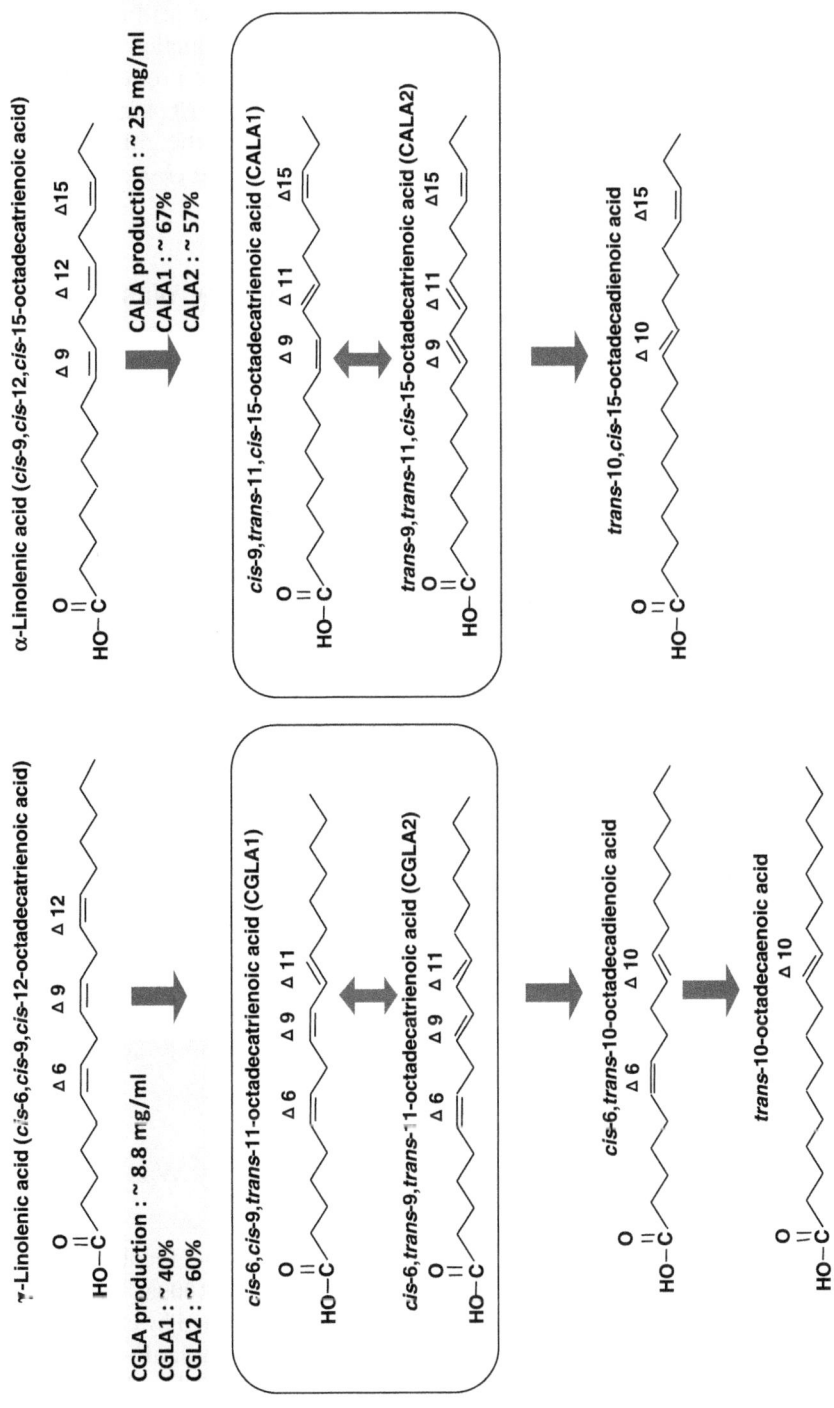

Figure 5.4 Proposed pathway of α-linolenic acid and γ-linolenic acid transformation by *Lactobacillus plantarum.*

These conjugated dienes were further saturated by this strain to the monoene of *trans*-10. Although the products derived from stearidonic acid and pinolenic acid could not be identified because of insufficient production amounts, on the basis of the above results, the three major fatty acids produced from stearidonic acid are surmised to be *cis*-6,*cis*-9,*trans*-11,*cis*-15–18:4, *cis*-6,*trans*-9,*trans*-11,*cis*-15–18:4, and *cis*-6,*trans*-10,*cis*-15–18:3, and the three major fatty acids produced from pinolenic acid are surmised to be *cis*-5,*cis*-9,*trans*-11–18:3, *cis*-5,*trans*-9,*trans*-11-18:3, and *cis*-5,*trans*-10–18:2.

5.5.2   Conjugated α-Linolenic Acid Production by Washed Cells of Lactic Acid Bacteria

Conjugated α-linolenic acid (CALA) was produced by the incubation of α-linolenic acid with the washed cells of *L. plantarum* AKU 1009a.[30] Washed cells exhibiting high levels of CALA productivity were obtained by cultivation in nutrient medium supplemented with 0.01% (w/v) α-linolenic acid. The cells at the late log phase showed significant productivity, but further incubation resulted in a rapid decrease in productivity. The CALA-producing reaction using the washed cells as the catalyst proceeded well under microaerobic conditions in 0.1 M potassium phosphate buffer (pH 6.5) at 37 °C with free α-linolenic acid mixed with bovine serum albumin in the ratio of 5:1 as the substrate. Under the optimum reaction conditions using 63 mg ml^{-1} of albumin-mixed α-linolenic acid as the substrate, the washed cells (33% [wet w/v]) produced 25 mg ml^{-1} of CALA in 72 h; thus, the molar yield from α-linolenic acid was 40% (Figure 5.4). The produced CALA comprised a mixture of the two isomers, *i.e.*, *cis*-9,*trans*-11,*cis*-15–18:3 (CALA1, 67% of total CALA) and *trans*-9,*trans*-11,*cis*-15–18:3 (CALA2, 33% of total CALA), and accounted for 48% of the total fatty acids obtained. Almost stoichiometric conversion was attained using 12 mg ml^{-1} of albumin-mixed α-linolenic acid as the substrate and washed cells (20% [wet w/v]) as the catalyst in 48 h. The 12 mg ml^{-1} of CALA produced consisted of 43% CALA1 and 57% CALA2, and accounted for 66% of the total fatty acids obtained. Of the CALA produced 40% and 60% were accumulated intracellularly (or associated with cells) and extracellularly, respectively, mainly as the free form.[30]

5.5.3   Conjugated γ-Linolenic Acid Production by Washed Cells of Lactic Acid Bacteria

Conjugated γ-linolenic acid (CGLA) was produced by the incubation of γ-linolenic acid with the washed cells of *L. plantarum* AKU 1009a.[31] Washed cells exhibiting high levels of CGLA productivity were obtained by cultivation in a nutrient medium supplemented with 0.03% (w/v) α-linolenic acid. The cells at the late log phase exhibited significant productivity, but further incubation resulted in a decrease in productivity. The CGLA-producing reaction using the washed cells as the catalyst proceeded well under microaerobic conditions in 0.1 M potassium phosphate buffer (pH 6.5) at 37 °C with free α-linolenic acid mixed with a detergent, *N*-heptyl-β-D-thio-glucoside, in the ratio of 5 : 1 as the substrate. Under the optimum reaction conditions using 13 mg ml^{-1} of detergent-mixed γ-linolenic acid as the substrate, the washed cells (32% [wet w/v]) produced 8.8 mg ml^{-1} of CGLA in 27 h; thus, the molar yield from γ-linolenic acid was 68% (Figure 5.4). The produced CGLA comprised a mixture of the two isomers, *i.e.*, *cis*-6,*cis*-9,*trans*-11–18 : 3 (CGLA1, 40% of total CGLA) and *cis*-6,*trans*-9,*trans*-11–18 : 3 (CGLA2, 60% of total CGLA), and accounted for 66% of the total fatty acids obtained. Of the CGLA produced 70% and 30% were accumulated intracellularly (or associated with cells) and extracellularly, respectively, mainly as the free form.[31]

5.6 Production of C20 Conjugated and Non-Methylene Interrupted Polyunsaturated Fatty Acids by Anaerobic Bacteria

An anaerobic bacteria, *Clostridium bifermentans*, saturated C20 poly-unsaturated fatty acids of arachidonic acid and eicosapentaenoic acid (EPA) into non-methylene interrupted polyunsaturated fatty acids, *i.e.*, *cis*-5,*cis*-8,*trans*-13-eicosatrienoic acid (20 : 3) and *cis*-5,*cis*-8,*trans*-13,*cis*-17-eicosate-traenoic acid (20 : 4), respectively, during anaerobic cultivation in the nutrient medium containing these polyunsaturated fatty acids (Figure 5.5). Similar transformations were observed with C18 and C20 polyunsaturated fatty acids with a ω6,ω9 *cis*,*cis*-diene system such as α-linolenic acid, γ-linolenic acid, and dihomo-γ-linolenic acid. The ω6,ω9 *cis*,*cis*-diene systems

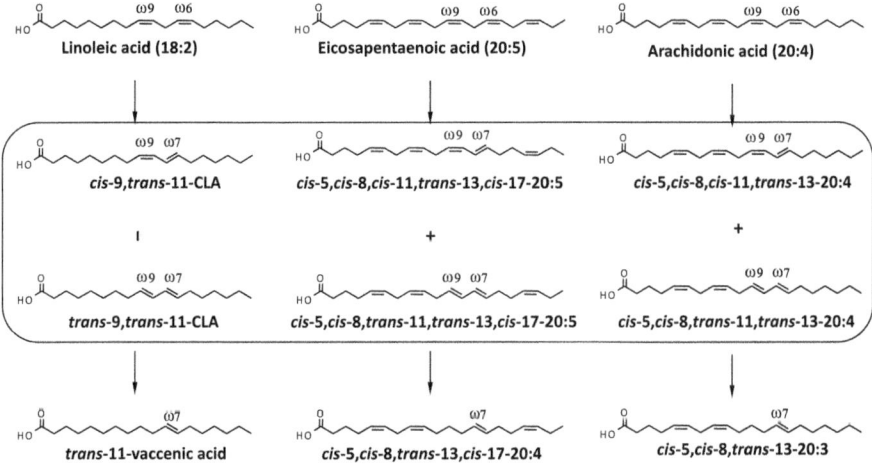

Figure 5.5   Proposed pathway of arachidonic acid and eicosapentaenoic acid transformation by *Clostridium bifermentans*.

were transformed into ω7 *trans*-monoene systems. When the EPA saturation to *cis*-5,*cis*-8,*trans*-13,*cis*-17–20 : 4 were analysed using the cell-free extracts of *C. bifermentans* under the anaerobic reaction conditions, conjugated isomers of EPA, *cis*-5,*cis*-8,*cis*-11,*trans*-13,*cis*-17-eicosapentaenoic acid (20 : 5) and *cis*-5,*cis*-8,*trans*-11,*trans*-13,*cis*-17–20 : 5 were identified as intermediates. Similarly, *cis*-5,*cis*-8,*cis*-11,*trans*-13–20 : 4 and *cis*-5,*cis*-8,*trans*-11,*trans*-13–20 : 4 were identified as intermediates of arachidonic acid saturation to *cis*-5,*cis*-8,*trans*-13–20 : 3 by *C. bifermentans*. On the basis of these results, we proposed the pathways of C20 polyunsaturated fatty acid transformation by *C. bifermentans* as shown in Figure 5.5, *i.e.*, ω6,ω9 non-conjugated *cis*,*cis*-diene saturation to ω7 *trans*-monoene *via* ω7,ω9 conjugated *cis*,*trans*- or *trans*,*trans*-diene as intermediates.

5.7 Metabolic Basis of Polyunsaturated Fatty Acid Biohydrogenation in Lactic Acid Bacteria

CLA producing metabolism was considered to be a part reactions of biohydrogenation (saturation of unsaturated fatty acids) that is detoxification metabolism of free polyunsaturated fatty acids in anaerobic bacteria. Biohydrogenation pathway of *L. plantarum* was found to consist of multiple reactions (Figure 5.6).[32–34] The first reaction of the biohydrogenation is hydration of the carbon–carbon double bond at Δ9 position to generate 10-hydroxy fatty acid catalyzed by a hydratase, CLA-HY. CLA-HY-required a redox cofactor, flavin adenine dinucleotide (FAD), for its activity and the activity was enhanced by a redox cofactor, nicotinamide adenine dinucleotide (NADH). The second reaction is dehydrogenation of the hydroxy group at C10 to generate 10-oxo fatty acid catalyzed by a dehydrogenase, CLA-DH, in the presence of NAD^+. The third reaction is isomerization of the carbon-carbon double bond at Δ12 to generate conjugated enone structure, 10-oxo-*trans*-11-fatty acid, catalyzed by an isomerase, CLA-DC. The fourth reaction is hydrogenation of the carbon–carbon double bond at Δ11 position to generate the carbon–carbon single bond catalyzed by an enone reductase, CLA-ER, in the presence of FAD/flavin mononucleotide (FMN) and NADH. The fifth reaction is hydrogenation of the oxo group at C10 to generate 10-hydroxy fatty acid catalyzed by CLA-DH in the presence of NADH. The last reaction is dehydration of hydroxy group at C10 to generate *cis*-9 or *trans*-10 monoenoic fatty acids catalyzed by CLA-HY in the presence of FAD and NADH. Through a branched pathway of this saturation metabolism, conjugated fatty acids are generated by combined actions of three enzymes, CLA-HY, CLA-DH, and CLA-DC (Figure 5.6). The branched pathway starts from hydrogenation of the oxo group at C10 in 10-oxo-*trans*-11-fatty acid to generate 10-hydroxy-*trans*-11-fatty acid catalyzed by CLA-DH. And the last reaction is dehydration of hydroxy group at C10 in 10-hydroxy-*trans*-11-fatty acid to generate *cis*-9,*trans*-11 and *trans*-9,*trans*-11 conjugated fatty acids catalyzed by CLA-HY (Figure 5.6). CLA-HY is a membrane bond protein and

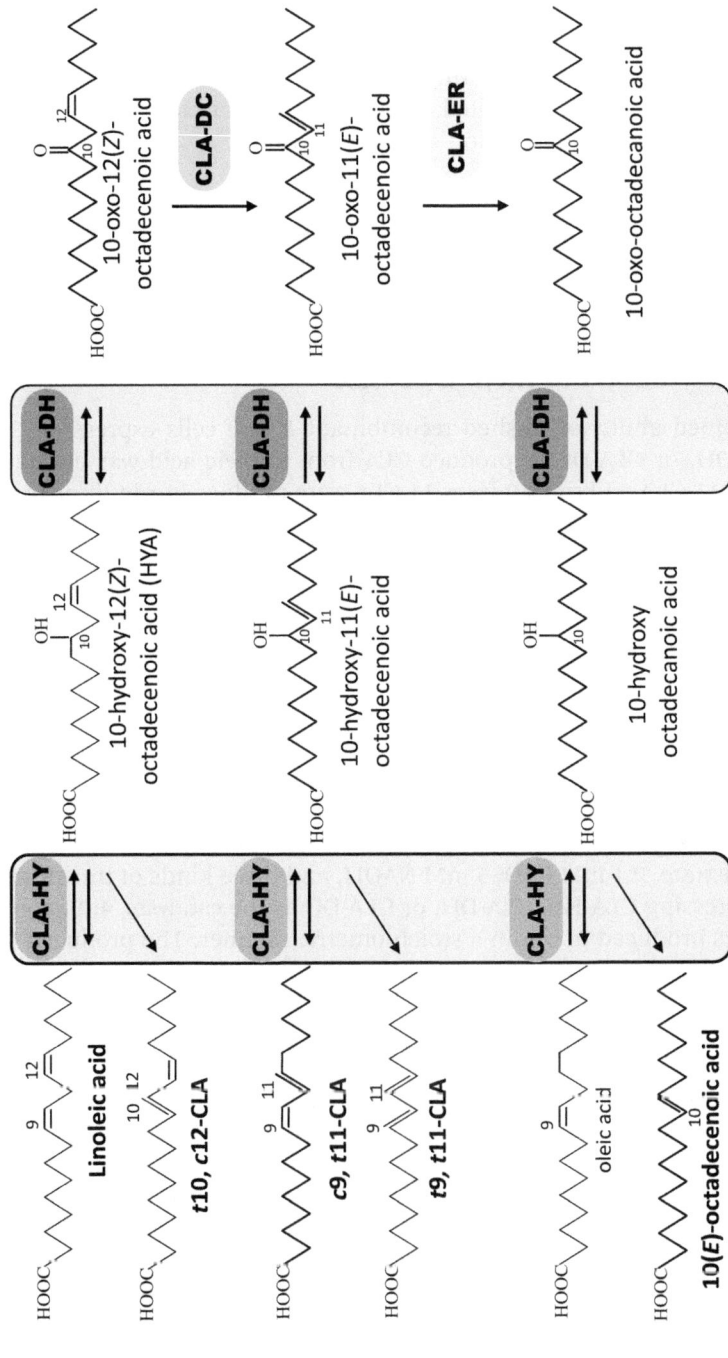

Figure 5.6   Proposed pathway of linoleic acid saturation, biohydrogenation, by *Lactobacillus plantarum*.

CLA-DH, CLA-DC, and CLA-ER are soluble proteins. The genes for CLA-DH, CLA-DC, and CLA-ER are tandemly located in the genome as a gene cluster. C18 fatty acids with *cis*-9,*cis*-12-diene system such as α-linolenic acid, γ-linolenic acid, and stearidonic acid undergo the same transformations in *L. plantarum* AKU 1009a, indicating that the corresponding intermediates, such as hydroxy, oxo, conjugated, and partially saturated fatty acids are produced by the combined action of these enzymes.[28,32–34]

5.8 Enzymatic Production of Conjugated Fatty Acids and Related Fatty Acids in Biohydrogenation Metabolism

5.8.1 Enzymatic Production of CLA

The combined ability of washed recombinant *E. coli* cells expressing CLA-HY, CLA-DH, or CLA-DC to produce CLA from linoleic acid was evaluated. *cis*-9,*trans*-11-CLA and *trans*-9,*trans*-11-CLA were produced from linoleic acid as well as 10-hydroxy-*cis*-12-octadecenoic acid in the presence of FAD and NADH. At an earlier reaction time, 10-hydroxy-*cis*-12-octadecenoic acid was produced from linoleic acid, and *cis*-9,*trans*-11-CLA and *trans*-9,*trans*-11-CLA gradually increased, followed by a decrease in the amount of 10-hydroxy-*cis*-12-octadecenoic acid. This reaction profile was exactly the same as the profile for CLA production using washed cells of *L. plantarum* AKU 1009a.[20,22] Based on these results, CLA-HY, CLA-DH, and CLA-DC were identified as the enzymes that catalyze CLA production in *L. plantarum* AKU1009a. The CLA production using washed cells of *E. coli* transformants proceeded much more efficiently than that using washed cells of *L. plantarum*. Under optimized reaction conditions with 4.0 mg ml^{-1} linoleic acid as the substrate, 0.1 mM FAD, 5 mM NADH, and three kinds of transformed *E. coli* expressing CLA-HY, CLA-DH, or CLA-DC as the catalysts, 4.0 mg ml^{-1} of CLA was produced in 5 h in a stoichiometric manner. The produced CLA was a mixture of CLA1 (18%) and CLA2 (82%).[32]

5.8.2 Enzymatic Production of Hydroxy Fatty Acid

The hydratase, CLA-HY, catalyzing the first step of polyunsaturated fatty acid saturation was overexpressed in *E. coli*, and the *E. coli* transformant was used as the catalyst for hydroxy fatty acid production. Under the optimized reaction conditions, about 30g L^{-1} of 10-hydroxy-18:0 was produced from oleic acid with more than 90% yield and with strict stereospecificity for *S*-isomer. The ability of hydration of double bonds in fatty acid was further screened in lactic acid bacteria, and it was found that *Pediococcus sp.* AKU 1080 produced a variety of hydroxyl fatty acids including a dihydroxy fatty acid (10-hydroxy, 13-hydroxy and 10,13-dihydroxy fatty acids) from C18 fatty acids with a *cis*-9,*cis*-12 diene system.[35] The growing cells of this strain were applied to the production of 13-hydroxy-*cis*-9–18:1. Under the optimum conditions, 2.3 mg ml^{-1} of 13-hydroxy-*cis*-9–18:1 was produced from 12.3 mg ml^{-1} of linoleic acid with 0.04 mg ml^{-1} 10-hydroxy-*cis*-12–18:1 and 0.05 mg ml^{-1} 10,13-dihydroxy-18:0 in the cultivation medium. Specific production of 13-hydroxy-*cis*-9–18:1, which is useful for the production of 13-oxo-fatty acids with anti-obesity activity, was attained using cell-free extracts of the strain as the catalyst. Under the optimum conditions, 0.4 mg ml^{-1} of 13-hydroxy-*cis*-9–18:1 was produced from 2.0 mg ml^{-1} of linoleic acid without 10-hydroxy-*cis*-12–18:1 and 10,13-dihydroxy-18:0.[35] Such regio- and stereo-selective introduction of hydroxyl group to unsaturated fatty acids by microorganisms is useful for production of hydroxy fatty acids which are valuable as starting materials for industrial chemicals, functional foods, and pharmaceuticals.

5.8.3 Enzymatic Production of Biohydrogenation Intermediates

Various combinations of *E. coli* cells expressing biohydrogenation enzymes, *i.e.*, CLA-HY, CLA-DH, CLA-DC, and CLA-ER, enable to produce various biohydrogenation intermediates (Figure 5.7). For example, CLA-DH transforms hydroxy fatty acids generated by CLA-HY to oxo fatty acids in the presence of NAD$^+$. CLA-DC transforms oxo fatty acids generated by CLA-DH to another kind of oxo fatty acids with enone structure, and the resulting enone-type oxo fatty acids are further saturated into partially saturated fatty acids (non-methylene-interrupted fatty acids) by CLA-ER in the presence of

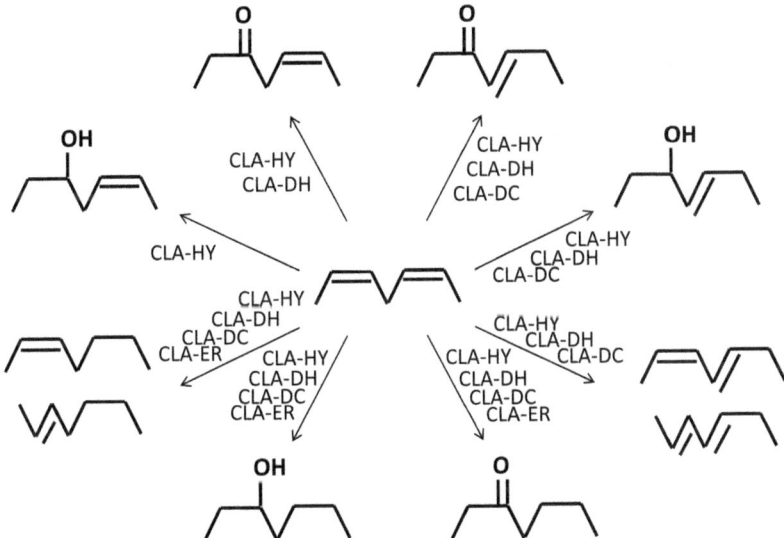

Figure 5.7 Scheme of fatty acid transformation by combined actions of enzymes involved in biohydrogenation.

FAD/FMN and NADH. Investigation of the physiological functions of these unique fatty acids of biohydrogenation intermediates produced by lactic acid bacteria will create novel functional lipids especially related to gastro-intestinal health control.

5.9 Discussion

Processes involving the washed (resting) cells of lactic acid bacteria as catalysts can help avoid the inhibitory effects of free polyunsaturated fatty acids (substrates) on cell growth and thus enable reactions with high substrate concentrations. The productivity obtained using the washed cell method is 10- to 100-fold higher than that obtained during cultivation.

Only two CLA isomers (CLA1 and CLA2) were produced from linoleic acid, ricinoleic acid or castor oil by lactic acid bacteria, suggesting that the biological CLA production processes are more isomer-selective than the chemical ones. However, it is still important to control the isomer production ratio for a more selective isomer synthesis. We investigated the factors affecting the isomer ratio in CLA production from linoleic acid, and found that it could be controlled by changing the reaction conditions. For example, the addition of L-serine, glucose, $AgNO_3$, or NaCl to the reaction mixture reduces the production of CLA2, resulting in selective production of CLA1 (about 75% selectivity).[21] CLA2 is produced with more than 97% selectivity, if the reaction is performed length of time using a low linoleic acid concentration.[20] Not only CLA but also conjugated trienoic acids were produced by the washed cells of lactic acid bacteria. The isomer selectivity of lactic acid bacteria is advantageous for the trienoic acid transformation, which is hard to control by chemical methods. The representative results of isomer specific production of conjugated C18 fatty acids are summarized in Figure 5.8.

CLA produced by lactic acid bacteria is a free fatty acid, however, triacylglycerol containing CLA is also interesting from physiological and nutritional viewpoints. The free fatty acid form of CLA can be transformed to the acylglycerol or ester form through lipase-catalysed reactions.[36] As an alternative method for triacylglycerol production, we attempted CLA production using molds that accumulate lipids as triacylglycerol. We found that some molds can produce CLA from *trans*-vaccenic acid (*trans*-11–18:1) through Δ9 desaturation,[37,38] as reported for mammalian cells.[15] Further optimization of fungal Δ9 desaturation with *Delacroixia coronata* enabled specific production of CLA1 containing triacylglycerol.[38]

During the production of CALA and CGLA by lactic acid bacteria from α-and γ-linolenic acid, respectively, other saturated fatty acids were also produced, indicating that the conjugated fatty acids are the intermediates of fatty acid saturation, as reported previously.[14] Saturation metabolism of polyunsaturated fatty acids is a representative mode of lipid metabolism by gastrointestinal microbes, such as lactic-acid bacteria, which reside in colon

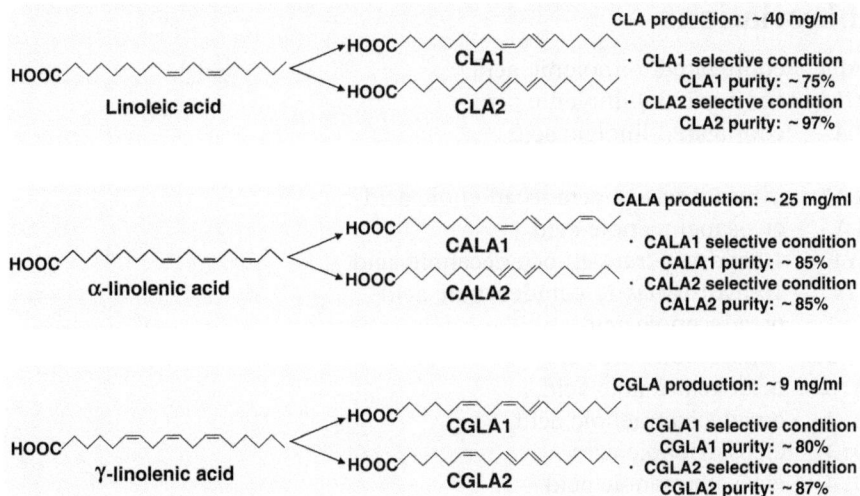

Figure 5.8   Isomer specific production of conjugated C18 fatty acids by *Lactobacillus plantarum.*

and intestine. Microorganisms in the gastrointestinal tract interact with their host in many ways and contribute significantly to the maintenance of host health.[39] Lipid metabolism by gastrointestinal microbes generates multiple fatty-acid species, such as conjugated fatty acids and *trans*-fatty acids that can affect host lipid metabolism.[13] In the representative gut bacterium *L. plantarum*, we identified genes encoding the enzymes involved in a novel saturation metabolism of polyunsaturated fatty acids, and revealed in detail the metabolic pathway that generates hydroxy fatty acids, oxo fatty acids, enone-type oxo fatty acids, conjugated fatty acids, and partially saturated *trans*-fatty acids (non-methylene interrupted fatty acids) as intermediates.

Furthermore, we observed these intermediates, especially hydroxy fatty acids, in host organs.[34] Levels of hydroxy fatty acids were much higher in specific pathogen-free mice than in germ-free mice, indicating that these fatty acids are generated through polyunsaturated fatty acids metabolism of gastrointestinal microorganisms. These findings suggested that lipid metabolism by gastrointestinal microbes affect the health of the host by modifying fatty-acid composition. Therefore, functional investigations of these fatty-acid intermediates of the polyunsaturated fatty acid saturation metabolism and probiotic effects of fatty acid-saturating microorganisms will provide new methods for improving our health by altering lipid metabolism related to the onset of metabolic syndrome. Exploration of the lipid metabolism of gastrointestinal microorganisms at the enzymatic and genetic levels, and integration of these findings with metagenomic information, might enable us to promote health by controlling intestinal lipid metabolism.

Abbreviations

CALA	conjugated α-linolenic acid
CGLA	conjugated γ-linolenic acid
CLA	conjugated linoleic acid
CLA1	*cis*-9,*trans*-11-octadecadienoic acid
CLA2	*trans*-9,*trans*-11-octadecadienoic acid
EPA	eicosapentaenoic acid
HY1	10-hydroxy-*trans*-12-octadecenoic acid
HY2	10-hydroxy-*cis*-12-octadecenoic acid
18:1	octadecenoic acid
18:2	octadecadienoic acid
18:3	octadecatrienoic acid
18:4	octadecatetraenoic acid
20:3	eicosatrienoic acid
20:4	eicosatetraenoic acid
20:5	eicosapentaenoic acid.

Acknowledgements

This work was partially supported by the Industrial Technology Research Grant Program in 2007 (No. 07A08005) and the Project for Development of a Technological Infrastructure for Industrial Bioprocesses on R&D of New Industrial Science and Technology Frontiers from the New Energy and Industrial Technology Development Organization (NEDO) of Japan, Grants-in-Aid for Scientific Research (No. 19780056, No. 16688004, and No. 18208009) and COE for Microbial-Process Development Pioneering Future Production Systems from the Ministry of Education, Culture, Sports, Science and Technology of Japan, and by the Bio-Oriented Technology Research Advancement Institution of Japan.

References

1. Y. L. Ha, J. Storkson and M. W. Pariza, *Cancer Res.*, 1990, **50**, 1097.
2. C. Ip, S. F. Chin, J. A. Scimeca and M. W. Pariza, *Cancer Res.*, 1991, **51**, 6118.
3. K. N. Lee, D. Kritchevsky and M. W. Pariza, *Atherosclerosis*, 1994, **108**, 19.
4. Y. Park, K. J. Albright, W. Liu, J. M. Storkson, M. E. Cook and M. W. Pariza, *Lipids*, 1997, **32**, 853.
5. Y. M. C. Silvia, P. V. H. John, G. B. Steven, A. L. Lisa and A. B. Martha, *J. Lipid Res.*, 1999, **40**, 1426.
6. O. A. Gudbrandsen, E. Rodríguez, H. Wergedahl, S. Mørk, J. E. Reseland, J. Skorve, A. Palou and R. K. Berge, *Br. J. Nutr.*, 2009, **102**, 803.
7. S. F. Chin, W. Liu, J. M. Storkson, Y. L. Ha and M. W. Pariza, *J. Food Comp. Anal.*, 1992, **5**, 185.

8. N. Sehat, M. P. Yurawecz, J. A. Roach, M. M. Mossoba, J. K. Kramer and Y. Ku, *Lipids*, 1998, **33**, 217.

9. N. Sehat, R. Rickert, M. M. Mossoba, J. K. Kramer, M. P. Yurawecz, J. A. Roach, R. O. Adlof, K. M. Morehouse, J. Fritsche, K. D. Eulitz, H. Steinhart and Y. Ku, *Lipids*, 1999, **34**, 407.

10. K. W. Wahle, S. D. Heys and D. Rotondo, *Prog. Lipid Res.*, 2004, **43**, 553.

11. M. W. Pariza, Y. Park and M. E. Cook, *Prog. Lipid Res.*, 2001, **40**, 283.

12. A. Philippaerts, S. Goossens, P. A. Jacobs and B. F. Sels, *ChemSus Chem*, 2011, **4**, 684.

13. J. M. Griinari and D. E. Bauman, Biosynthesis of conjugated linoleic acid and its incorporation into meat and milk in ruminants, in *Advances in Conjugated Linoleic Acid Research*, ed. M. P. Yurawecz, M. M. Mossoba, J. K. G. Kramer, M. W. Pariza and G. J. Nelson, AOCS Press, Champaign, 1999, vol. 1, pp. 180–200.

14. C. R. Kepler, K. P. Hirons, J. J. Mcneill and S. B. Tove, *J. Biol. Chem.*, 1966, **241**, 1350.

15. R. O. Adolf, S. Duval and E. A. Emken, *Lipids*, 2000, **35**, 131.

16. J. Jiang, L. Bjorck and R. Fonden, *J. Appl. Microbiol.*, 1998, **85**, 95.

17. J. C. Andrade, K. Ascenção, P. Gullón, S. M. S. Henriques, J. M. S. Pinto, T. A. P. Rocha-Santos, A. C. Freitas and A. M. Gomes, *Int. J. Dairy Tech.*, 2012, **65**, 467.

18. E. F. O'Shea, P. D. Cotter, C. Stanton, R. P. Ross and C. Hill, *Int. J. Food Microbiol.*, 2012, **152**, 189.

19. J. Ogawa, S. Kishino, A. Ando, S. Sugimoto, K. Mihara and S. Shimizu, *J. Biosci. Bioeng.*, 2005, **100**, 355.

20. S. Kishino, J. Ogawa, Y. Omura, K. Matsumura and S. Shimizu, *J. Am. Oil Chem. Soc.*, 2002, **79**, 159.

21. S. Kishino, J. Ogawa, A. Ando, T. Iwashita, T. Fujita, H. Kawashima and S. Shimizu, *Biosci. Biotechnol. Biochem.*, 2003, **67**, 179.

22. J. Ogawa, K. Matsumura, K. Kishino, Y. Omura and S. Shimizu, *Appl. Environ. Microbiol.*, 2001, **67**, 1246.

23. S. Kishino, J. Ogawa, A. Ando, Y. Omura and S. Shimizu, *Biosci. Biotechnol. Biochem.*, 2002, **66**, 2283.

24. A. Ando, J. Ogawa, S. Kishino and S. Shimizu, *J. Am. Oil Chem. Soc.*, 2003, **80**, 889.

25. A. Ando, J. Ogawa, S. Kishino and S. Shimizu, *Enzyme Microb. Technol.*, 2004, **35**, 40.

26. M. Igarashi and T. Miyazawa, *Cancer Lett.*, 2000, **148**, 173.

27. R. Suzuki, R. Noguchi, R. Ota, M. Abe, K. Miyashita and T. Kawada, *Lipids*, 2001, **36**, 477.

28. S. Kishino, J. Ogawa, K. Yokozeki and S. Shimizu, *Lipid Technology*, 2009, **21**, 177.

29. S. Kishino, J. Ogawa, K. Yokozeki and S. Shimizu, *Appl. Microbiol. Biotechnol.*, 2009, **84**, 87.

30. S. Kishino, J. Ogawa, A. Ando and S. Shimizu, *Eur. J. Lipid Sci. Technol.*, 2003, **105**, 572.

31. S. Kishino, J. Ogawa, A. Ando, K. Yokozeki and S. Shimizu, *J. Appl. Microbiol.*, 2010, **108**, 2012.
32. S. Kishino, S. B. Park, M. Takeuchi, K. Yokozeki, S. Shimizu and J. Ogawa, *Biochem. Biophys. Res. Commun.*, 2011, **416**, 188.
33. S. Kishino, J. Ogawa, K. Yokozeki and S. Shimizu, *Biosci. Biotechnol. Biochem.*, 2011, **75**, 318.
34. S. Kishino, M. Takeuchi, S.-B. Park, A. Hirata, N. Kitamura, J. Kunisawa, H. Kiyono, R. Iwamoto, Y. Isobe, M. Arita, H. Arai, K. Ueda, J. Shima, S. Takahashi, K. Yokozeki, S. Shimizu and J. Ogawa, *Proc. Natl. Acad. Sci. USA*, 2013, **110**, 17808.
35. M. Takeuchi, S. Kishino, K. Tanabe, A. Hirata, S. B. Park, S. Shimizu and J. Ogawa, *Eur. J. Lipid Sci. Technol.*, 2013, **115**, 386.
36. T. Nagao, Y. Shimada, Y. Yamaguchi-Sato, T. Yamamoto, M. Kasai, K. Tsutsumi, A. Sugihara and Y. Tominaga, *J. Am. Oil Chem. Soc.*, 2002, **79**, 303.
37. A. Ando, J. Ogawa, S. Kishino, T. Ito, N. Shirasaka, E. Sakuradani, K. Yokozeki and S. Shimizu, *J. Am. Oil Chem. Soc.*, 2009, **86**, 227.
38. A. Ando, J. Ogawa, S. Sugimoto, S. Kishino, E. Sakuradani, K. Yokozeki and S. Shimizu, *J. Appl. Microbiol.*, 2009, **106**, 1697.
39. J. L. Round and S. K. Mazmanian, *Nat. Rev. Immunol.*, 2009, **9**, 313.

CHAPTER 6

Recent Advances in the Production of CLA and Conjugated Vegetable Oils: Production of CLA and Conjugated Vegetable Oils via Metal Catalysis

K. BELKACEMI,* N. CHORFA AND S. HAMOUDI

Department of Soil Sciences and Agri-Food Engineering, Paul-Comtois Building, Laval University, Quebec, G1V 0A6, Canada
*Email: khaled.belkacemi@sga.ulaval.ca

6.1 Introduction

The interest for polyunsaturated fats isomerization to produce Conjugated Linoleic Acid (CLA) has increased since the last years due to the growing interest for the biologically activities of (9-*cis*,11-*trans*) and (10-*trans*,12-*cis*) CLA isomers.[1] Conjugated dienes formation can be catalysed by alkalines bases, enzymes, photo-irradiation, or by supported metals as well as organometallic complexes.[2,3] This chapter deals with a non-exhaustive literature review of conjugated isomerization of linoleic acid and vegetable oils catalysed by transition metals in homogeneous and heterogeneous systems. Even if the reactions seemed to be similar, however, the distribution of the

RSC Catalysis Series No. 19
Conjugated Linoleic Acids and Conjugated Vegetable Oils
Edited by Bert Sels and An Philippaerts
© The Royal Society of Chemistry 2014
Published by the Royal Society of Chemistry, www.rsc.org

products and the mechanisms involved could be different, and for thoses reasons we opted to distinguish between studies conducted with linoleic acid and its derivatives and thoses conducted with vegetable oils. This review constitutes an update of more recent reviews covered for this important topic.

6.2 Homogeneous Catalytic Approaches for Conjugated Isomerization

Homogeneous catalysts for CLA production are based on transition metal of group VI (Chromium) and group VIII ("Iron Metal", and noble metals such as Rh, Ru, Ir, Os or Pt). Overviews of their catalytic performances for the isomerization of linoleic acid and its esters as well as vegetable oils are given below.

6.2.1 Catalysts Activities

6.2.1.1 Conjugated Isomerization of Linoleic Acid and Its Derivatives

Conjugated isomerization is a dual reaction occurring during the hydrogenation process of fats and oils in both homogeneous and heterogeneous catalysis.[4] Carbonyl complexes of chromium,[5] iron[6] and ruthenium,[7] are homogeneous hydrogenation catalysts characterized also to be active for geometic and positional isomerization activities promoted by the operating conditions (Table 6.1). Methyl linoleate (ML) isomerization with methyl-, benzene- and cycloheptatriene-$Cr(CO)_3$ can be achieved with or without solvent, and produces 65–78% conjugated dienes at 175–185 °C after 2–6 h.[5] Polyunsaturated fats heated at 180–185 °C with $Fe(CO)_5$ formes iron tricarbonyl complexes, and then the diene-$Fe(CO)_3$ complex is decomposed by ferric chloride $FeCl_3$ in ethanol giving high yields of conjugated fatty acids (80–95%).[6] Basu and Sharma[7] investigated methyl linoleate isomerization with carbonyl clusters of group VIII metals ([$M_3(CO)_{12}$] where $M = Fe$, Ru or Os). The best conditions are obtained with [$Ru_3(CO)_{12}$] (1 mol.% Ru) in isopropanol at 80 °C during 4 h where a conversion of 77%, selectivity towards conjugated products of 89%, and 11% for methyl oleate hydrogenation product issued from hydrogen transfer were obtained. The use of octane as solvent increased the selectivity towards conjugated dienes. Ruthenium alkoxide complexes prepared *insitu* with ruthenium trichloride $RuCl_3$–$3H_2O$ in boiling hexanol during 30 min, leads to 61% methyl linoleate conversion, a selectivity of 75% conjugated methyl linoleate and 25% hydrogenated products and polymer.[8,9] High temperatures favored the production of *trans-trans* CLAs and polymers. The formation of monoenes and conjugated trienes highlighted the hydrogenation and dehydrogenation side reactions.

The most efficient organometallic complexes for conjugated isomerization of linoleic acid and its esters under mild conditions in homogeneous system are based on rhodium and Wilkinson catalyst $RhCl(PPh_3)_3$

Table 6.1 Production of conjugated linoleic acids isomers in homogeneous catalytic systems.

Starting material	Catalyst	Solvent	T (°C)	t (h)	Conversion (wt%)	TOF[a] (h^{-1})	CLAs[b] (wt%)	Ref.
methyl linoleate	Methyl benzoate-Cr(CO)$_3$	n-hexanes	175	5	100	1.30	65	5
methyl linoleate	Benzene-Cr(CO)$_3$	n-hexanes	185	6	100	1.06	63.7	5
methyl linoleate	Cycloheptatriene-Cr(CO)$_3$	n-hexanes	185	5	99.9	1.57	78.4	5
methyl linoleate	Cr(CO)$_6$	n-hexanes	195	2	100	3.32	66.4	5
methyl linoleate	Fe(CO)$_5$/FeCl$_3$	petroleum ether/ethanol	180	4	90	0.10	89.8	6
methyl linoleate	Ru(η^6-naphthalene)(η^4-cycloocta-1,5-diene)/ CH$_3$CN	hexane	60	24	100	3.90	95	14
methyl linoleate	[RhCl(C$_8$H$_{12}$)$_2$]$_2$/(p-CH$_3$C$_6$H$_5$)$_3$P/SnCl$_2$·2H$_2$O	ethanol	60	24	100	36.29	87.1	12
linoleic acid	[(p-CH$_3$C$_6$H$_4$)$_3$P]$_3$RhCl/SnCl$_2$·2H$_2$O	—	140	6	100	13.5	90	30
methyl linoleate	RhCl$_3$ tris(triphenylphosphine)	methanol	65	4	85	7.17	95	10
methyl linoleate	cis-Cl$_2$[(C$_6$H$_5$)$_3$P]$_2$Pt-SnCl$_2$	n-butyl carbitol	180	5	60	2.6	24	33
methyl linoleate	RuHClCO(PPh$_3$)$_3$	BMI · NTf$_2$[c]	80	24	68	3.33	68	20
methyl lincleate	RhCl(PPh$_3$)$_3$/SnCl$_2$	ethanol + BMI · NTf$_2$[c]	60	24			85	20
methyl lincleate	RhCl(PPh$_3$)$_3$/SnCl$_2$	ethanol + NBu$_4$ · Br[d]	60	24			99	20

[a] Turn over frequencies: Moles of conjugated dienes/(moles of catalyst * h).
[b] CLA yield.
[c] BMI · NTf$_2$ = 1-n-butyl-3-methylimidazolium bis(trifluoromethylsulfonyl)imidate.
[d] NBu$_4$ · Br = Tetrabutylammonium bromide.

(Table 6.1): $H_2RhCl[(C_6H_5)_3P]_2$,[10] $H-Rh[(p-CH_3C_6H_4)_3P]_2$,[11,12] cationic complexes $[Rh(dien)L_2]^+ClO_4^-$ (dien = NBD or COD, L = PPh$_3$ or L$_2$ = dppe),[13] and catalysts based on ruthenium Ru(η^6-naphthalene)(η^4-cycloocta-1,5-diene) and Ru(η^6-arene)-(η^4-C$_8$H$_{12}$) (arene = benzene and *p*-cymene).[14]

The active hydrogenation catalyst $H_2RhCl[(C_6H_5)_3P]_2$ is formed from RhCl(PPh$_3$)$_3$ by the elimination of one molecule of triphenylphosphine and addition of two hydrogen atoms (under H$_2$ pressure).[15] Dejarlais and Gast[10] reported that methanol at 65 °C can replace hydrogen for quantitative conversion of methyl linoleate where a selectivity up to 95% towards conjugated dienes in confguration *cis-trans* and *cis-cis* can be reached. Larock group had made a significant improvement of this isomerization process by changing the ligand and introducing a Lewis acid to minimize the catalyst deactivation.[11,12] They use 0.1 mol.% [RhCl(C$_8$H$_{12}$)$_2$]$_2$, 0.4 mol.% (*p*-CH$_3$C$_6$H$_4$)$_3$P, and 0.8 mol.% SnCl$_2$.2H$_2$O in absolute ethanol at 60 °C during 24 h. The obtained isomerization product yield was 87.1% where the main CLA isomers were 9-*cis*,11-*trans* (33.8%) and 10-*trans*,12-*cis* (45.1%). Andjelkovic *et al.*[12] reported that the intensity of the bisallylic proton decreased with increasing conversion in the first 2 h of reaction and slowed over the next 22 h. An analysis of the reaction mechanism revealed that the reaction followed approximately first-order kinetics during the initial stages of the reaction. Possible reason for the catalyst deactivation could be the catalyst decomposition under the reaction conditions.[16] Others cationic complexes based on rhodium as $[Rh(NBD)(PPh_3)_2]^+ClO_4^-$ revealed a significant conjugated isomerization activity of methyl linoleate (93%) in isopropanol at 80 °C.[13] Most of the conjugated dienes are in *cis-trans* configurations, and there is no hydrogenation product or polymer.

The Ru(η^6-naphthalene)(η^4-cycloocta-1,5-diene) complex with acetonitrile in aprotic solvent (hexane) at 60 °C, leads to a full conversion of methyl linoleate after 24 h with a high selectivity for conjugated products (95%).[14] This Ru-complex can be recycled and it is characterized by the lability of η^6-naphthalene-Ru bond in acetonitrile leaving coordination sites more accessible for the incoming reagent.[17,18]

Although the homogeneous conjugated isomerization of methyl linoleate with rhodium and ruthenium organometallic complexes is quantitative and selective, the noble metal catalysts recover and re-use processes are still problematic. A possible solution to ease the separation of the catalyst from the product mixture is the use of ionic liquids (ILs). Ionic liquids are salts which have a melting point below 100 °C because the anions and cations are very bulky and asymmetric so they can crystallize only with difficulty (Figure 6.1). The high polarity and low volatility of the ILs can serve as *supported ionic liquid-phase catalysis.*[19]

Consorti *et al.*[20] formulated biphasic catalytic systems for selective isomerization of ML with Ru and Rh organomettallic complexes such as RuHCl(CO)(PPh$_3$)$_3$ and RhCl(PPh$_3$)$_3$-SnCl$_2$ dissolved in BMI.NTf$_2$ at 80–60 °C (Table 6.1). After 24 h, ML is quantitatively converted into conjugated isomers 80–85%. An ionophilic ligand is employed as ancillary ligand to

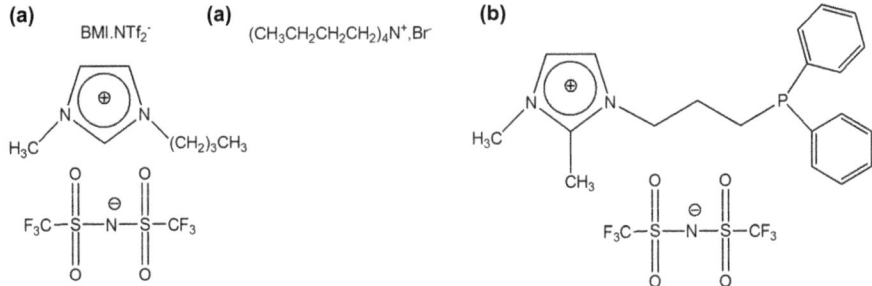

Figure 6.1   (a) Ionic liquids namely 1-*n*-butyl-3-methylimidazolium (BMI) and tetra-*n*-butylammonium cations associated with bromide and bis(trifluoromethylsulfonyl)imide (NTf$_2^-$) anions, and (b) ionophilic phosphine ligand.[20]

immobilize Ru and Rh complexes in the ionic liquid phase, and the upper product phase is collected by decantation.

6.2.1.2   Conjugated Isomerization of Vegetable Oils

The majority of CLAs products are commercially available in isomeric mixtures of free fatty acids or esters. However, recent studies have shown that triglycerides forms are more appropriate for incorporation into functional foods and bioavailability of CLA isomers.[21]

Frankel group[5,6,22,23] observed that chromium and iron carbonyls catalysts for hydrogenation process have also an interesting isomerization activity (Table 6.2). Soybean and safflower oils are conjugated at 73% and 48% with benzoate-Cr(CO)$_3$.[5] The reaction of polyunsaturated vegetable oils with iron pentacarbonyl Fe(CO)$_5$ formed an iron tricarbonyl complexe Fe(CO)$_3$-Diene that is decomposed by ferric chloride FeCl$_3$ to give high yield of conjugated oil (90–97%).[6] However, the iron tricarbonyl complexe is totally decomposed and lost at the end of the process. Frankel and Metlin[23] reported the decomposition of Fe(CO)$_3$-diene complexes under high carbon monoxide pressure to produce conjugated oil and iron pentacarbonyl Fe(CO)$_5$ recoverable and reusable. The iron tricarbonyl complex of soybean oil heated at 180 °C under 3600 psi CO produced 82% conjugated oil in all-*trans* configuration and 84% Fe(CO)$_5$ are recovered. Although, the yield of conjugated oil is 11–12% lower than that of the method using ferric chloride FeCl$_3$,[6] the Fe(CO)$_3$-Diene conversion into reusable Fe(CO)$_5$ makes the CO procedure more economic.

Ruthenium complexes are known catalysts for olefins isomerization,[24] and their association with Brönsted acid (HCOOH) or Lewis acid (SnCl$_2 \cdot 2H_2O$) can promote the double bond migration of polyunsaturated vegetable oils. Sleeter[25,26] patented the isomerization of linseed oil catalysed by [Ru$_3$(CO)$_{12}$]/HCOOH and [RuCl$_2$(PPh$_3$)$_3$]/HCOOH at 180 °C, and the final product contained 75–90% conjugated dienes and trienes. Unfortunately, the authors did not give any information on the catalytic action of formic acid and the conjugation was determined from the consumption of the unconjugated reagent. Moreover, the reactions are carried out under air which promote the polymerization of conjugated dienes and trienes, and also increases the deactivation of the ruthenium catalysts. Larock et al.[11] developped one of the most efficient homogeneous catalyst, RuHCl(CO)(PPh$_3$)$_3$, for the conjugated isomerization of vegetable oils (Table 6.2). After 24 h, in benzene at 60 °C, soybean oil is quantitatively converted into its conjugated isomers. Although, more Ru catalyst is required compared to [RhCl(C$_8$H$_{14}$)$_2$]$_2$/(p-CH$_3$C$_6$H$_4$)$_3$P/SnCl$_2$.2H$_2$O and PtCl$_2$(PPh$_3$)$_2$/SnCl$_2 \cdot$2H$_2$O catalysts, the advantages of the Ru system relates to the lower cost of ruthenium compared to rhodium and platinum, and the lewis acid SnCl$_2 \cdot$2H$_2$O is not required. Nevertheless, the use of benzene, in order to dissolve the catalyst, is a clear drawback. Krompiec et al.[27] reported that the use of RuHCl(CO)(PPh$_3$)$_3$ without any solvent produced 54% conjugated soybean oil, but much higher temperatures (212–226 °C) are required.

The cationic complexes based on iridium [Ir(NBD)(PPh$_3$)$_2$]$^+$ClO$_4^-$, and rhodium [Rh(NBD)(PPh$_3$)$_2$]$^+$ClO$_4^-$, in a boiling mixture of methanol with dichloromethane or acetone during 12 h, produced 64% of conjugated isomers in linseed oil and 96% in safflower oil.[13,28] After evaporation of solvents, the catalyst was precipitated in pentane and recovered at 95% after a filtration step. Andjelkovic et al.[12] developed a Wilkinson catalyst for the synthesis of highly conjugated vegetable oils under mild conditions (24 h at 60 °C in ethanol). According to ^{1}H NMR analysis of the *bis*-allylic protons, 54% linoleic acid present in soybean oil was quantitatively converted into its conjugated isomers. The two major products were (9-*cis*,11-*trans*) and (10-*trans*,12-*cis*) CLA isomers which accounted for 81% of all CLAs in the mixture. Andjelkovic et al.[12] reported that the highest fraction of biologically active CLA isomers (9-*cis*,11-*trans*) and (10-*trans*,12-*cis*), was obtained with oils that contain an intermediate amount of both linoleic and linolenic acids. However, if the linolenic acid amount is in excess (linseed oil) or in deficit (safflower oil), the thermodynamically more stable (9-*trans*,11-*trans*) CLA isomer was favored.

The isomerization processes of vegetable oils in a homogeneous system can be achieved without solvent if the temperatures are higher than 100–200 °C necessary to dissolve the organometallic complexes, which are very sensitive to oxygen and water.[29] For instance, the conjugated isomerization of soybean oil as reported by Andjelkovic et al.[12] can be completed without ethanol but at high temperature (180 °C) yielding a quantitative conversion into conjugated isomers.[30,31] Krompiec et al.[27,32] described the isomerization of pure vegetable oils catalysed by ruthenium complexes, RuClH(CO)(PPh$_3$)$_3$, at high temperatures (221–238 °C). The ruthenium complexes were removed from the conjugated products by sorption onto diatomaceous earth. The final conversion and level of conjugation is highly depended on the oil source. For instance, 30% (rape), 54% (soybean and sunflower), 50% (linseed) conversions can be achieved with a small amount of polymers (less than 10%) and less than 1 ppm of ruthenium. Finally, also Pt-complexes can be used to

Table 6.2 Production of conjugated polyunsaturated oils in homogeneous catalytic systems.

Oils	Catalyst	Solvent/gas	T (°C)	t (h)	TOF[a] (h^{-1})	C[b] (wt%)	H[c] (wt%)	CLAs[d] (wt%)	Ref.
soybean[e]	Fe(CO)$_3$	CO (259 bar)	198	2	0.39	95		83.1	23
soybean[e]	Fe(CO)$_3$/FeCl$_3$	petroleum ether/ethanol	180	4	0.23	90–95	4.1	96.6	6
linseed[e]	Fe(CO)$_3$/FeCl$_3$	petroleum ether/ethanol	185	2	0.46	90–95	4.8	95.8	6
safflower[e]	Fe(CO)$_3$/FeCl$_3$	petroleum ether/ethanol	185	4	0.23	90–95	2.3	97.2	6
safflower	Methyl benzoate-Cr(CO)$_3$	N$_2$	175	6	1.22	100		73.3	5
soybean	Methyl benzoate-Cr(CO)$_3$	N$_2$	175	6	1.22	100		72.9	5
linseed	Methyl benzoate-Cr(CO)$_3$	N$_2$	175	6	0.80	100		48.2	5
soybean	[RhCl(C$_8$H$_{12}$)$_2$]$_2$/(p-CH$_3$C$_6$H$_5$)$_3$P/ SnCl$_2$ · 2H$_2$O	ethanol	60	24	41.25	100	—	99	12
soybean	Cl$_2$PtL$_2$-SnCl$_2$	—	180	24				19	33
linseed	Cl$_2$PtL$_2$-SnCl$_2$	—	180	24				21	33
soybean	RhCl(PPh$_3$)$_3$/SnCl$_2$	ethanol + MBPy · NTf$_2$[f]	60	24	7.07	87	—	83	20
soybean	RhCl(PPh$_3$)$_3$/SnCl$_2$	Ethanol + NBu$_4$ Br[g]	60	24	9.50	100	—	97	20
soybean	RhCl(PPh$_3$)$_3$/SnCl$_2$	Ethanol + NBu$_4$ Br[g] + ionophilic ligand	60	24	6.90	87	—	81	20
castor bean	Tungstic acid + zinc powder	—	250	24	0.21	99.5	—	34.4	35
soybean	PtCl$_2$(PPh$_3$)$_2$/SnCl$_2$ 2H$_2$O	1 atm H$_2$	100	24	25.57	100	—	89	11
soybean	RuHCl(CO)(PPh$_3$)$_3$	benzene	60	24	27.30	100	—	95	11

[a] Turn over frequencies: Moles of conjugated dienes/(moles of catalyst * h).
[b] Conversion.
[c] Hydrogenation yield.
[d] CLA yield based on total linoleate and linolenate in original oils and GLC analyses: safflower, 76.5%; linseed, 62.6%; soybean, 59.1%.
[e] Iron complex oil expressed as octadecadienoate oil-Fe(CO)$_3$.
[f] MBPy · NTf$_2$ = N,N-methyl-n-butylpyrrolidinium bis(trifluoromethylsulfonyl)imide.
[g] NBu$_4$ · Br = Tetrabutylammonium bromide.

conjugate pure vegetable oils (Table 6.2). The platinum complexes Cl_2PtL_2 can be dissolved in soybean oil at 180 °C during 24 h, producing 19% and 1% conjugated dienes and trienes respectively.[33] Larock *et al.*[11] significantly improved this process by addtion of a Lewis acid to the platinum complexe, $PtCl_2(PPh_3)_2/SnCl_2 \cdot 2H_2O$, and by carrying out the reaction under hydrogen. Up to 89% conversion to conjugated soybean oil was obtained after 24 h at 100 °C.

Ionic liquids can serve as a *supported ionic liquid-phase for catalysed* conjugated isomerization of vegetable oils in a biphasic system.[20] The homogeneous isomerization of soybean oil with $RhCl(PPh_3)_3/SnCl_2$ in ethanol and $NBu_4.Br$ produced 97% of conjugated isomers without transesterification side reaction (Table 6.2). The suppression of this side reaction may be the result of the elimination of the Lewis acidic SnX_2 species *via* formation of anionic SnX_3^- species from the reaction of $SnCl_2$ with $NBu_4.Br$. The use of ionophilic phosphines decreased metal leaching into the product phase (Figure 6.1).

Dehydration of castor bean oil mostly composed of ricinoleic acid (85–90% 12-hydroxy-9-*cis* C18:1), can lead to conjugated allyphatic lipids under Brönsted or Lewis acidic media with metals oxides as catalysts at high temperatures, and zinc powder as an antipolymerizing agent.[34] Villeneuve *et al.*[35] used tungstic acid (Table 6.2). After 24 h, 6% residual ricinoleic acid was present, 34% CLA isomers and 50% linoleic acid were produced. The isomer (9-*cis*,11-*trans*) CLA was the major product (59.8%).

6.2.2   Mechanisms

The complexation of the double bond with the metal center is the key step in the conjugated isomerization, and two different isomerization mechanisms have been identified. The *alkyl mechanism* occurring especially with unsaturated metal hydride complexes, such as Rh(I)-, Ru(II)-, Ni(II)-, Pd(II)-, Pt(II)-hydride complexes.[19] A typical specie is $H\text{-}Rh((p\text{-}CH_3C_6H_4)_3P)_2$ developped by Andjelkovic *et al.*[12] for conjugated isomerization of linoleic acid and its esters and polyunsaturated vegetable oils. The metal alkyl specie is formed by addition of the metal-hydride to the double bond, and then a hydrogen atom of the alkyl group is eliminated to lead to isomerization and form again a metal hydride.[19] The *allylic mechanism* is typical of metals that preferably form π-allyl complexes such as Cr, Fe, Ni, Pd or Rh. The first step in the mechanism is an association of a double bond to ML_n catalyst, and the migration of an H atom to form metal-η^3-allyl-hydride specie. An 1,3-hydrogen migration leads to the formation of a new metal alkene complex, and then the isomerized alkene is cleaved leading to the back-formation of complex ML_n.[19] The alkene isomerization is frequently not a single reaction, but rather occurs in the context of other alkene reactions. Indeed, the *cis,cis*-1,4-pentadiene, part of linoleate is involved in autoxidation, alkali conjugation and catalytic hydrogenation.[36,37] Frankel[5] described the formation of allyl/H-Cr(CO)$_3$ complexes with either the Δ^9 or Δ^{12} double bond of linoleate, and the 1,3-hydrogen shift yielding (9-*cis*,11-*trans*) and (10-*trans*,12-*cis*) CLA isomers as main initial products (Figure 6.2). Depending on the temperature

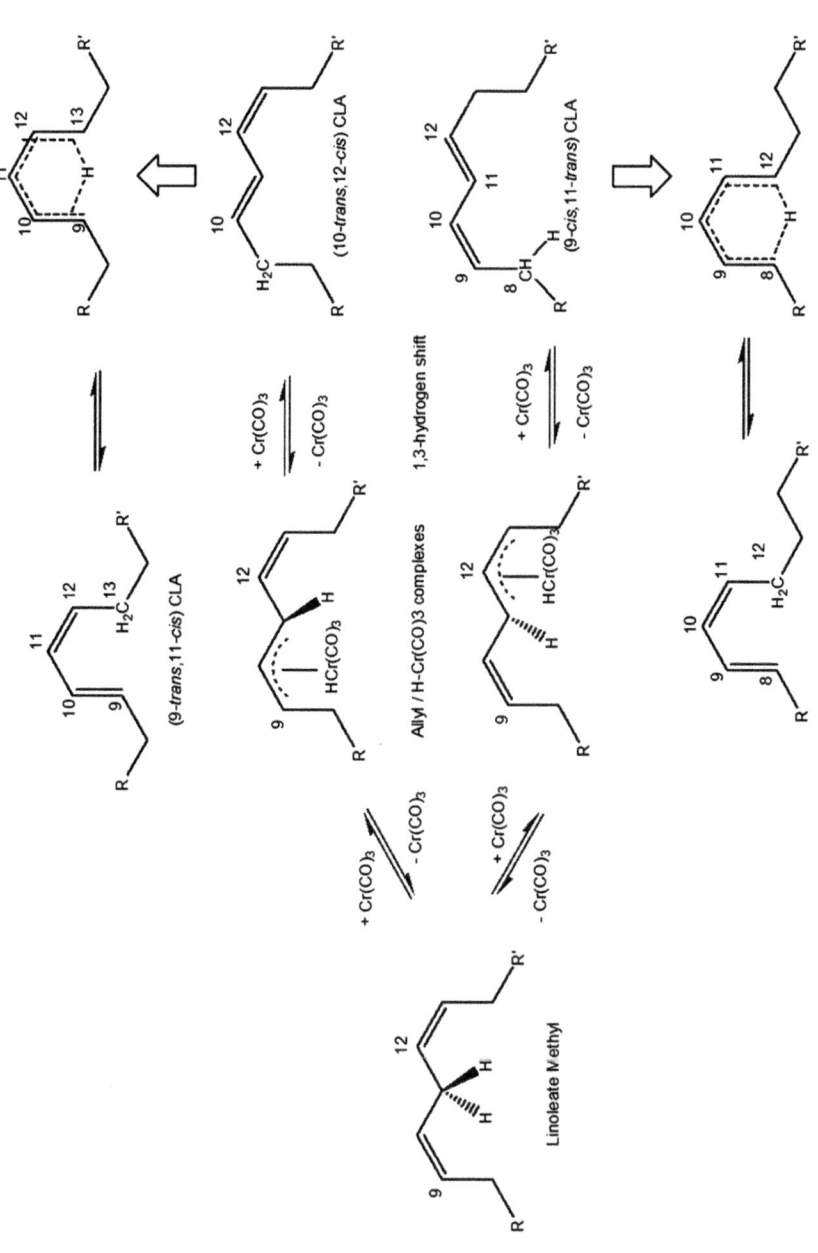

Figure 6.2 Formation of allyl/H-Cr(CO)₃ complexes with either the Δ^9 or Δ^{12} double bond of linoleate, and the 1,3-hydrogen shift yielding CLA isomers, and then sigmatropic rearrangement to produce others CLA isomers.[38]

and the reaction time, these two main compounds can undergo an intra-molecular sigmatropic rearrangement to produce (8-*trans*,10-*cis*) and (9-*trans*,11 *cis*) CLA isomers (Figure 6.2).[38]

6.3 Heterogeneous Catalytic Approaches for Conjugated Isomerization

It is known that conjugation and *cis–trans* isomerizations of polyunsaturated fats are competing parallel reactions with hydrogenation in heterogeneous system. Consequently, evoking conjugated isomerization catalysts description is untimely related to that of hydrogenation catalysts in terms of chemical and textural properties, as well as active compounds.

6.3.1 Catalysts' Properties

Noble metals, such as Pd, Rh, Pt and Ru, are known as very active hydrogenation/isomerization catalysts since they have vacant *d*-orbitals that can interact with π-bounds of fatty acids and activate an adjacent C–H bond, a necessary reaction event for double bound migration or saturation.

The different catalysts used to carry out vegetable oil hardening and/or conjugated isomerization are prepared with common synthesis techniques such as co-precipitation or pore volume impregnation, which allow the dispersion of the active metals on amorphous, weakly structured, or highly structured supports, such as amorphous silicas, aluminas, magnesium oxides, activated carbon, natural earths or nanostructured silicas. The magnitude of BET surface area, total pore volume, pore sizes distribution of the catalyst supports, and degree of the metal dispersion and crystallites sizes are the major catalyst key-parameters conferring high activity and significant selectivities for the desired products.

Commercial catalysts such as rhodium and ruthenium (5 wt%) deposited on activated carbon with a surface area of 1100–1050 m^2/g were used for the heterogeneously conjugated isomerization of polyunsaturated fats.[39] Bernas *et al.*[40–45] have conducted selective conjugation reactions of linoleic acid dissolved in *n*-decane over different catalysts based on ruthenium, nickel, palladium, rhodium, iridium and osmium supported on mesoporous materials: Al_2O_3, SiO_2, MCM-41 and MCM-22. Kreich and Claus[46] have described a new selective methodology for CLA production by using Ag/SiO_2 catalysts, activated under a constant flow rate of hydrogen.

An overview of various heterogeneous catalytic systems for CLA formation from linoleic acid and its methyl ester was also carried out by Philippaerts *et al.*[3,47] These authors evaluated various Ru-loaded zeolites, varying in Ru loading, zeolite topology (ZSM-5, β, Y), Si/Al ratio, and charge balancing cations (H^+, Na^+, Cs^+) in *n*-decane in absence of a hydrogen source.[47] Other researchers reported that supported gold can also be applied for conjugating linoleic acid under hydrogen flow.[48,49] Table 6.3 summarizes some

Table 6.3   Textural parameters of some metal-based heterogeneous catalysts for CLAs production.

Catalysts	Metal (wt%)	BET surface area (m^2 g^{-1})	Pore size (nm)	Ref.
Rh/C	5	1050		39
Ru/C	5	841	10.0–100.0	41
Ru/Al$_2$O$_3$	5	156	10.0–100.0	41
Ni/H-MCM-41	5	1051	1.0–1.5	41
Pd/H-Y	5	594	10.0–100.0	41
Ag/SiO$_2$	7.7	600		46
Ru/C	5	1050		46
Ru/zeolites	0.5	ND	<1.0	47
Ru/H-USY	0.5	ND	<1.0	47
Ru/Na-USY	0.5	ND	<1.0	47
Ru/Cs-USY	0.5	ND	<1.0	47
Au/C	1	ND	ND	48
Au/SiO$_2$	1	ND	ND	48
5Au/ZrO$_2$-DPU-C350	5	ND	ND	49
Au/TiO$_2$-DPU-C350	5	ND	ND	49
0.5Au/TiO$_2$-SI-C300	0.5	ND	ND	49
10Au/Al$_2$O$_3$-DPU-C200	10	ND	ND	49
2Au/Fe$_2$O$_3$-SI-C300	2	ND	ND	49
2Au/MnO$_2$-DPU-C300	2	ND	ND	49
2Au/TS-1-DP-C400	2	ND	ND	49

properties of these catalysts. Details on the isomerization activity of these catalysts are presented in the next sections of the chapter.

The team of Professors Belkacemi and Hamoudi[50–53] showed that the hydrogenation of vegetable oils, with catalysts based on noble metals supported on mesostructured silica SBA-15, are a very interesting alternative to the nickel-based catalysts, because it results in a low content of saturated and *trans* (<10%) fatty acids, harmful to health and an important *cis*-monoenes (>40%) content, beneficial to health.[54–56] Therefore, this type of mesostructured catalysts was tested during the hydrogenation process of vegetable oils to direct the secondary conjugation at the expense of *cis–trans* isomerization. The acidity of SBA-15 can be increased by replacing part of its silicon atoms by aluminum, which has Lewis-type acidic properties because of the presence of an empty orbital in its electronic structure, allowing its coordination to the double bonds of fatty acids and thus promote their isomerization. Chorfa *et al.*[57] have attained 80% yields of Al in SBA-15 by using Al tri-*sec*-butoxide as Al-source.

Chorfa *et al.*[58] showed that conjugation activity of the Rh/SBA-15 catalyst was favored when 1 ppm of sulfur is added to safflower oil. To determine an interaction between the sulfur and rhodium at the origin of this reactivity, a catalyst based on rhodium (1% by mass) and sulfur (0.02% by mass, or 1 ppm in 200 g oil) impregnated on mesoporous silica SBA-15 was prepared and tested during the process of hydrogenation/directed isomerization of safflower oil.[58,59] The textural properties of mesoporous silica SBA-15 and

Table 6.4   Textural parameters of the heterogeneous catalyst supports and catalysts for hydrogenation/directed conjugated isomerization of safflower oil.

Supports/ Catalysts	Metal (wt%)	BET surface area $(m^2 \cdot g^{-1})$	Pore volume $(cm^3 \cdot g^{-1})$	Pore size (nm)	Ref.
Cab-O-Sil	—	200	0.84	33.0	77
Ag/Cab-O-Sil	7.7	181	1.00	18.5–33.0	77
SBA-15	1	931	1.20	6.2	57
Al_SBA(100)	1	1006	1.18	4.7–7.4	57
Rh/SBA-15	1	651	0.95	3.7–6.3	59
S-Rh/SBA-15	1 (Rh) + 0.02 (S)	635	0.90	3.5–5.8	58

these catalysts are presented in Table 6.4. It is noted that the specific surface area and pore volume of the catalysts are less important than those of the SBA-15, owing to the presence of Rh clusters inside the pores. The addition of sulfur accentuates this phenomenon.

6.3.2   Catalyst Activities

6.3.2.1   Conjugated Isomerization of Linoleic Acid and Its Derivatives

Although some heterogeneous processes for isomerization of linoleic acid or methyl linoleate have already been described in literature, low productivity is the main drawback. Different metals and supports have been tested for the production of CLA, including: Ru/C,[39,60,61] Rh/C,[39] and Ru dispersed on different supports (γ-Al$_2$O$_3$, SiO$_2$.Al$_2$O$_3$, C and MgO) and in combination with Ni.[62] However, besides isomerization also hydrogenation (formation of oleate, elaidate and stearate), polymerization and coke formation were observed.[63]

Bernas *et al.*[40–45] screened Ru, Ni, Pd, Pt, Rh, Ir, Os, and bimetallic Pt-Rh supported by activated carbon, Al$_2$O$_3$, SiO$_2$.Al$_2$O$_3$, MCM-22, H-MCM-41, H-Y and H-β. In a pre-activation step the catalyst surface is first saturated with hydrogen and then the isomerization of linoleic acid to CLA occurs under a N$_2$ atmosphere at high temperature. Conjugation in highly protic solvents, such as methanol and isopropanol, showed very high activity but also high hydrogenation selectivity, while high isomerization selectivity was obtained in aprotic solvents such as hexane or cyclohexane.

Very recently, Phillipaerts *et al.*[3,47] reviewed the formation of CLAs from linoleic acid and its alkyl esters using heterogeneous catalysts. To summarize their work, the authors stated that the productivity of most heterogeneous catalytic systems are generally orders of magnitude lower than the industrial processes with homogeneous base, with two exceptional systems, *i.e.* non-activated commercial Ru/C, and Ru/Cs-USY perform very well. These catalytic systems show very promising productivities of 0.9 and 0.7 g CLA/L · solvent^{-1} · min^{-1}, respectively; much higher than the most homogeneous metal systems. Moreover, Ru/Cs-USY requires less Ru to achieve high productivity, indicating that Ru in this catalytic system is very active. They concluded that the most important criteria to obtain high CLA productivity and isomer selectivity are (1) absence of a hydrogen donor, (2) absence of catalyst acidity, (3) high metal dispersion, and (4) highly accessible pore architecture.

Other metals were reported to have a certain catalytic conjugated isomerization activity. For example, Kreich and Claus[46] described a highly selective method for the synthesis of CLAs over heterogeneous silver catalysts and in the constant presence of hydrogen. Similarly, the use of heterogeneous gold catalysts were tested under a constant hydrogen flow, and depending on the Au catalyst used, linoleic acid isomerization or hydrogenation is favored.[48,49] The activity of these catalysts is presented in Table 6.5.

6.3.2.2 Dual Reactions Hydrogenation/Isomerization of Vegetable Oils

The catalytic transfer hydrogenation with alcohols and organic acids as hydrogen source can produce conjugated linoleic acids during hydrogenation process of vegetable oils over commercial nickel catalysts.[64-69] The major advantage is to use safe hydrogen donors as an alternative for dangerous gaseous hydrogen. Ju *et al.*[70] reported that the CLA contents in the hydrogenated soybean oil with nickel catalyst (N-545), in 2.5% ethanol for 150 min, were 120.4 mg/g oil. Hydrogenation time and alcohol content also greatly influenced the isomeric distribution of CLAs.

Mossoba *et al.*[71] and Banni *et al.*[72] found some *cis-trans* conjugated linoleic acids isomers in hydrogenated soybean oil and margarine from hydrogenation processes using commercial Nickel catalyst. Based on those observations, Jung *et al.*[73,74] reported the effects of catalyst types, catalyst amount, reaction temperature, agitation rate, hydrogen pressure, and different oils on the quality and quantity of CLAs of hydrogenated soybean oil with commercial Ni catalysts. The authors concluded that the gas–liquid mass transfer limitation conditions of low hydrogen pressure, low stirring rate, high selective nickel catalyst content, and high reaction temperature favored the production of CLAs in vegetable oils during hydrogenation. Moreover, the qualitative and quantitative effects of sulfur addition on the formation of conjugated linoleic acids in vegetable oils during hydrogenation with nonselective type Ni catalyst showed that sulfur addition generally increased the CLAs content.[75] However, the higher sulfur above the optimum level decreased the CLAs formation, and the sulfur action was not described.

Recently, Chorfa *et al.*[59] showed that it is technically possible to favor the conjugated isomerization activity of Rh/SBA-15 bifunctional catalyst, at the expense of *cis–trans* isomerization activity, during the partial hydrogenation of safflower oil, by optimizing the operating conditions. High reaction temperature (180 °C), low hydrogen pressure (0.275 bar) and low agitation

Table 6.5 Production of conjugated linoleic acid isomers in heterogeneous catalytic systems.

Starting Material	Catalyst	Solvent	T (°C)	time (h)	TOF[a] (h^{-1})	C[b] (wt%)	H[c] (wt%)	CLAs[d] (wt%)	Ref.
methyl linoleate	Ni/C	—	170	6	—	—	—	61	63
methyl linoleate	Ru-Ni/Al$_2$O$_3$	—	200	4	—	87	10	40	62
methyl linoleate	Rh/C	hexane	230	4	3.12	53.1	15.54	37.37	39
methyl linoleate	Rh/C	cyclo-hexane	230	4	2.16	40.2	5.7	34.12	39
methyl linoleate	Rh/C	isopropyl alcohol	230	4	0.53	78.65	73.29	4.29	39
methyl linoleate	Ru/C	hexane	250	4	1.62	34.82	3.46	30.64	39
methyl linoleate	Ru/C	cyclo-hexane	250	4	2.04	42.61	9.14	31.38	39
methyl linoleate	Ru/C	isopropyl alcohol	250	4	0.33	99.83	96.44	2.2	39
linoleic acid	Ru/Al$_2$O$_3$[e]	n-decane	120	6	0.15	44	16	28	41
linoleic acid	Ru/C[e]	n-decane	120	6	0.49	77	24	53	41
linoleic acid	Ni/H-MCM-41[e]	n-decane	120	6	0.13	51	15	37	41
linoleic acid	Pt/H-Y-zeolite[e]	n-decane	120	6	0.28	40	11	29	41
linoleic acid	Pd/H-Y-zeolite[e]	n-decane	120	6	0.17	78	60	17	41
linoleic acid	Ag/SiO$_2$[e]	n-decane + 100 ml/min H$_2$	165	2	0.46	69	12	81	46
methyl linoleate	Ru/Cs-USY	n-decane	165	2	117.59	94	—	67	47
linoleic acid	Au/C	n-decane	150	4	0.004	4	0	4	48

[a]Turn over frequencies: Moles of conjugated dienes/(moles of catalyst * h).
[b]Conversion.
[c]Hydrogenation yield.
[d]CLA yield.
[e]Catalyst preactivated with H$_2$.

Table 6.6 Production of conjugated polyunsaturated oils in heterogeneous catalytic systems during partial hydrogenation/directed conjugated isomerization processes.

Oils	Catalyst	Solvent	H_2 (bar)	T (°C)	Time (h)	TOF^a (h^{-1})	C^b (wt%)	H^c (wt%)	$CLAs^d$ (wt%)	Ref.
soybean	Nickel catalyst (N-545)	ethanol	—	210	6	—	6	3	97	70
soybean	Nickel catalyst (SP-7, Engelhard)	—	0.49	210	1	—	19	21	79	73
safflower	Rh/Al_SBA(100)	—	0.28	180	5	66.6	17	25	34	57
safflower	Rh/SBA-15	—	0.28	180	5	84.7	21	7	35	59
safflower	S-Rh/SBA-15	—	0.28	180	5	157.3	39	15	35	58

[a]Turn over frequencies: Moles of conjugated dienes/(moles of catalyst×h).
[b]Conversion.
[c]Hydrogenation yield.
[d]CLA yield based on total linoleate and linolenate in original oils and GLC analyses: safflower 76.5% and soybean 59.1%.

rate (300 rpm), allowed to obtain partially hydrogenated oil which may contain up to 70 mg·g^{-1} oil of total CLA isomers where 60% of them are biologically active and health-beneficial (9-*cis*,11-*trans*) and (10-*trans*,12-*cis*) CLA isomers, and only a limited level of undesired *trans*-C18:1 reaching up to 8% of total fatty acids.

Finally, Chorfa *et al.*[58] investigated the sulfur effect on CLA isomers formation during the dual hydrogenation/directed isomerization of safflower oil over Rh/SBA-15 by direct addition of increased concentrations of 3-mercapto-1,2-propanediol to the reaction medium or by doping Rh/SBA-15 with the same sulfur-based compound yielding the sulfur-doped Rh-catalyst (S-Rh/SBA-15). These catalysts exhibited interesting activity, stability and recyclability. The maximum CLA content obtained during the dual reactions with Rh/SBA-15 and 0, 0.2, 1, 2, 5 and 10 ppm sulfur addition was 73, 99, 131, 110, 105 and 68 mg CLA/g oil, respectively, while the amount of harmful *trans* monoenes remained below 8%. The safflower oil partially hydrogenated under the same operating conditions over S-Rh/SBA-15 catalyst, without sulfur addition, produced up to 100 mg CLA/g oil. These results showed clear evidence of sulfur promotion effect on CLA formation during the dual hydrogenation/directed isomerization of safflower oil.

All these results show that it is possible to tailor characteristics of the hydrogenation catalyst in such a way to confer it bifunctional activity: hydrogenation and conjugated isomerization. The activity of these catalysts is presented in Table 6.6.

6.3.3 Mechanisms

The hydrogenation/isomerization mechanism of polyunsaturated fats in heterogeneous catalysis under hydrogen has been determined by Horiuti

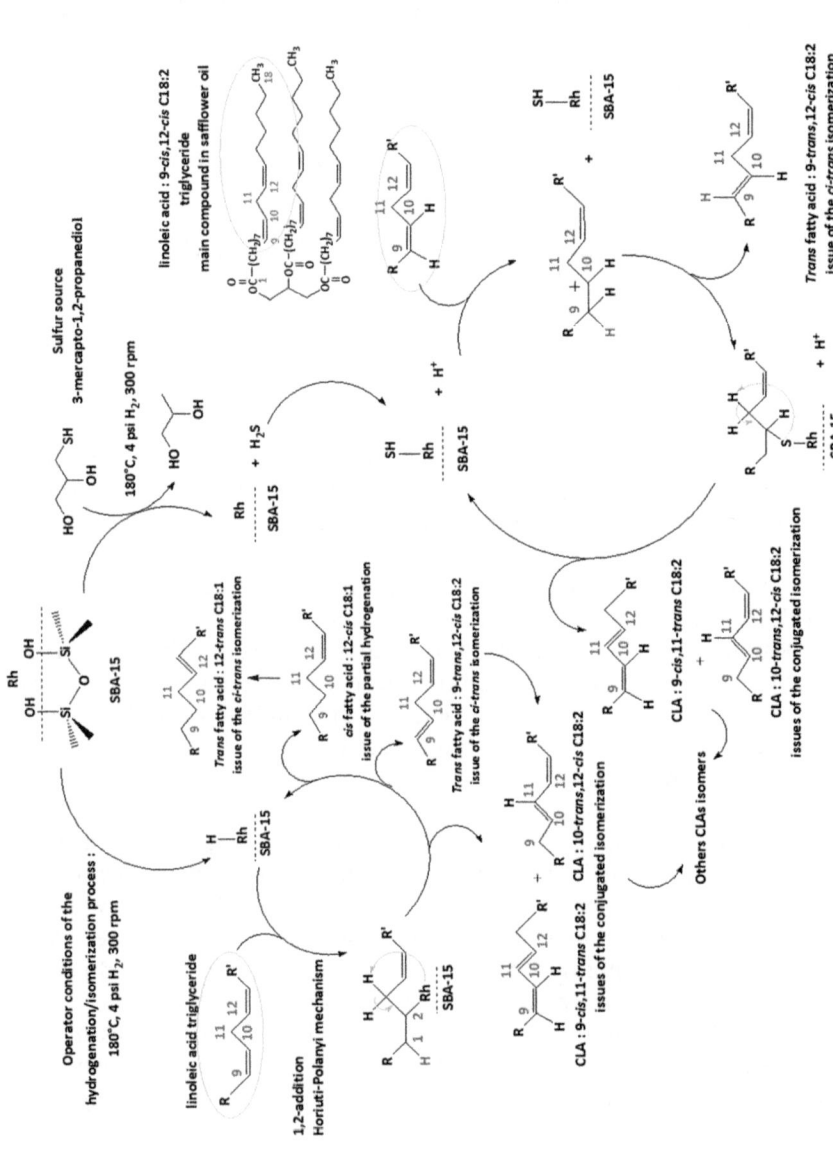

Figure 6.3 Trilinolein (triacylglycerol of linoleic acid) hydrogenation/directed conjugated isomerization proposed mechanism using Rh/SBA-15 with sulfur promotion (Chorfa *et al.* 2013[58]).

and Polanyi.[76] It involves many steps where each of the double bonds of the fatty acids can be transferred and adsorbed at the catalyst surface to react with hydrogen. An instable partially hydrogenated double bond intermediate is then formed and can either react with hydrogen to complete saturation of the double bond, or eliminate hydrogen to create a new unsaturation. Finally, the saturated or unsaturated bond is desorbed from the catalyst surface. Therefore, the Horiuti Polanyi mechanism leads to saturated fat, but also to geometric and/or positional isomers of unsaturated fatty acids. The selectivity of these reactions is controlled by the hydrogen concentration at the catalyst surface. Kreich and Claus[46] showed that strongly adsorbed hydrogen on ruthenium catalyst favors the hydrogenation activity while weakly adsorbed hydrogen in the case of silver catalyst promotes the iso-merization activity. Deshpande *et al.*[39] reported that the hydrogenation/isomerization activities of Rh/C and Ru/C catalysts are influenced by hydrogen derived from the solvent. When the metal hydride M-H concen-tration is important at the catalyst surface (it is the case when isopropyl alcohol is used as a solvent), a second hydrogen atom is added to the double bond leading to hydrogenated products. On the other hand, if the concen-tration of M-H species is low at the catalyst surface (such as the case of cyclohexane solvent), there is abstraction of a hydrogen atom by the metal from an adjacent carbon atom leading to a migration of the double bond.

Reporting for the first time the promotion effect of sulfur during the conjugated isomerization of vegetable oils over noble metal-based catalysts, Chorfa *et al.*[58,77] explained mechanistically the sulfur promotion of S-Rh/SBA-15 conjugated isomerization activity by the formation of the more nucleophilic rhodium sulfide intermediate (HS-Rh/SBA-15) rather than the rhodium hydride (H-Rh/SBA-15) (Figure 6.3).

6.4   Concluding Remarks

From this overview of catalysis approaches and catalysts developed for the conjugated isomerization of linoleic acid and its derivatives as well as vegetable oils, we have shown that presently the two approaches of homo-geneous and heterogeneous catalysis can lead to CLA production with interesting yields and productivities.

A quantitative conversion of linoleic acid and its esters as well as vegetable oils can be obtained in homogeneous system under mild operating con-ditions with organometallic complexes based on rhodium or Wilkinson catalysts, and ruthenium. The H-Rh[(p-CH$_3$C$_6$H$_4$)$_3$P]$_2$ complex was revealed as an a highly efficient and selective homogeneous catalyst for the iso-merization of vegetables oils to CLAs where the two biologically active CLA isomers (9-*cis*,11-*trans*) and (10-*trans*,12-*cis*) were obtained as the major CLAs.[11,12] Similarly, the Ru(η^6-naphthalene)(η^4-cycloocta-1,5-diene) complex with acetonitrile, in aprotic solvent (hexanes) leads to a full conversion of methyl linoleate with a high selectivity for conjugated products (95%).[14] Although the homogeneous conjugated isomerization with organometallic complexes is quantitative and selective, the noble metal catalysts recover and re-use processes are still problematic. Ionic liquid can serve as supported ionic liquid-phase catalysis for conjugated isomerization of vegetable oils in a biphasic system.[19,20]

A sustained interest in the use of very small loading of noble metal dispersed on new inorganic highly structured supports as new generation of catalysts can be viewed as an alternative to Ni-catalysts. Rhodium and ruthenium-based catalysts are considered the most active and selective metal catalyst towards CLA formation.

Conjugated and *cis–trans* isomerizations of polyunsaturated fats are competing parallel reactions with hydrogenation in heterogeneous systems. Conjugation in highly protic solvents, such as methanol, showed very high activity but also high hydrogenation selectivity, while high isomerization selectivity was obtained in aprotic solvents such as hexane. In *n*-decane, Ru/ Cs-USY is a highly selective catalyst for the synthesis of conjugated methyl linoleate with a TOF of about 118 h^{-1}.[47] Recently, Chorfa *et al.*[57–59] showed that is technically possible to favor the conjugated isomerization activity of Rh/SBA-15 bifunctional catalyst, at the expense of *cis–trans* isomerization activity, during the partial hydrogenation of safflower oil, by optimizing the operating conditions without any solvent.

References

1. A. Bhattacharya, J. Banu, M. Rahman, J. Causey and G. Fernandes, *J. Nutr. Biochem.*, 2006, **17**, 789.
2. S. Krompiec, R. Penczek, M. Krompiec, T. Pluta, H. Ignasiak, A. Kita, S. Michalik, M. Matlengiewicz and M. Filapek, *Curr. Org. Chem.*, 2009, **13**, 896.
3. A. Philippaerts, S. Goossens, P. A. Jacobs and B. F. Sels, *Chemsuschem*, 2011, **4**, 684.
4. E. N. Frankel, T. L. Mounts, R. O. Butterfield and H. J. Dutton, *Adv. Chem. Ser.*, 1968, 177.
5. E. N. Frankel, *J. Am. Oil Chem. Soc.*, 1970, **47**, 33–36.
6. E. N. Frankel, E. A. Emken and V. L. Davison, *J. Am. Oil Chem. Soc.*, 1966, **43**, 307.
7. A. Basu and K. R. Sharma, *J. Mol. Catal.*, 1986, **38**, 315.
8. D. Mukesh, C. S. Narasimhan, K. Ramnarayan and V. M. Deshpande, *Ind. Eng. Chem. Res.*, 1989, **28**, 1261.
9. C. S. Narasimhan, K. Ramnarayan and V. M. Deshpande, *J. Mol. Catal.*, 1989, **52**, 305.
10. W. J. Dejarlai and L. E. Gast, *J. Am. Oil Chem. Soc.*, 1971, **48**, 21.
11. R. C. Larock, X. Y. Dong, S. Chung, C. K. Reddy and L. E. Ehlers, *J. Am. Oil Chem. Soc.*, 2001, **78**, 447.
12. D. D. Andjelkovic, B. Min, D. Ahn and R. C. Larock, *J. Agric. Food Chem.*, 2006, **54**, 9535.
13. A. Basu and T. G. Kasar, *J. Am. Oil Chem. Soc.*, 1986, **63**, 1444.

14. P. Pertici, V. Ballantini, S. Catalano, A. Giuntoli, C. Malanga and G. Vitulli, *Mol. Catal., A: Chemical*, 1999, **144**, 7.
15. J. A. Osborn, F. H. Jardine, J. F. Young and G. Wilkinso, *J. Chem. Soc., A: Inorg. Physic. Theor*, 1966, 1711.
16. M. C. Baird, C. J. Nyman and G. Wilkinso, *J. Chem. Soc., A: Inorg. Phys. Theor*, 1968, 348.
17. P. Pertici, V. Ballantini, P. Salvadori and M. A. Bennett, *Organometallics*, 1995, **14**, 2565.
18. P. Pertici, G. U. Barretta, F. Burzagli, P. Salvadori and M. A. Bennett, *J. Organomet. Chem.*, 1991, **413**, 303.
19. A. Behr and P. Neubert, *Applied Homogeneous Catalysis*, Wiley-CH, 2012.
20. C. S. Consorti, G. L. P. Aydos, G. Ebeling and J. Dupont, *Appl. Catal., A:-General*, 2009, **371**, 114.
21. C. E. Fernie, I. E. Dupont, O. Scruel, Y. A. Carpentier, J. L. Sebedio and C. M. Scrimgeour, *Eur. J. Lipid Sci.Technol.*, 2004, **106**, 347.
22. E. N. Frankel, E. P. Jones, H. M. Peters and H. J. Dutton, *J. Am. Oil Chem. Soc*, 1964, **41**, 186.
23. E. N. Frankel and S. Metlin, *J. Am. Oil Chem. Soc.*, 1967, **44**, 37.
24. D. Bingham, D. E. Webster and P. B. Wells, *J. Chem. Soc.,-Dalton Trans.*, 1974, 1519.
25. R. T. Sleeter, *Patent*, 1996, E.P. 0736593 B0736591.
26. R. T. Sleeter, *Patent*, 1998, US 08/472,919.
27. S. Krompiec, J. Jerzy, J. Majewski and J. Grobelny, *Pol. J. Appl. Chem.*, 1998, **42**, 43.
28. A. Basu, S. Bhaduri and T. G. Kasar, *Patent*, 1985, EP0160544A0160542.
29. J. E. Lyons, *J. Org. Chem.*, 1971, **36**, 2497.
30. H. Singer, R. Seibel and U. Mees, *Fette Seifen Anstr.*, 1977, **79**, 147.
31. H. Singer, W. Stein and H. Lepper, *Fette Seifen Anstr. V.*, 1972, **74**, 193.
32. S. Krompiec, J. Suwinski, J. Majewski and J. Grobelny, *Pol. J. Appl. Chem.*, 1997, **41**, 35.
33. W. J. Dejarlai and L. E. Gast, *J. Am. Oil Chem. Soc.*, 1971, **48**, 157.
34. W. C. Forbes and H. A. Neville, *Ind. Eng. Chem.*, 1940, **32**, 555.
35. P. Villeneuve, R. Lago, N. Barouh, B. Barea, G. Piombo, J. Y. Dupre, A. Le Guillou and M. Pina, *J. Am. Oil Chem. Soc.*, 2005, **82**, 261.
36. E. N. Frankel, E. Selke, C. D. Evans, H. J. Dutton and D. G. McConnell, *J. Org. Chem.*, 1961, **26**, 4663.
37. P. L. Nichols, S. F. Herb and R. W. Riemenschneider, *J. Am. Chem. So.*, 1951, **73**, 247.
38. A. Saebo, J. L. Sebedio, W. W. Christie and R. Adolf, *Advances in Conjugated Linoleic Acid Research*, Champaign, Illinois, 2003.
39. V. M. Deshpande, R. G. Gadkari, D. Mukesh and C. S. Narasimhan, *J. Am. Oil Chem. Soc.*, 1985, **62**, 734.
40. A. Bernas, N. Kumar, P. Maki-Arvela, B. Holmbom, T. Salmi and D. Y. Murzin, *Org. Proc. Res. Dev.*, 2004, **8**, 341.
41. A. Bernas, N. Kumar, P. Maki-Arvela, N. V. Kul'kova, B. Holmbom, T. Salmi and D. Y. Murzin, *Appl. Catal., A: Gen.l*, 2003, **245**, 257.

42. A. Bernas, N. Kumar, P. Maki-Arvela, E. Laine, B. Holmbom, T. Salmi and D. Y. Murzin, *Chem. Commun.*, 2002, 1142.
43. A. Bernas, P. Laukkanen, N. Kumar, P. Maki-Arvela, J. Vayrynen, E. Laine, B. Holmbom, T. Salmi and D. Y. Murzin, *J. Catal.*, 2002, **210**, 354.
44. A. Bernas, P. Maki-Arvela, N. Kumar, B. Holmbom, T. Salmi and D. Y. Murzin, *Ind. Eng. Chem. Res.*, 2003, **42**, 718.
45. A. Bernas and D. Y. Murzin, *React. Kinet. Catal. Lett.*, 2003, **78**, 3.
46. M. Kreich and P. Claus, *Angew. Chem. Int. Ed.*, 2005, **44**, 7800.
47. A. Philippaerts, S. Goossens, W. Vermandel, M. Tromp, S. Turner, J. Geboers, G. Van Tendeloo, P. A. Jacobs and B. F. Sels, *ChemSusChem*, 2011, **4**, 757.
48. O. A. Simakova, A. R. Leino, B. Campo, P. Maki-Arvela, K. Kordas, J. P. Mikkola and D. Y. Murzin, *Catal. Today*, 2010, **150**, 32.
49. P. Bauer, P. Horlacher and P. Claus, *Chem. Eng. Tech.*, 2009, **32**, 2005.
50. K. Belkacemi, N. Kemache, S. Hamoudi and J. Arul, *Int. J. Chem. React. Eng.*, 2007, 5.
51. K. Belkacemi and S. Hamoudi, *Ind. Eng. Chem. Res.*, 2009, **48**, 1081.
52. K. Belkacemi, A. Boulmerka, S. Hamoudi and J. Arul, *Int. J. Chem. Reac. Eng.*, 2005, 3.
53. K. Belkacemi, A. Boulmerka, J. Arul and S. Hamoudi, *Top. Cata.s*, 2006, **37**, 113.
54. M. Plourde, Hydrogénation des huiles végétales: nouvelle approche catalytique minimisant la production des acides gras trans et saturés, MSc thesis, Université Laval, 2003.
55. A. Boulmerka, Hydrogénation catalytique des huiles de tournesol et canola minimisant la production des acides gras trans et saturés : étude des paramètres de procédé, MSc thesis, Université Laval, 2006.
56. N. Kemache, *Hydrogénation des huiles végétales en présence de catalyseurs bimétalliques à base de Pd et monométalliques à base de Pd doppé au soufre et supportés sur une silice mésoporeuse*, PhD Thesis, Université Laval, 2010.
57. N. Chorfa, S. Hamoudi and K. Belkacemi, *Appl. Catal., A: Gen.*, 2010, **387**, 75.
58. N. Chorfa, S. Hamoudi, J. Arul and K. Belkacemi, *Chemcatchem*, 2013, **5**, 1917.
59. N. Chorfa, S. Hamoudi, J. Arul and K. Belkacemi, *Can. J. Chem. Eng.*, 2012, **90**, 41.
60. D. Mukesh, S. Narasimhan, R. Gadkari and V. M. Deshpande, *Ind. Eng. Chem. Prod.t R. D.*, 1985, **24**, 318.
61. S. Narasimhan, D. Mukesh, R. Gadkari and V. M. Deshpande, *Ind. Eng. Chem. Prod. s. D.*, 1985, **24**, 324.
62. D. Mukesh, C. S. Narasimhan, V. M. Deshpande and K. Ramnarayan, *Ind. Eng. Chem. Res.*, 1988, **27**, 409.
63. S. B. Radlove, H. M. Teeter, W. H. Bond, J. C. Cowan and J. P. Kass, *Ind. Eng. Chem.*, 1946, **38**, 997.

64. O. Arkad, H. Wiener, N. Garti and Y. Sasson, *J. Am. Oil Chem. Soc.*, 1987, **64**, 1529.

65. H. N. Basu and M. M. Chakrabarty, *J. Am. Oil Chem. Soc.*, 1966, **43**, 119.

66. M. M. Chakrabarty, D. Bhattacharyya and A. K. Basu, *Am. Oil Chem. Soc.*, 1972, **49**, 510.

67. K. Mondal and S. B. Lalvani, *J. Am. Oil Chem. Soc.*, 2000, **77**, 1.

68. M. Naglic, A. Smidovnik and T. Koloini, *J. Am. Oil Chem. Soc.*, 1998, **75**, 629.

69. T. Tagawa, T. Nishiguchi and K. Fukuzumi, *J. Am. Oil Chem. Soc.*, 1978, **55**, 332.

70. J. W. Ju, W. S. So, J. H. Kim, B. J. Bae, E. N. Choi, Y. H. Kwon, I. M. Chung, S. H. Yoon and M. Y. Jung, *J. Food Sci.*, 2003, **68**, 1915.

71. M. M. Mossoba, R. E. McDonald, D. J. Armstrong and S. W. Page, *J. Chromatogr. Sci.*, 1991, **29**, 324.

72. S. Banni and J. C. Martin, *Conjugated Linoleic Acid and Metabolites*, The Oil Press edn., Dundee, 1998.

73. M. O. Jung, J. W. Ju, D. S. Choi, S. H. Yoon and M. Y. Jung, *J. Am. Oil Chem. Soc.*, 2002, **79**, 501.

74. M. O. Jung, S. H. Yoon and M. Y. Jung, *Agric. Food Chem.*, 2001, **49**, 3010.

75. J. W. Ju and M. Y. Jung, *J. Agric. Food Chem.*, 2003, **51**, 3144.

76. H. Pines, *The Chemistry of Catalytic Hydrocarbon Conversions*, New York, 1981.

77. N. Chorfa, Isomérisation et hydrogénation des huiles végétales pour la production des acides linoléiques conjugués, PhD thesis, Université Laval, 2011.

CHAPTER 7

Analysis of Conjugated and Other Fatty Acids

PIERLUIGI DELMONTE,*[a] ALI REZA FARDIN-KIA,[a]
NOELIA ALDAI,[b] MAGDI M. MOSSOBA[a] AND
JOHN K. G. KRAMER*[c]

[a] Office of Regulatory Science, Center for Food Safety and Applied
Nutrition, Food and Drug Administration, 5100 Paint Branch Pkwy,
College Park, MD 20740, USA; [b] Department of Pharmacy and
Food Sciences, Faculty of Pharmacy, University of the Basque
Country (UPV/EHU), 01006, Vitoria-Gasteiz, Spain; [c] Guelph Food
Research Center, Agriculture and Agri-Food Canada, Guelph, ON,
Canada (retired)
*Email: Pierluigi.Delmonte@fda.hhs.gov; jkgkramer@rogers.com

7.1 Introduction

There are several different types of fatty acids (FAs) occurring in natural and
synthetic products which include saturated-, branched-chain-, polyunsatur-
ated-, conjugated-, hydroxy-, cyclic-, keto-, and furan-FAs, just to name a few.
Increased interest in any of these FA groups is generally associated with the
finding of specific chemical and/or biological properties. This also applies to
conjugated fatty acids (CFAs) that have experienced renewed interest since
the discovery of their anti-carcinogenic,[1] anti-oxidative[2] and cytotoxic[3,4]
properties. In this chapter we will not address the evidence for these findings but only review the many different analytical techniques used for their
analysis. Gas chromatography (GC) is by far the most effective method for

RSC Catalysis Series No. 19
Conjugated Linoleic Acids and Conjugated Vegetable Oils
Edited by Bert Sels and An Philippaerts
© The Royal Society of Chemistry 2014
Published by the Royal Society of Chemistry, www.rsc.org

the routine analysis of the FA composition of any matrix, including CFAs. Mass spectrometry (MS), nuclear magnetic resonance (NMR), and Fourier transform infrared (FTIR) spectroscopy are valuable techniques for characterization and confirmation of the structure of CFAs, but they found limited applicability in routine analysis. These complementary techniques are outside the scope of this chapter, as are the methods to determine the positions and geometric configurations of double bonds.

Although the main emphasis in this chapter is the analysis of CFAs, the other FAs that are generally analysed together cannot be ignored. For instance, in natural ruminant and oil seed products, the measurement of the other FAs, especially *trans* FAs is also required. The measurement of CFAs may also be required for determining the content of *trans* FA for labeling purposes in some countries, since most CFAs contain a *trans* double bond. In biological systems CFA are synthesized from *trans* FA precursors and metabolized to other CFA metabolites. They can also occur as minor products in partially hydrogenated vegetable oils (PHVOs).

In GC separations of FAMEs, the aim should always be a complete resolution of all the CFA and other FAs, because many FAs have different biological activities. Improvements in recent years have been attributed to the availability of very long fused capillary GC columns with increased efficiency to separate fatty acid methyl esters (FAMEs). Non-resolution of FAMEs should not be an option, but performing the separation of all occurring FAMEs in a single analysis remains the challenge. Currently, a major limitation has been the availability of authentic standards to systematically examine the chromatographic behavior of FAMEs on most GC columns. For this reason, the complete series of synthetic C18 conjugated dienoic acids (CDA) and *cis* (*c*) and *trans* (*t*) monounsaturated fatty acids (MUFAs) prepared by Delmonte and his colleagues have been very valuable to evaluate the selectivity of different GC columns.[5-10] In this chapter we wish to review the different chromatographic techniques needed to resolve conjugated and non-conjugated FAMEs and provide supporting evidence from other analytical techniques.

7.2 Chemical Structure

A CFA is defined as any FA that contains two or more double bonds of which at least two of them are conjugated (–C–C–C=C). They differ from common polyunsaturated fatty acids (PUFAs) in which the isolated double bonds are separated by one or more methylene groups, which are referred to as methylene interrupted (MI) or non-methylene interrupted (NMI), respectively. The double bonds in naturally occurring MI PUFAs is in the *cis* configuration, while the two conjugated double bonds form a rigid flat four-carbon chain unit that can occur in four different geometric configurations, *i.e.*, *c,t*, *t,c*, *c,c*, or *t,t*. Therefore, a C18 CDA can theoretically occur in 54 possible geometric and positional isomers. A total of sixteen C18 CDA isomers were identified in cheese lipids using three silver-ion high performance liquid chromatography (Ag$^+$-HPLC) columns in series.[11] If the CFA contains one additional double bond in the molecule, it can be either isolated from the conjugated system giving rise to a mono-conjugated trienoic acid (MCTA) such as *c*9,*t*11,*c*15-18:3, or conjugated to the existing conjugated dienoic system giving rise to a di-conjugated trienoic acids (DCTA) such as *c*9,*t*11,*c*13-18:3.[12] Both of these CFAs can occur in many isomeric forms. Some MCTAs are found in ruminant fats,[13,14] and several DCTAs are found in oils of plants seeds.[15–19] In addition to C18 CFAs, several elongated and desaturated C20 CFA metabolites have been identified in tissues of rats fed a synthetic 18:2 CDA mixture marketed as conjugated linoleic acid (CLA).[20–22] In aquatic plants, CFAs of 16 to 22 carbon chain-length have been identified, which includes dienes, trienes and tetraenes.[23] Some natural CFAs are listed in Table 7.1, but in this chapter we will consider primarily the analysis of C18 CDAs, and briefly mention MCTA and DCTAs.

The presence of a conjugated double bond system in the molecule gives rise to unique ultraviolet (UV) and infrared (IR) absorptions, both of which have been used to detect and quantitate CFAs in the presence of other non-conjugated FAs. The separation of the individual CFA isomers by chain-length and positional/geometric composition sometimes requires more than one method. Each will be evaluated in turn by discussing their respective benefits and limitations. Several combinations of methods are listed at the end of this chapter. A brief section is also included to stress the importance of preparing lipid extracts and their conversion to FAME derivatives. These methods need to preserve the native FA composition of any matrix investigated.

7.3 Ultraviolet and Infrared Absorption of Conjugated Fatty Acids

7.3.1 Ultraviolet

The presence of a two conjugated double bond system in a molecule gives rise to a UV absorption at ~233 nm. Slight differences occur in the UV absorption between *t,t*-18:2 (230 nm), *c,t*- or *t,c*-18:2 (232 nm), and *c,c*-18:2 (234 nm) CDA isomers.[7] A three double bond conjugated system gives rise to a triplet with absorptions at ~262, 272 and 283 nm.[15,16] Precise UV maximum absorptions vary between detectors, but not the relative order of *t,t*- before the *c,t*- and *t,c*- isomers and these before *c,c*-18:2. Stand alone UV spectroscopy is extensively used to detect[15,16] and quantitate CFA in the presence of other methylated products, and as detection technique for Ag$^+$-HPLC[11,29,30] and reverse phase (RP)-HPLC separations.[8,30,31] A typical separation of the FAMEs prepared from bovine milk fat shows the UV absorption of several CDAs at 232.1 nm and DCTAs eluting in the CDA region at 266.3 nm (Figure 7.1).[7] By selecting specific absorption wavelengths with the use of a diode array detector (DAD), CFA can be analysed in the presence of

Table 7.1 Structures and names of *trans*-containing conjugated fatty acids found in partially hydrogenated vegetable oils, oil seeds and ruminant fats.

Structure and Abbreviations	Common name	Source
Dienoic acid (DA)		
Conjugated DA (CDA)		
$t7,c9$-18:2	Yurawic acid[12]	Ruminant fats
$c9,t11$-18:2	Rumenic acid[24]	Ruminant fats
$t10,c12$-18:2		Ruminant fats
$t10,t12$-18:2		Ruminant fats
$t11,c13$-18:2		Ruminant fats
$t9,t11$-18:2	Mangold acid[17]	Ruminant fats
$c11,t13$-20:2		Synthetic[11]
8,12,14-20:3		In liver lipids of rats fed CFA[20]
5,8,12,14-20:4		In liver lipids of rats fed CFA[20]
5,8,11,13-20:4		In liver lipids of rats fed CFA[20]
Trienoic acid (TA)		
Mono-conjugated TA (MCTA)		
$c9,t11,c15$-18:3	Rumelenic acid[13]	Ruminant fats
$c9,t13,c15$-18:3	Iso-rumelenic acid[13]	Ruminant fats
$c9,t11,t15$-18:3[14]		Ruminant fats
$t9,t11,c15$-18:3[25]		Ruminant fats
Di-conjugated TA (DCTA)		
$t2,t4,c6$-10:3		Latex of poinsettia[26]
$c8,t10,c12$-18:3	Jacaric acid[17]	*Jacaranda mimosifolia*
$t8,t10,c12$-18:3	α-Calendic acid[17]	Pot marigold
$t8,t10,t12$-18:3	β-Calendic acid[27]	Pot marigold
$c9,t11,c13$-18:3	Punicic acid[17]	Pomegranate, snake gourd
$c9,t11,t13$-18:3	α-Eleostearic acid[17]	Tung, bitter gourd
$t9,t11,c13$-18:3	Catalpic acid[17]	*Catalpa ovata*
$t9,t11,t13$-18:3	β-Eleostearic acid[17]	Tung, bitter gourd, catalpa
$c5,t7,t9,c14$-20:4		*Acanthophora spicifera* (red algae)[23]
$t5,t7,t9,c14$-20:4		*Acanthophora spicifera* (red algae)[23]
$c5,t7,t9,c14,c17$-20:5		*Acanthophora spicifera* (red algae)[23] *Ptilota filicina* (red algae)[23]
$t5,t7,t9,c14,c17$-20:5		*Acanthophora spicifera* (red algae)[23] *Ptilota filicina* (red algae)[23]
Tetraenoic acid		
$c9,t11,t13,c15$-18:4	α-Parinaric acid	*Impatiens edgeworthii*[23]
$t9,t11,t13,t15$-18:4	β-Parinaric acid	*Impatiens edgeworthii*[23] Makita tree (*Parinarium laurinum*)[28]
$c5,c8,t10,t12,c14$-20:5	Bosseopentaenoic acid	*Bossiella orbingnian* (red algae)[23]
$c4,c7,t9,t11,c13,c16,c19$-20:7	Stellaheptaenoic acid	*Anadyomene stellata* (green algae)[23]

CFA, conjugated fatty acid.

other methylated products such as FAMEs and dimethylacetals (DMA).[32] CFAs can also be analysed as their free fatty acids (FFA) by Ag$^+$-HPLC using UV detection.[33] UV has also been used after RP-HPLC separations.[8,30,31,34]

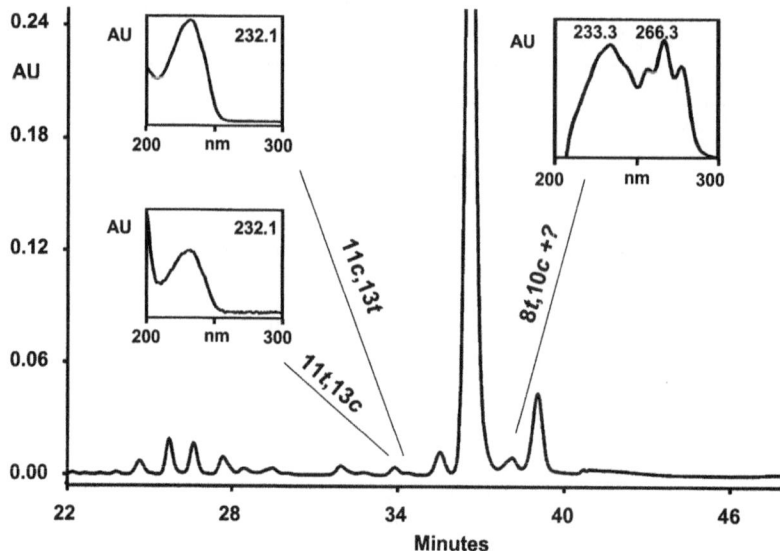

Figure 7.1   A partial Ag$^+$-HPLC separation of bovine milk FAMEs with UV detection
of C18 CDA at ∼232 nm and of MCTAs at ∼266 nm (triplet).[7]
Reproduced with permission of *J. AOAC Intern.* and the authors.

7.3.2   Infrared

The presence of a two or three double bond conjugated system in the molecule gives rise to an IR absorption with a doublet for *c/t* isomers at ∼988
and 949 cm^{-1}, a singlet for the all *trans* isomers at ∼993 cm^{-1}, and a
doublet for the all *cis* isomers at 3037 and 3005 cm^{-1}.[16,35–37] The work by
Mossoba and colleagues showed the immense potential to combine GC
separations using 100 m 100% cyanopropylsiloxane (CPS) columns with
direct deposition-FTIR for the analysis and confirmation of numerous geometric isomers of CFA (Figure 7.2) and other unsaturated FAs.[11,35–39]

7.4   Methylation of Lipids Containing Conjugated
Fatty Acids

The FAs contained in oils, fats and foods are most commonly analysed by GC
after their extraction and conversion to their volatile FAME derivatives using
either acid or base catalysts. CFAs are easily isomerized under acidic condition in which the *cis* double bond in the CFA system is converted to *trans*.[40]
Therefore, acid-catalysed methylations should be avoided when CFA are
being analyzed. In addition, acid-catalysed methylation of C18 CDA leads to
the formation of methoxy artifacts that elute slightly after the C18 CDA using
100 m 100% CPS columns.[40,41] On the other hand, the positions and
geometric configurations of the CFA double bonds are not altered under

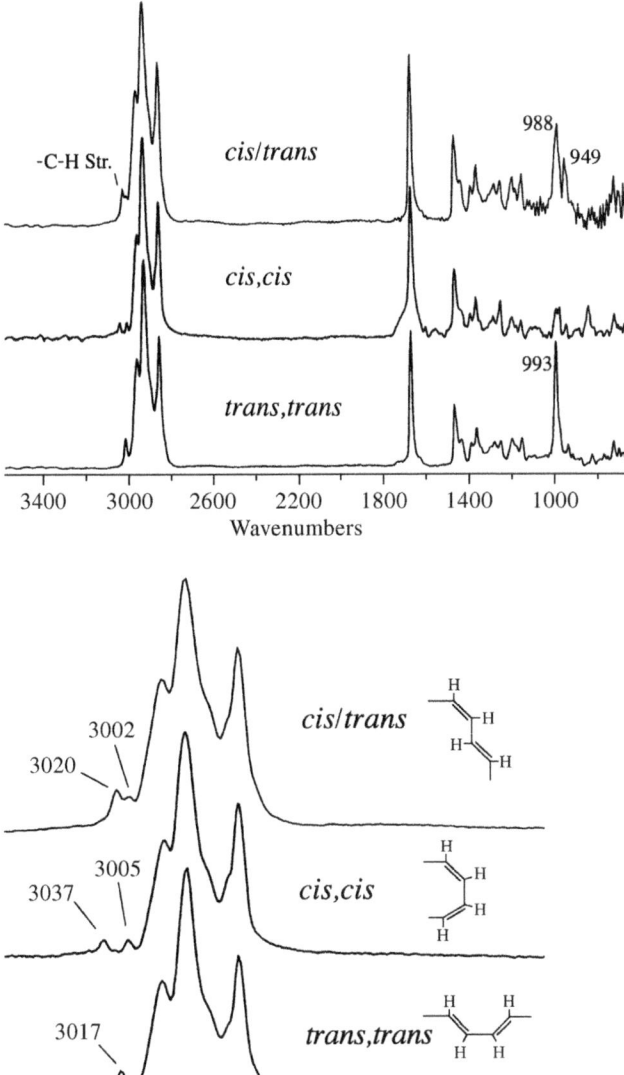

Figure 7.2 GC-direct deposition-FTIR spectroscopy of the 4,4-dimethyloxazoline (DMOX) derivatives of *c/t-*, *c,c-* and *t,t-*C18 conjugated dienoic acids after GC separation on a 50 m CP-Sil 88 column.[38]
Reproduced with permission of AOCS Press and the authors.

base-catalysed conditions.[42] However, it should be noted that base catalysts do not methylate FFAs, plasmalogens that contain an alk-1-enyl ether bond, and sphingomyelin with *N*-acylated FA.[29,30,42–44] This becomes a challenge when analysing samples that contain both acid labile and base resistant lipids. Alkali saponification is no solution since the base resistant compounds are also not hydrolysed under these conditions. To overcome this problem, a two-step methylation procedure has been proposed, in which samples are first methylated using a base catalyst, followed by a brief acid catalysed methylation at ~ 50 °C.[29,40,45] An alternative would be to conduct separate base and acid catalysed methylations.[29,42-44] This approach combined the results from two sample preparations: acid-catalysed methylation provides the quantitative analysis of all lipid classes, while base-catalysed methylation corrects the quantification of specific FAME regions and identifies artifacts such as methoxy FAMEs and products of plamalogenic lipids produced during acid-catalysed methylation.[29,30,42-44] Animal tissues contain plasmalogens, which after acid-catalysed methylation yield both FAMEs and DMAs; the latter are derived from the alk-1-enyl ether linkages in the plasmalogens (Figure 7.3).[46] During GC analysis, some DMAs lose a methanol group in the hot injection port to yield alk-1-enyl methyl ethers (AMEs) depending on the temperature of the injector (Figure 7.3B).[29,30,44] The FAMEs and DMEs can be separated by thin layer chromatography (TLC) using 1,2-dichloroethane as developing solvent (Figure 7.3C and D).[29,30,44,47] The intact DMAs can be analysed by conversion to the more stable cyclic acetals (Figure 7.3E).[20,30,44]

7.5   Fatty Acid Methyl Ester Standards

7.5.1   Conjugated Fatty Acid

Progress in the evaluation and analysis of CFAs has been limited by the availability of authentic CFA reference materials. Only a few C18 CDAs are commercially available such as c9,t11-, t10,c12- and t11,c13-18:2 from Matreya Inc. (Pleasant Gap, PA). In an attempt to obtain more CDA standards, several researchers have used the mixture UC-59M available from Nu-Chek Prep Inc. (Elysian, MN) that contains primarily four positional C18 CDA isomers, t8c10-, c9t11, t10c12-, and c11t13-18:2; the corresponding c/c and t/t isomers are present in minor amounts. All possible c/t isomers of these four positional CDAs can be easily prepared by treating the UC-59M mixture with a dilute solution of iodine.[6,48] This isomerized UC-59M reference mixture contained sufficient components to establish an elution pattern for the C18 CDAs on GC and Ag^+-HPLC columns.[49] However, authentic standards of an additional number of common C18 CDAs generally present in natural products are not available, such as the geometric isomers of 6,8- 7,9-, 12,14- and 13,15-18:2. Direct synthesis was required to obtain these FAs.[5-7,10,50] Delmonte and colleagues systematically prepared a complete series of C18 CDA isomers from 6,8- to 13,15-18:2 by partial hydrazine reduction of known PUFAs followed by conjugation of the generated dienes with KOH in ethylene glycol, and finally isolating the desired C18 CDA by semi-preparative Ag^+-HPLC.[5-7,10] This series of C18 CDAs was then used to evaluate their elution on different GC,[5,6,10,52-54] Ag^+-HPLC,[5-8,51] and RP-HPLC[8,51] columns.

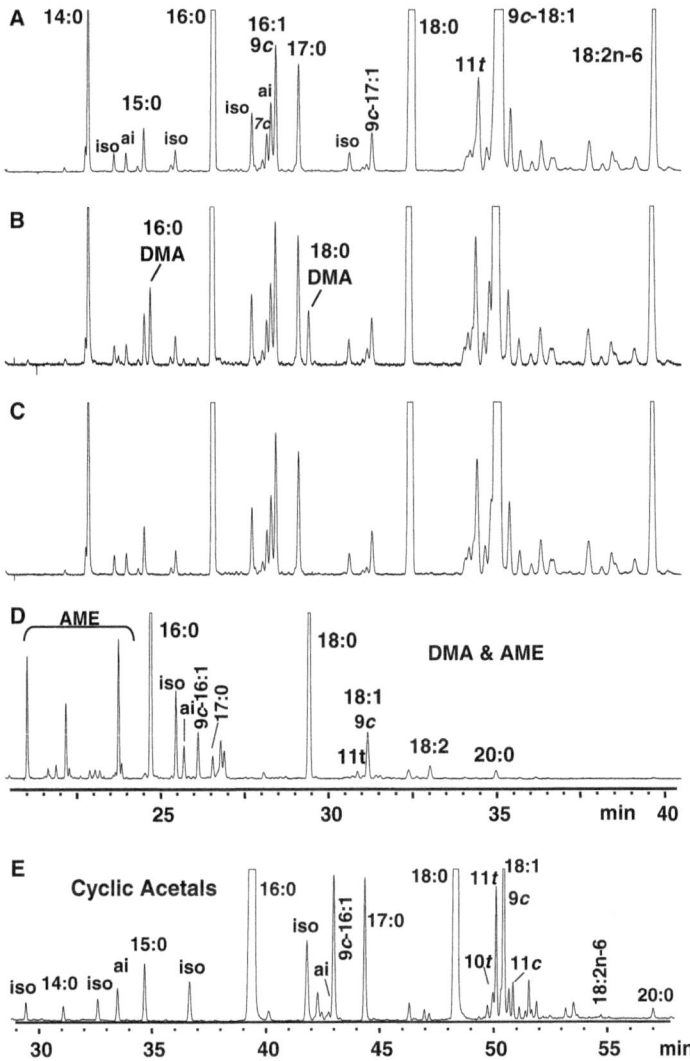

Figure 7.3 Partial GC separation of total sheep lipids methylated using a base-(A) or an acid-catalysed procedure (B).[30] The acid-catalysed methylated mixture (B) was separated by TLC (developing solvent, 1,2-dichloroethane) into its FAMEs (C) and dimethylacetals (DMAs) (D). The DMAs were converted to stable cyclic acetals (E). All GC operations were conducted using a 100 m CP-Sil 88 column.
Reproduced with permission of *Lipids* and the authors.

7.5.2 Other FAME

Several excellent GC reference FAME mixtures are available from commercial suppliers containing most of the non-conjugated FAs occurring in natural products (*i.e.*, Nu-Chek Prep Inc.; Sigma-Aldrich Ltd, St. Louis, MO; Matreya

Inc.; Larodan Fine Chemical, Malmö, Sweden). When choosing a GC standard mixture select one with as many representative FAMEs from all regions of the chromatogram as possible. In addition, select one in which the FAMEs have different concentrations such as in GLC-463 (Nu-Chek Prep Inc.). If the concentration of all FAME components is the same, it becomes increasingly difficult to identify the various FAMEs in the GC chromatograms, and can easily lead to misidentification. It would appear that this occurred in the assignment of 20:3n-3 and 20:4n-6 on a BPX-90 column, where the authors claimed a reversal in the elution order of these two FAMEs compared to those obtained using other BPX columns.[55a] If necessary, complement these GC standard mixtures by adding specific FAMEs of interest that may not be in the reference mixture. For example, the commercial GC reference mixture GLC-463 from Nu-Chek Prep Inc. contains 63 FAMEs and it can be enriched with the four positional C18 CDA isomer mix UC-59M and the long-chain saturated FAMEs 21:0, 23:0 and 26:0. The GC reference mixtures should be regularly injected to monitor the performance of the GC column throughout the analyses. With the age of the column there may be slight changes in the retention times and relative elution orders of FAMEs that can be easily spotted with frequent analyses of GC standards.

A number of specific *trans*- and *cis*-18:1 methyl esters are available from Sigma-Aldrich Inc. A complete mixture of most *trans* and *cis* isomers of 14:1 to 22:1 can also be obtained by isolating larger amounts of the *trans* and *cis* MUFAs from natural sources such as ruminant fats and PHVOs by Ag^+-TLC,[29,56–58] Ag^+-solid phase extraction (Ag^+-SPE),[59,60] or Ag^+-HPLC.[53,54] Mixtures of *trans* and *cis* MUFAs from 14:1 to 20:1 were also synthesized by repeated bromination/debromination of commercially available FAME isomers.[9,10]

Mixtures containing all geometric isomers of linoleic (c9,c12-18:2 or 18:2n-6) and α-linolenic acids (c9,c12,c15-18:3 or 18:3n-3) are commercially available. Geometric isomers of PUFAs can be easily prepared by isomerization of their commercial sources using *p*-toluenesulfinic acid as catalyst.[9,10,61] Most of the NMI FAME standards are not commercially available and need to be synthesized.[12,62]

Several quantitative GC mixtures are also commercially available to establish correction factors for FAMEs, and these mixtures come with a certificate of analysis. One needs to be cautious using these quantitative mixtures, since the purity of the FAMEs in these certified mixtures might be assessed using GC columns with less resolution capabilities than the one used by the analyst. In such cases the determined response factor (area/amount) for the FAME may be questionable.[36] Attention must also be paid to the storage and preparation of these reference mixtures, since short chain FAMEs tend to evaporate and long chain PUFAs tend to oxidize. The improper use of experimentally determined flame ionization detector (FID) response factors may result in a larger error than using theoretically calculated response factors.

7.6 Gas Chromatographic Methods for Fatty Acid Analysis

Gas chromatography is the technique of choice to separate FAMEs by chain-length, number and geometry of double bonds, and FAMEs with CFA systems in a molecule. The FID provides sensitive response and allows the prediction of the theoretical response factors for FAMEs not available as reference materials. However, the separation of geometric and positional isomers of unsaturated FAMEs requires polar separation columns. The presence of a conjugated system in a molecule results in a later elution of these FAMEs on all GC columns compared to FAMEs having the same total number of carbon and double bonds whether they are MI or NMI. This applies both to CDA,[61,63,64] MCTA,[13,14] and DCTA.[65] A major problem in the GC analysis of CFAs is the coelution with other FAMEs in the same eluting region. Therefore, every effort must be made to identify the coeluting FAMEs to avoid misidentification or overestimation of the CFAs. In one study, MS was used to identify 21:0 (Figure 7.4A) and several isolated 20:2 (Figure 7.4B) FAMEs eluting among the C18 CDA using a 100 m CP-Sil 88 column.[66,67] However, which FAMEs elute in the separation region of a specific CFA group depends on the type of GC stationary phase, column dimensions (length and diameter), and elution temperature that are selected. Increasing the column length improves the resolution because of an increase in separation column theoretical plates. The capillary column chromatographic resolution is proportional to the square root of the column number of theoretical plates.[54] Lowering the elution temperature is associated with improved separation of homologous FAME isomers,[59,68] but will also result in a change in the relative elution pattern of FAMEs with different number and geometric configuration of double bonds.[59,69–71]

Several types of GC columns have been used in the analysis of CFAs. It is not our intent here to review the separation provided by every GC column ever used for FAME analysis, but to provide a critical assessment of selected columns of each type. For ease of comparison, the GC columns in this study were arbitrarily grouped based on where 18:3n-3 eluted relative to the C20 FAMEs. Based on this criterion, there are four major types of GC columns, those with medium polarity (*i.e.*, Supelcowax-10, DB-Wax, Omegawax) in which 18:3n-3 elutes before 20:0,[61,72,73] those with intermediate polarity (BPX-70, CP Select for FAME, HP-88) in which 18:3n-3 elutes just before or after 20:0, polar columns (*i.e.*, SP-2560, CP-Sil 88. Rt-2560) in which 18:3n-3 elutes after the *cis*-20:1 isomers,[59,61,69] and highly polar columns (SLB-IL111) in which 18:3n-3 elutes after all the *cis*-20:1 isomers and just before 22:0.[52–54] It is well known that not all GC columns used for FAME analysis fall specifically into these four groups, but rather there is a continuum of polarities. Often GC columns of intermediate polarity were selected as a compromise between their capability of separating positional/geometric FAME isomers and their thermal stability (*i.e.*, BPX-90, HP-88, CP Select for

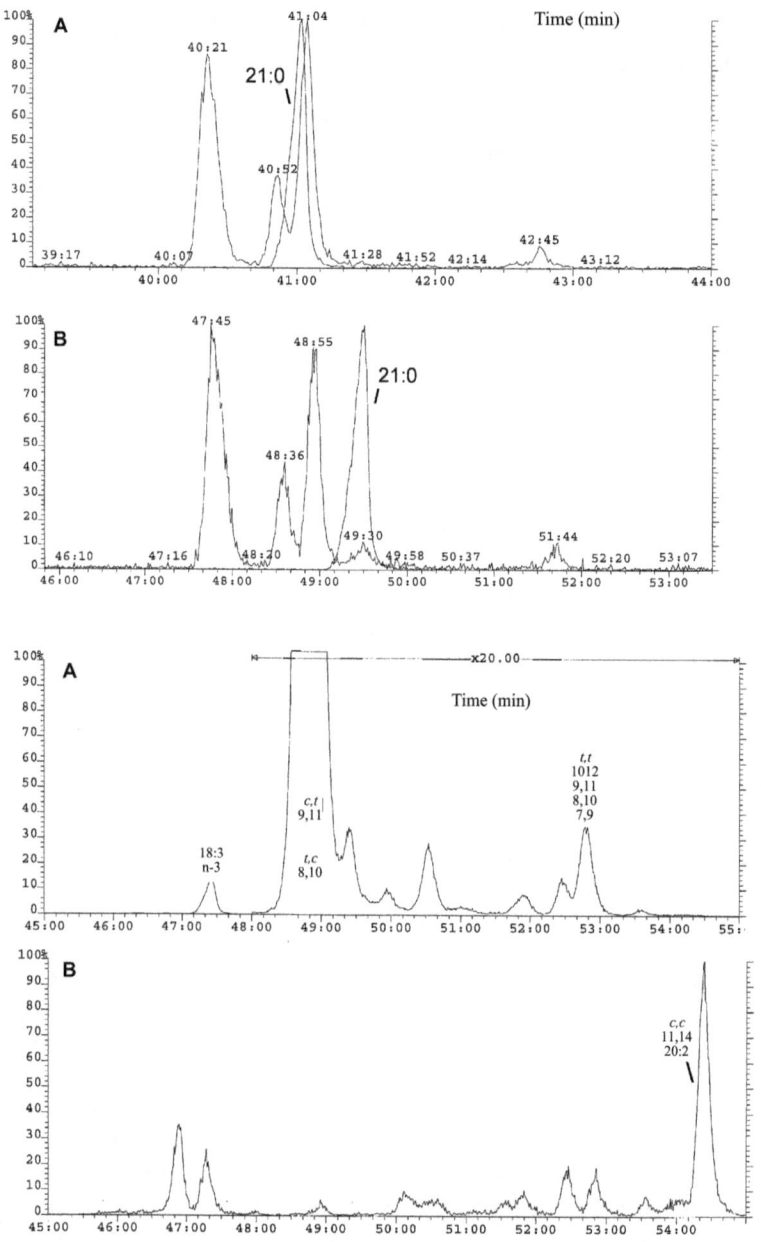

Figure 7.4 A low-resolution reconstructcd ion profiles of the molecular ions of the C18 CDAs in mixture UC-59M (Nu-Chek Prep Inc.) recorded at m/z 294, and for 21 : 0 at m/z 340.[66] The GC was operated at 180 °C (A) and 170 °C (B) (upper graphs). High-resolution selected ion profiles of m/z 294.2559 (A) and m/z 322.2872 (B) (lower graphs) recorded for C18 CDA and 20 : 2 in total cheese FAME, respectively. The signals recorded at m/z 322 were 1% of those recorded at m/z 294.

Reproduced with permission of *Lipids* and the authors.

FAME). In each case, the inclusion of substituents in the stationary phase structure to stabilize CPS phases resulted in a decreased polarity and an earlier elution of 18 : 3*n*-3 compared to the 100% CPS columns.

The main limitation to test the different GC columns has been the availability of authentic standards. It is unfortunate that during the 1960s and 1970s when Gunstone's group synthesized numerous series of *cis* and *trans* mono- and di-unsaturated FAMEs, the very long fused silica capillary GC columns we are currently using were not available. In recent years, Delmonte's group has synthesized a series *cis* and *trans* mono- and conjugated FAMEs that have been successfully used to evaluate the polar and highly polar ionic liquid GC columns. Each of these types of GC columns will be discussed in turn, and not only regarding their ability to separate CFAs but also *cis* and *trans* MUFAs and other FAMEs that are present in natural products.

7.6.1 Separation of FA on GC Columns with Medium Polarity

7.6.1.1 Column Properties

The popularity of GC columns of medium polarity, mainly the polyethylene glycol (PEG) type columns, appears to be related to the ease of identifying complex mixtures of PUFAs, since all the MI and NMI unsaturated FAs (but not CFAs) of chain length X elute before the next saturated fatty acid (SFA) containing two more carbon atoms in the acyl chain [(X + 2):0]. Therefore, 18 : 4*n*-3 elutes before 20 : 0, 20 : 5*n*-3 elutes before 22 : 0, and 21 : 5*n*-3 before 23 : 0.[30,72,74] However, in the case of the C22 unsaturated FAs, while 22 : 5*n*-3 always elutes before 24 : 0, 22 : 6*n*-3 elutes before 24 : 0 only under some conditions,[74,75] and most often between 24 : 0 and 24 : 1.[72,75,76] No other exceptions of this elution order rule were observed using PEG columns. However, one needs to be careful when evaluating new GC phases. For instance, the new ionic liquid column SLB-IL60, which is characterized by a similar selectivity as a PEG column, was shown to provide an altered elution order with 22 : 6 eluting before 22 : 5.[77] An additional benefit of the medium polarity GC columns is their stability. Several of these phases are bonded and cross-linked (Supelcowax-10, Durabond-Wax, *etc.*), which results in low bleed, and low minimum and high upper temperature limits.[72] As a result of these reasons, the PEG type columns have been the GC columns of choice in the official methods for the analysis of marine oils.[78]

7.6.1.2 FAME

Even though PEG type columns provide a clear separation of PUFAs with different chain length, less interference of the most abundant PUFAs with other FAMEs, and a separation of SFAs and MUFAs, they provide little separation among the geometric and positional isomers of MUFAs (Figure 7.5A), which is a serious disadvantage.[61,76] There was no separation among the

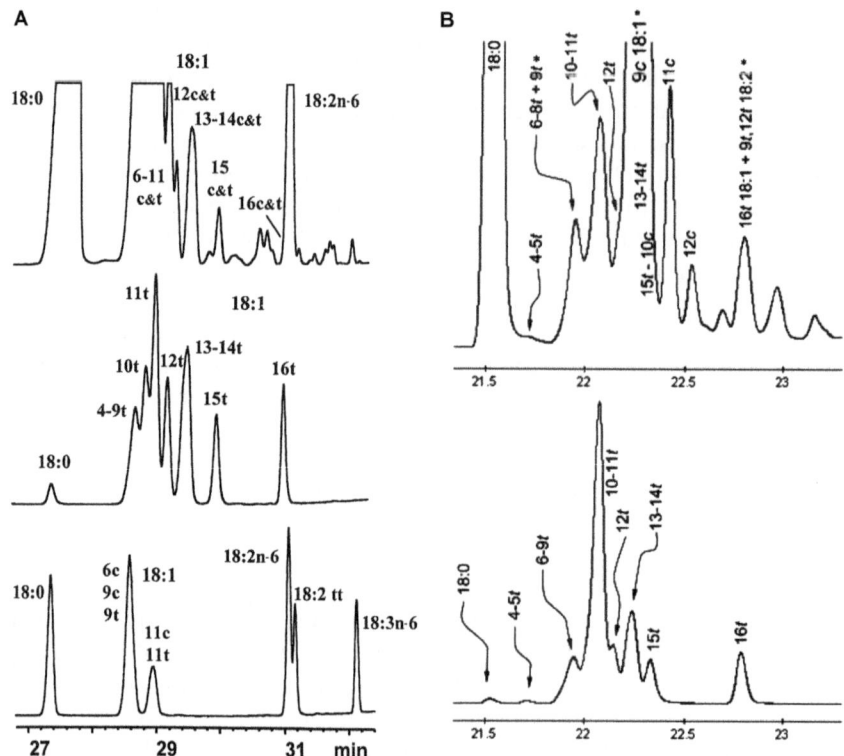

Figure 7.5 Partial GC separation of the 18:1 isomers region using a 60 m Supelcowax-10 column with a temperature program from 65 °C to 240 °C (left side),[61] and a 60 m BPX-70 column with a temperature program from 60 °C (5 min) to 165 °C at 15°C/min (held for 1 min) and then to 225 °C at 2°C/min (held for 4 min) (right side).[73] Below each separation is the *trans* fraction isolated using Ag+-TLC. In the case of the Supelcowax-10 column, the separation of the FAMEs in GLC-463 eluting in this region is shown.
Reproduced with permission of *Lipids* and *Eur. J. Lipid Sci. Technol.*, respectively, and the authors.

geometric isomers of MUFAs on a typical PEG column such as the 60 m Supelcowax-10 (Figure 7.5A) unless an isothermal condition at 130 °C (or lower) was used (not shown), in which case the *cis* MUFAs eluted just ahead of the *trans* MUFAs.[61] The geometric isomers of PUFAs were better resolved.[61] Figure 7.5A shows the elution of the isolated *trans*-18:1 isomers compared to the total milk 18:1 FAMEs on a 60 m Supelcowax-10 column.[61]

7.6.1.3 Conjugated Dienoic acid (CDA)

The separation of CDA isomers on the PEG type column is rather limited. The C18 CDAs contained in the UC-59M mixture were separated into two pairs: *c*9,*t*11- eluted with *t*8,*c*10-18:2 first followed by *c*11,*t*13- with

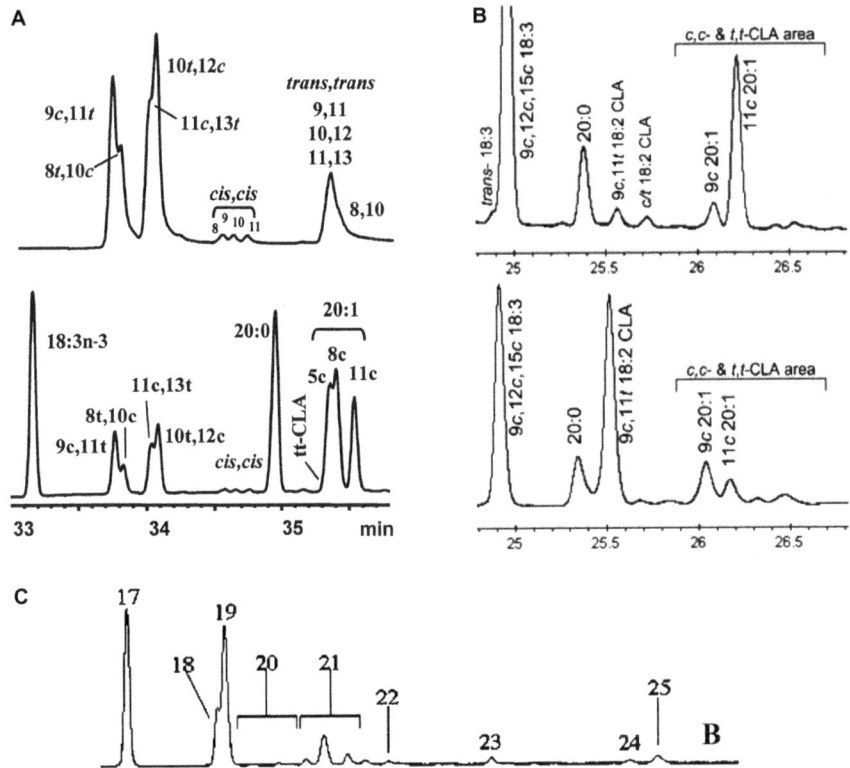

Figure 7.6 Partial GC separation of the C18 CDA isomers on a 60 m Supelcowax-10 column[61] (A), a 60 m BPX-70 column[73] (B), and a 120 m BPX-70 column[13] (C). The FAMEs in (C) were: 17, 18:3n-3; 18, 20:0, 19, c9,t11-18:2; 20, c/t-18:2 CDA; 21, c,c-18:2 CDA, 20:1 and 21:0; 22, t,t-18:2 CDA; 23, c9,t11,c15-18:3; 24, 20:2n-6; 25, unidentified. Each separation includes other FAMEs eluting in the C18 CDA region. Reproduced with permission of *Lipids, Eur. J. Lipid Sci. Technol.*, and *J. Dairy Sci.*, respectively, and the authors.

t10,c12-18:2 (Figure 7.6A).[61] In addition, only a partial separation of the four positional c,c- was observed, but none of the t,t-C18 CDA isomers separated. Furthermore, there was extensive overlap between the C18 CDAs and 20:0 and the t- and c-20:1 isomers. On the 60 m Supelcowax-10 column the c/t- and c,c-18:2 CDAs eluted before 20:0, while the t,t-18:2 CDA isomers eluted among the t- and c-20:1 positional isomers (Figure 7.6A, bottom).[61]

7.6.1.4 Diconjugated Trienoic Acid (DCTA).

The applied methylation procedure is very critical to preserve the intact isomeric composition of the DCTAs.[18] The DCTAs were generally resolved using low or medium polarity GC columns.[18,19,27,79] However, these GC columns did not completely resolve all the isomers present in the different natural products; only seven C18 DCTA isomers are listed in Table 7.1 of the many possible isomers. To identify the coeluting DCTA isomers, ^{13}C-NMR[18] and MS[28] were used, although a GC solution would certainly be preferred for routine analysis.

7.6.1.5   Limitations

- The main limitation of the PEG columns is their inability to resolve the geometric and positional isomers of MUFAs and CDAs. This is particularly critical when analysing ruminant products and biological samples that contain *trans* FAs and CDAs.
- Increasing the column length showed little improvements in the separation of the 18:1 and CDA isomers.[61]
- Because of the lack of separation among the geometric and positional isomers of unsaturated FAMEs the PEG type columns are not recommended for the analysis of CDAs and *trans* FAs.

7.6.2   Separation of FA on GC Columns with Intermediate Polarity

7.6.2.1   Column Properties

The GC columns of intermediate polarity combine thermal stability of the chemically bonded phases with the improved separation of FAME isomers compared to PEG columns, but inferior to the 100% CPS columns. Two of the more common of these columns is the BPX-70 (bis-cyanopropyl-silphenylene) column, and the CP Select for FAME column of unspecified phase composition. These columns are reported to be 100% chemically bonded and stable up to 290 °C.[80] In both of these columns 18:3*n*-3 elutes close to 20:0 and 18:4*n*-3 elutes among the 20:1 FAMEs.[73]

7.6.2.2   FAME

The BPX-70 column, being slightly more polar than the PEG columns, provides very limited separation of the geometric isomers of MUFAs with the *trans* eluting before *cis* isomers (Figure 7.5B).[73] A thorough evaluation of FAME separations on the BPX-70 column was reported.[73] The separation of the 18:1 isomers was not much improved by using a 120 m BPX-70 column.[13]

On a 100 m CP Select for FAME column 18:3*n*-3 eluted just before 20:0, and after 20:0 on a 200 m column, while 18:4*n*-3 eluted among the 20:1 FAMEs.[80,81] A column length of 200 m helped to increase the resolution of the geometric and positional isomers of 18:1 (Figure 7.7),[36,80,82] such that the separation of the 18:1 FAMEs appeared comparable to that observed with a 100 m 100% CPS type GC column.[36,82] However, a critical evaluation of the different FAME regions of this column has not been reported. This includes the resolution of the C18 CDA isomers, which FAMEs interfere in the CDA elution region, and the 18:3/20:1 region. Based on the limited

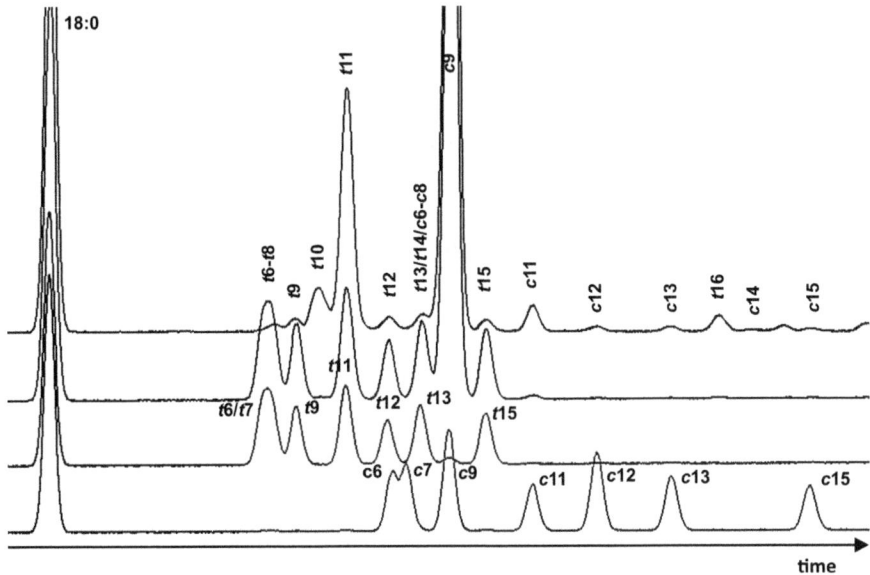

Figure 7.7   Partial GC separation of the 18 : 1 region using a 200 m CP Select column operated using an isothermal condition at 180 °C.[36] The chromatograms include total milk fat FAME and three different mixtures of 18 : 1 isomers. Chromatogram kindly provided by Dr G. Jahreis and Mr P. Möckel, University of Jena, Germany.

information available,[80,82] there appears to be extensive overlaps in the 16 : 1, 20 : 1 and 22 : 1 regions, which may be one reason why the 200 m CP Select for FAME column has not been extensively used.[82,83]

7.6.2.3 Conjugated Di- and Trienoic Acids

On the 60 m BPX-70 column all the C18 CDA isomers eluted just after 20 : 0. An extensive overlap occurs between most C18 CDA isomers and the *t*- and *c*-20 : 1 positional isomers (Figure 7.6B).[73] There was no improvement in the separation by using a 120 m BPX-70 column since now *c*9,*t*11-18 : 2 coeluted with 20 : 0 (Figure 7.6C).[13] However, the 120 m BPX-70 column did provide a good separation of the MCTA rumelenic acid (*c*9,*t*11,*c*15-18 : 3) from the other FAMEs.[13]

7.6.2.4 Limitations

- The GC columns of intermediate polarity (BPX and CP Select for FAME) are less efficient in separating unsaturated FAME isomers than polar or highly polar ionic liquid GC columns of equal length. To improve the separations, longer column lengths are generally used (120 m BPX-70, 200 m CP Select for FAME).
- Increasing the length of the GC columns of intermediate polarity will improve the separation of FAME isomers somewhat (proportional

to the square root of the column number of theoretical plates), but it does not change the basic characteristic selectivity of the GC column.

- To increase the potential of the 200 m CP Select for FAME column the chromatographic properties of this column will needs to be thoroughly investigated with authentic standards.

7.6.3   Separation of FA on Polar GC Columns

7.6.3.1   Column Properties

The current choice of GC columns for the analysis of *trans* FAs (plus all the other FAMEs) in fats and oils are the 100 m 100% CPS type columns marketed as CP-Sil 88 (Agilent Technologies), SP-2560 (Supelco Inc.), and Rt-2560 (Restek Corp.).[36,84] There are two other less common used polar columns that are sometimes included among these polar columns which include the BPX-90 column (corresponding to 90% cyanopropyl substituted phase on a polysilphenylene-siloxane backbone),[55a,55b] and the HP-88 that has a bis(cyanopropylsiloxane)-co-methylsilarylene phase corresponding to a CPS content of $\sim$88%.[85] The non-cyanopropyl components in these phases provide greater thermal stability to the columns at the expense of slightly reduced polarity. This was evident in the elution of 18:3n-3 just before c11-20:1 using the latter columns[55,73,85] instead of behind c13-20:1 using the 100% CPS type columns. The 100% CPS columns are not bonded and therefore elution temperatures above 225 °C and rinsing with organic solvents should be avoided.

7.6.3.2   FAME

The great advantage of these polar columns is their ability to separate geometric and positional isomers of FAMEs, but that has also proven to be the greatest challenge. The increased polarity has resulted in marked improvements in the separation of the 16:1,[56,57,59,71] 18:1,[10,57,59,61,71] 18:2,[59,70,71] 20:1-18:3,[59,69,71,86] and other PUFA regions.[71,76] However, it has also led to many overlapping FAMEs of different chain lengths, degrees of unsaturation, and geometric configuration. These cannot be resolved in a single chromatographic separation on the 100 m 100% CPS columns. To obtain their separation two main approaches have been used: (a) changing the elution temperature of the column, and (b) using Ag$^+$ separations to help in the identification and confirmation of the individual FAMEs.

The easiest approach is to change the elution temperature of the column. The success of this approach is based on the well-recognized fact that the selectivity of polar columns is affected by the elution temperature. Changing the column temperature, changes the relative separation of members within and between homologous series of FAMEs.[59,70,84] We are not aware of any elution reversals within members of a homologous series of FAMEs, which apparently has been reported for alkyl phosphates.[87] However, changing the temperature changes the elution times of different homologous series of FAMEs differently. If the temperature differences selected are big enough, it will lead to recognizable differences in separations and/or coelutions.[59,70,84] This technique was successfully used to resolve branched-chain FAME from coeluting 16:1 isomers (Figure 7.8),[56] 11-cyclohexylundecanoic acid in among the dienoic FAMEs in milk fats,[70] the *trans*-18:3 FAMEs from the co-eluting 20:1 FAMEs in PHVOs (Figure 7.9),[69] better separations of the four *trans*-18:1 isomer peaks t6/t7t8-, t9-, t10- and t11-18:1 in ruminant fats (Figure 7.10),[59] and identifying the different PUFAs in marine oil samples.[71,76]

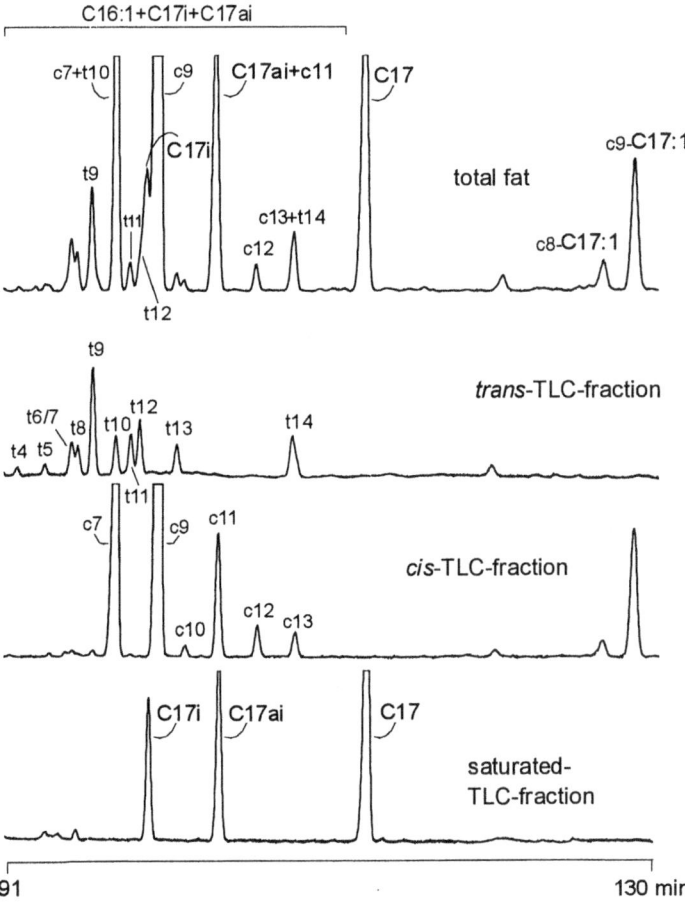

Figure 7.8 Partial GC chromatograms of the 16:1 region comparing the total milk FAMEs with the *trans*, *cis* and SFA fractions obtained by Ag$^+$-TLC. A 100 m CP-Sil 88 column was used operated at an isothermal condition of 120 °C.[56]
Reproduced with permission of the *Eur. J. Lipid Sci. Technol.* and the authors.

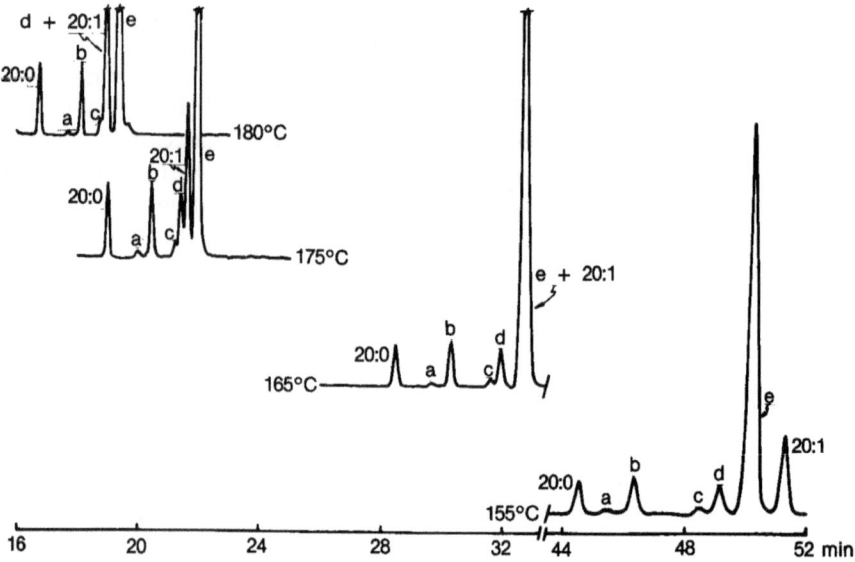

Figure 7.9 Separation of the 20:1/18:3 isomer region of a commercial deodorized rapeseed oil using a 50 m CP-Sil 88 column and operated at different isothermal operating conditions from 155 °C to 180 °C. For identification of letters see reference 69.
Reproduced with permission of *Am. Oil Chem. Soc.* the and authors.

However, one needs to be careful since the change in column temperatures selected may not result in a separation of the desired FAMEs. For example, the separation of the *trans* positional isomer pairs $t11/t12$-16:1, $t13/t14$-18:1, $t15/t16$-20:1 and $t17/t18$-22:1 cannot be resolved on 100 m 100% CPS columns using isothermal or temperature program conditions between 175 °C and 150 °C.[59,68] However, lowering the column temperature to 120 °C will separate these pairs of FAMEs (Figure 7.11),[56–60,68] but as shown in that figure the resolution between these isomer pairs slightly diminishes for the higher chain-length MUFAs.[60] On the other hand, one may deal with unresolved FAMEs that are not separated even if the column temperature is lowed to 120 °C. Take for example the FAMEs $t6/t7$-16:1,[56,59,60] $t6/t7t8$-18:1,[56,58–60,68] $t5$ to $t9$-20:1,[57,59,60] and $t5$ to $t11$-22:1.[60] These FAMEs remain unresolved at 175 °C and at 120 °C (Figure 7.11).[59,60,68] As will be discussed later, the more polar ionic liquid column SLB-IL111 provide better separation of these FAMEs;[52–54] see Section 7.6.4 below.

Another area that has received very little attention is the 18:2 region in ruminant fat and PHVOs, which has many FAMEs that have not been definitely characterized. A simple comparison of the 18:2 separation region obtained from different elution temperatures on the same GC column shows different separation patterns,[59,68,70] However, these 18:2 FAMEs will first need to be identified with authentic standards as well as isolated and characterized by chemical (hydrazine reduction), MS, and/or NMR

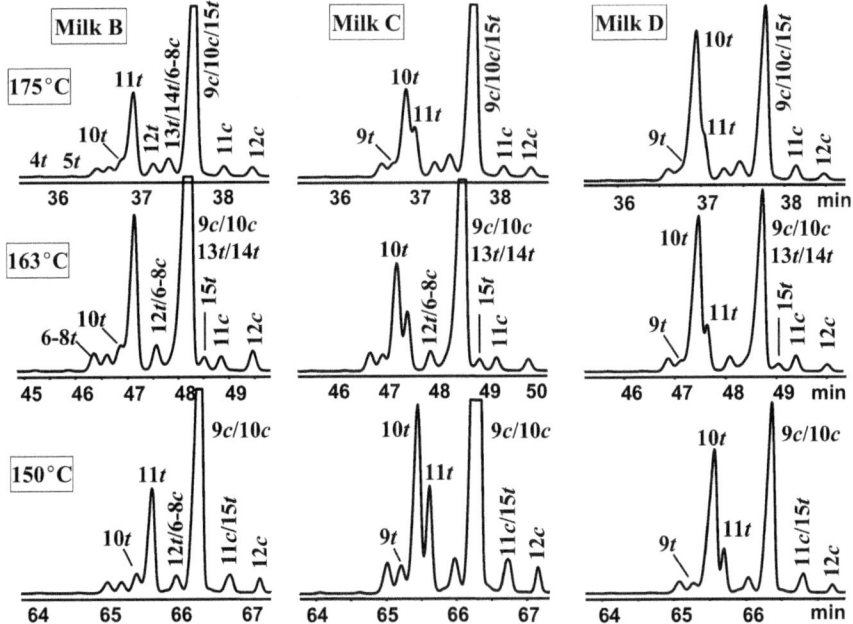

Figure 7.10 Partial GC chromatogram of three milk fats (B-D) are shown from *t*4-
to *c*12-18:1 containing different levels of *t*10- and *t*11-18:1. Separations were performed using a temperature program with a plateau at
175 °C (top row), 163 °C (middle row) or 150 °C (bottom row) and a
100 m SP-2560 column.[59]
Reproduced with permission of *Lipids* and the authors.

techniques. Therefore, when dealing with unresolved, or poorly resolved
FAMEs, one needs to be prepared not only to try different elution temperatures but also examine alternative GC columns such as the 100 m (or 200 m)
SLB-IL111 ionic liquid column and other techniques.

Methods involving Ag$^+$ chromatography have been extensively used to
complement GC separations.[60] Overlapping geometric isomers and FAMEs
with different degrees of unsaturation can be effectively addressed using
prior Ag$^+$ fractionations with Ag$^+$-TLC,[56–58,68,86,88] Ag$^+$-HPLC,[53,54] or Ag$^+$-
SPE.[59,60,76] The value of Ag$^+$ fractionation is the effectiveness of separating
FAMEs based on the number and geometric configuration of the double
bonds in the molecule. It not only resolves overlapping FAMEs but also
provides valuable structural support for the identification of FAMEs based
on the Ag$^+$ fraction in which they elute. Figure 7.12 shows a partial GC
separation of the 18:1/18:2 region from total milk fat FAME obtained using
Ag$^+$ SPE fractionation. The *trans*- and *cis*-18:1 and dienoic FAMEs eluting in
the 18:1/18:2 retention time region are clearly identified.[59] It cannot be
stressed enough that the identification of the different geometric and positional isomers of unsaturated FAMEs should always be confirmed. For the
analysis of the many minor PUFAs in a marine oil it was found that prior

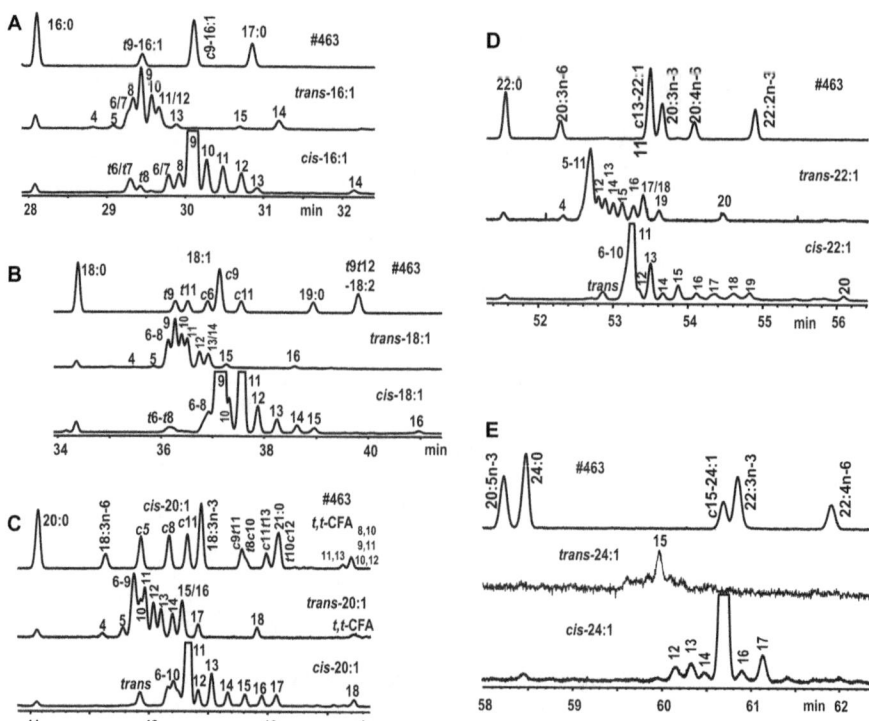

Figure 7.11 Partial GC separations of the 16:1 (A), 18:1 (B), 20:1 (C), 22:1 (D) and 24:1 (E) retention time regions. Each region contains the corresponding FAMEs present in GC reference mixture GLC-463 (upper), and the *trans* (center) and *cis* (bottom) MUFA fractions derived from a Ag⁺-SPE separation of a partially hydrogenated fish oil.[60] A 100 m SP-2560 column was used operated using a temperature program with a plateau at 175 °C. All PUFA were labeled using the n nomenclature, while the MUFAs are labeled with their Δ nomenclature.
Reproduced with permission of *Lipid Technol.* and the authors.

Ag⁺-SPE[76] and Ag⁺-HPLC[53,54] fractionation was not sufficient. It required further GC-MS in the positive chemical ionization mode (CI⁺) with isobutane as ionization reagent to identify minor PUFAs as well as other components in the marine oil;[54] see Section 7.6.4 on ionic liquid columns below.

It should be noted that 50 m columns with the same polarity (100% CPS) are not recommended because they provide limited separation of the geometric 18:1 isomers.[88] A typical separation of total milk FAME using a 50 and a 100 m CP Sil-88 column was previously reported.[88] It was estimated that ~35% of the total *trans*-18:1 content in dairy products was underestimated in the TRANSFAIR study [89,90] because they used a 50 m CP-Sil 88 column.[90] On the other hand, we are not aware of any reports in which 100% CPS columns longer than 100 m were tested.

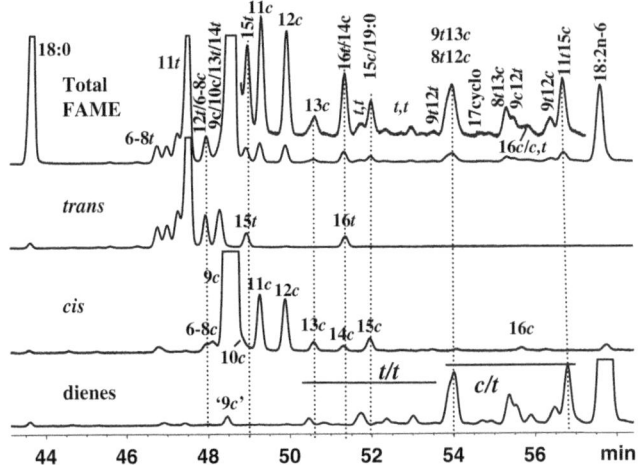

Figure 7.12 Partial GC chromatogram of a milk fat from 18:0 to 18:2n-6 using a GC temperature program with a plateau at 163 °C. The insert shows an enlargement of the *c*11-18:1 to *t*11*c*15-18:2 region. The *trans*, *cis* and diene fractions were isolated from total milk fat FAMEs using a Ag⁺-SPE cartridge.[59]
Reproduced with permission of *Lipids* and the authors.

7.6.3.3 Conjugated Dienoic Acid (CDA)

It is somewhat surprising that the only additional separation of C18 CDA a 100 m 100% CPS column was able to achieve compared to a GC column of medium polarity (Figure 7.6) was the additional separation of *c*11,*t*13- from *t*10,*c*12-18:2 and *t*11,*t*13- from the remaining three *t,t*-18:2 isomers in the UC-59M mixture (Figure 7.13B).[61] Even operating the column temperatures at 120 °C did not significantly improve the separation of the C18 CDA isomers present in the UC-59M mixture (Figure 7.13C).[91] However, the 100 m 100% CPS columns do separate the two *c/t* isomers of 9,11- and 10,12-18:2, which are not resolve by Ag⁺-HPLC.[58] This finding proved to be a critical separation to measure these two C18 CDA isomers in complex mixtures since it required combining the GC results from the 100% CPS column and the results from the Ag⁺-HPLC separation (Figure 7.14).[58]

An additional advantage of the 100 m CPS columns is the elution of the C18 CDAs between 18:3n-3 and 20:2n-6, a part of the GC region in which only a few minor interfering FAMEs elute (Figure 7.13A).[11,32,61] In this region of the GC chromatogram only 21:0 and a few minor 20:2 isomers where shown to elute using a 100 m 100% CPS column, as demonstrated by high-resolution selected-ion MS (Figure 7.4).[66,67] It should be noted that the elution of 21:0 relative to the C18 CDA isomers differs between different 100 m 100% CPS columns, even from the same supplier, and the elution pattern may also change with the age of the column (Figure 7.15).[29,44] For this reason, one should always monitor the GC column performance by

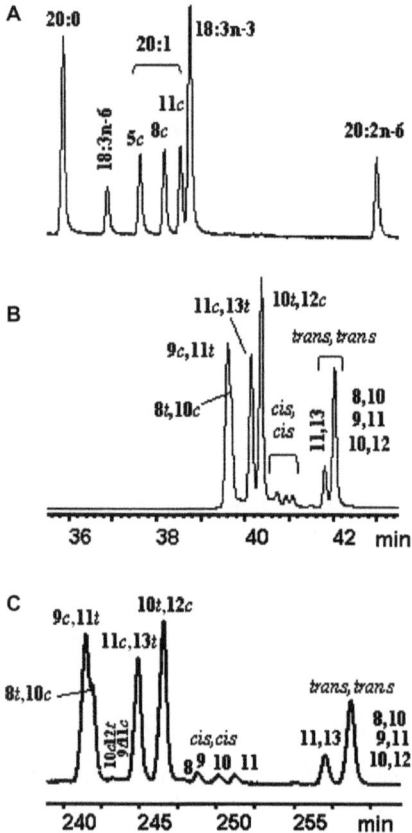

Figure 7.13   Partial GC separation of the FAMEs in reference mixture GLC-463 from 20:0 to 20:2n-6 (A). The C18 CDA isomers present in UC-59M are separated in (B). A 100 m CP-Sil 88 column was operated using a temperature program with a plateau at 175 °C. In panel (C), the UC-59M mixture was resolved using the same GC column and an iso-thermal condition at 120 °C.[61,91]
Reproduced with permission of *Lipids* and the authors.

analyzing a reference standard and include 21:0 in that standard if it is not present.

Figure 7.16 shows the elution sequence of all the synthesized *c/t* C18 CDA isomers from 6,8- to 13,15-18:2 using a 100 m CP-Sil 88[6] or SP-2560 column.[10] To more accurately compare the elution of C18 CDAs, the relative retention time (RRT) of each isomer was determined after adjusting for the solvent front and expressing the retention times relative to that of methyl γ-linolenate (GLA; 18:3n-6) as follows: $RRT_{GLA} = (RT_{isomer} - RT_{solvent})/(RT_{GLA} - RT_{solvent})$.[6] These RRT values are reported in Table 7.2. A few general rules apply. First, the retention time (RT) sequence for each positional C18 CDA isomer was: *c/t* < *c,c* < *t,t*. Second, *c,t* C18 CDA isomers eluted before the *t,c* isomers, except for the two 6,8-18:2 isomers which coeluted.

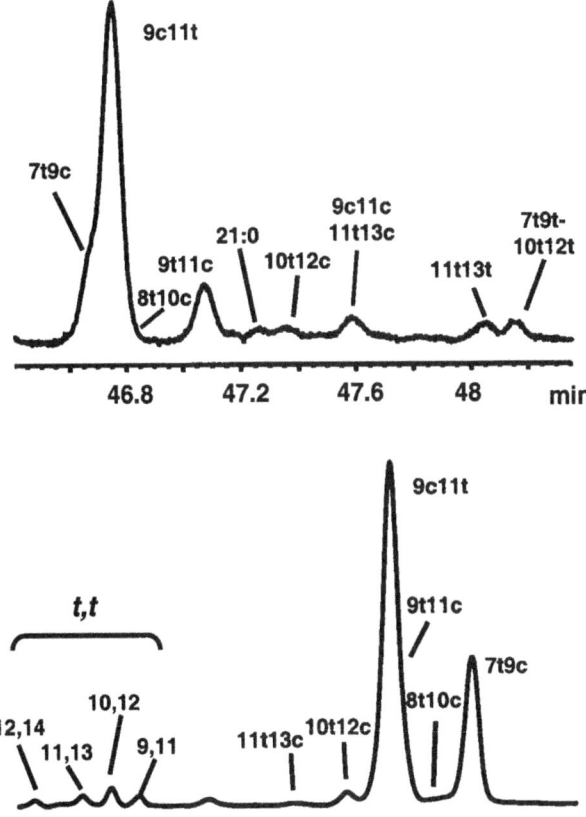

Figure 7.14 Partial GC separation of the C18 CDA region of a milk fat FAME mixture using a 100 m CP-Sil 88 column operated using a temperature program with a plateau at 175 °C (upper panel). The lower panel shows the separation of the same milk fat FAME mixture using three Ag+-HPLC cartridges in series and a mobile phase consisting of 0.1% acetonitrile, 0.5% tertiary butyl ether and 99.4% hexane.[29,58] Redrawn with permission of *J. AOAC Intern.*, and the authors.

Third, the RT of the *c,t* C18 CDA isomers increased as the Δ value of the *cis* double bond increased, except for the two 6,8-18:2 isomers which eluted after *c*7,*t*9-18:2. Fourth, the RT of the *c,c* C18 CDA isomers increased with increased Δ values. Finally, the RT of *t,t* C18 CDA isomers increased with increased Δ values, but the *t,t* isomers from 6,8- to 10,12-18:2 coeluted. Unfortunately, samples from natural or industrial sources are often more difficult to analyse since the isomer distribution is not in equal concentrations as it is in the UC-59M reference mixture or the iodine isomerized mixture. When the isomer distribution is unequal, such as in ruminant products, it leads to unresolved peaks. MS has been used to identify closely eluting or co-eluting C18 CDA isomers. In this way *t*7,*c*9-18:2 was identified in ruminant fats eluting just ahead of rumenic acid (*c*9,*t*11-18:2)

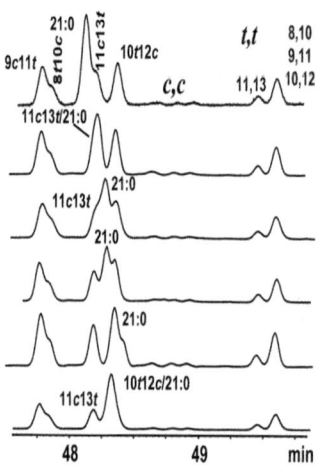

Figure 7.15   Partial GC chromatogram of the C18 CDA FAME region on three separate
100 m CP-Sil 88 columns and different ages. The standard CDA FAME
mixture UC-59M (Nu-Chek Prep Inc.) contained four positional C18 CDA
isomers and was spiked with FAME 21:0. The GC separations were
sorted by increased elution times of 21:0 relative to the CDA isomers.[29]
Reproduced with permission of AOCS Press and the authors.

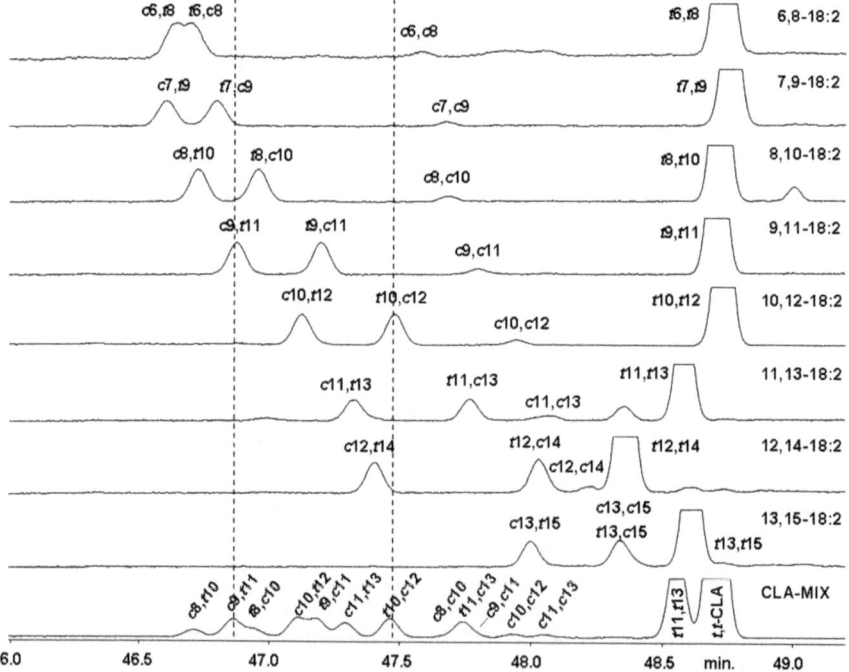

Figure 7.16   GC separation of the individual synthetic geometric and positional
isomers of the CDAs from 6,8- to 13,15-18:2 using a 100 m SP-2560
column and a temperature program with a plateau at 175 °C.[10] An
isomerized mixture of UC-59M was included on the bottom.
Reproduced with permission of *J. AOAC Intern.*, and the authors.

Table 7.2 GC relative retention times (RRT) of C18 CDA FAME isomers.[6]

Isomer	RRT/GLA	Isomer	RRT/GLA
c7,t9	1.083	c9,c11	1.137
c8,t10	1.085	c10,c12	1.145
c6,t8	1.086	c13,t15	1.147
t6,c8	1.086	t12,c14	1.149
t7,c9	1.091	c11,c13	1.151
c9,t11	1.091	c12,c14	1.159
t8,c10	1.097	t13,c15	1.166
c10,t12	1.104	c13,c15	1.166
t9,c11	1.109	t12,t14	1.171
c11,t13	1.114	t11,t13	1.184
c12,t14	1.118	t13,t15	1.184
t10,c12	1.122	t9,t11	1.191
c6,c8	1.128	t8,t10	1.191
c8,c10	1.131	t10,t12	1.192
c7,c9	1.131	t6,t8	1.192
t11,c13	1.136	t7,t9	1.193

GLA, γ-linolenic acid; RT, retention time.

(Figure 7.17A).[39] The coelution of several *t,t*-C18 CDA isomers and their elution order was also established on the 100m 100% CPS column by GC-MS (Figure 7.17B).[66,92] Even though MS makes it possible to identify co-eluting C18 CDA isomers, this is not a practical method for routine analysis. A direct GC separation would be preferred, but so far it has not been possible by GC to match the superior separation of the C18 CDA isomers obtained by Ag^+-HPLC;[11,49,92] see Section 7.8 below.

7.6.3.4 Mono- and Diconjugated Trienoic Acid (MCTA and DCTA)

Only a few MCTAs have been identified using 100 m CPS columns. The MCTA isomer *c*9,*t*11,*t*15-18:3 eluted after the *t,t*-18:2 CDA isomers and before 20:2*n*-6, while *c*9,*t*11,*c*15-18:3 eluted between 20:2*n*-6 and 22:0.[14,93] It is not known whether there are other naturally occurring MCTAs, but that may be partly due to a lack of standards. If so, every effort should be made to ensure that the measured MCTAs are real and not the product of isomerization during analysis. To our knowledge, the DCTAs have not been analysed using long-chain polar GC columns, except in one case in which a 50 m CP-Sil 88 CB column was used.[18] A preliminary investigation has shown that the DCTAs elute in the 24:0 to 24:1 region using a 100 m SP-2560 column and a temperature program with a plateau at 175 °C (JKGK unpublished data).

7.6.3.5 Limitations

- There is a lack of separation between the shorter-chain SFAs from C10 to C13 and the MUFAs with one less carbon, which is evident in the analysis of milk fats. Changing the column temperature did not resolve these FAMEs, but GC columns with medium polarity provide good

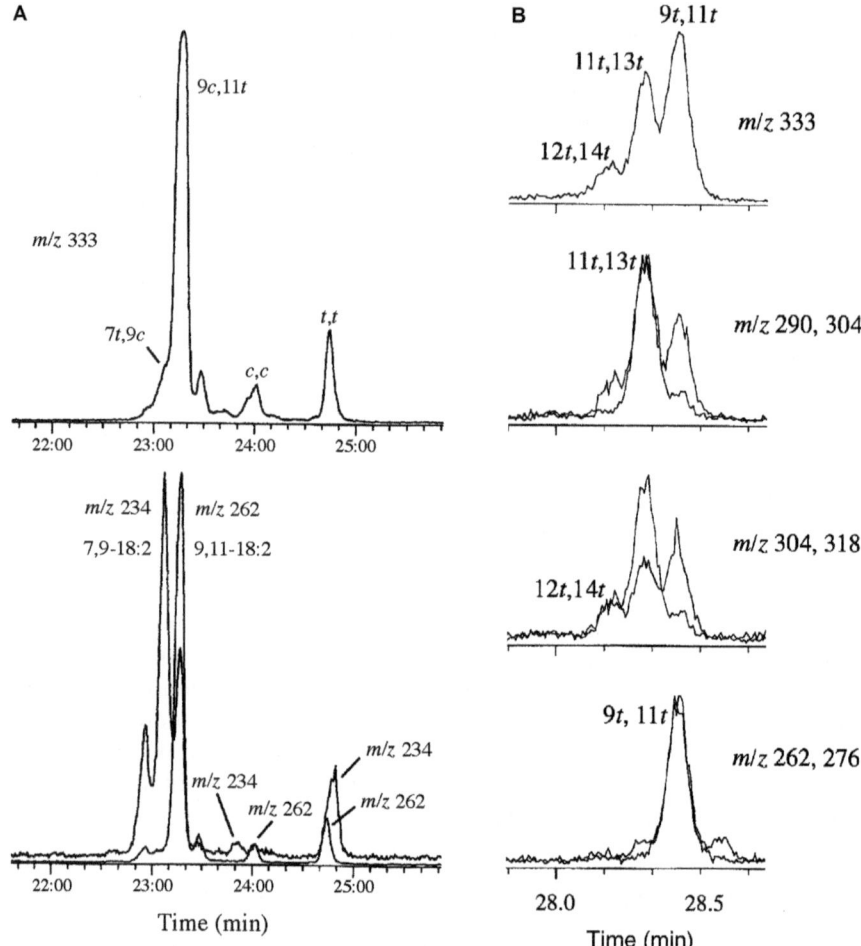

Figure 7.17   Reconstructed ion chromatograms of the DMOX derivatives of the C18 CDAs from human adipose tissue (A) and the *t,t*-CDA isomers isolated from cheese lipids by Ag$^+$-HPLC (B). The ion profile at *m/z* 333 is shown for all C18 CDAs (A, top), and at the bottom of (A) at *m/z* 234 for the two 7,9-18:2 isomers of *c*7,*t*9-18:2 and *t*7,*c*9-18:2, and at *m/z* 262 for *c*9,*t*11-18:2.[39] The reconstructed ion profile for all *t,t*-C18 CDA isomers in cheese lipids are detected at *m/z* 333 (B, top), and the allylic ions (*m* + 2) and *m* + 3 ion are shown for the 11,13-18:2 (B, 2nd from top), 12,14-18:2 (B. 3rd from top) and 9,11-18:2 (B, bottom) isomers.[66]
       Reproduced with permission of *Lipids* and the authors.

separations,[61] as well as highly polar ionic liquid columns;[53], see Section 7.6.4 below.

- On the 100 m 100% CPS column, the 16:1 region consists of partially resolved *t*- and *c*-16:1 isomers, C17 branched-chain FAs, and phytanic acid isomers. The individual FAMEs can be partially measured by operating the GC column at different temperatures,[59] but a prior Ag$^+$ separation and subsequent GC analysis of the resulting fractions is more reliable (Figure 7.8).[56] The dietary consumption of t9-16:1 isomer has been reported to be associated with a number of diseases,[94,95] but this finding may need to be confirmed with a pure isomers and a combination of appropriate analytical methods.

- Complete resolution of the *t*- and *c*-18:1 isomers remains a problem using the 100 m 100% CPS columns. Conducting GC analysis using two different temperature programs will provide identification of most 18:1 isomers, except the separation of t13- from t14-18:1, of c10- from the predominant c9-18:1, and of t16- from c14-18:1, which often coelute (Figures 7.10 and 7.12).[59] A prior Ag$^+$ separation is recommended for a definitive GC identification (Figure 7.12).[56,57,59,88] Separating the t10- and t11-18:1 remains a challenge when these two FAMEs are present in uneven amounts, in which case a lower temperature program at 150 °C provides a better separation (Figure 7.10).[59]

- There is extensive overlap between the 18:3 and 20:1 isomers. This problem was partially resolved for PHVO samples by lowering the column temperature from 180 °C to 155 °C (Figure 7.9).[69] This solution was successful largely because PHVOs contain only a few 20:1 FAMEs. On the other hand, ruminant fats contain appreciable amounts of *t*- and *c*-20:1 and 18:3 FAMEs, and they cannot be separated using temperature programs that plateau at either 175 °C (Figure 7.18A) or 150 °C (Figure 7.18B).[59] It is doubtful whether any temperature setting will separate these isomers in ruminant fats on 100 m 100% CPS columns.

- The identification of eicosapentaenoic acid (EPA; 20:5n-3) and 24:0 can be confusing since these two FAMEs elute very closely together on 100 m 100% CPS columns. The elution order may reverse between different columns, temperature programs, and with the age of the column. The GC reference mixture GLC-463 is not too helpful since these two FAMEs are both present at 2% of total FAME, unless that mixture is spiked with an additional amount of 24:0. To differentiate between these two FAMEs one could also use a different GC reference standard in which 20:5n-3 and 24:0 have different concentrations, such as the Supelco 37 Component FAME Mix (Sigma-Aldrich).

- Use different elution temperatures to test for coelution of FAMEs or improve their separation. However, if that fails, complement the separation using the very highly polar SLB-IL111 ionic liquid column.

- It is impossible to measure the two most abundant C18 CDAs t7,c9- and c9,t11-18:2 in ruminant fats using a 100 m 100% CPS column without the use of complementary techniques that resolves these isomers (Figure 7.14). In most reports the t7,c9-18:2 isomer is ignored, or its presence is acknowledged without providing an actual separation from c9,t11-18:2. Others have determined the ratio of these two CDAs in a

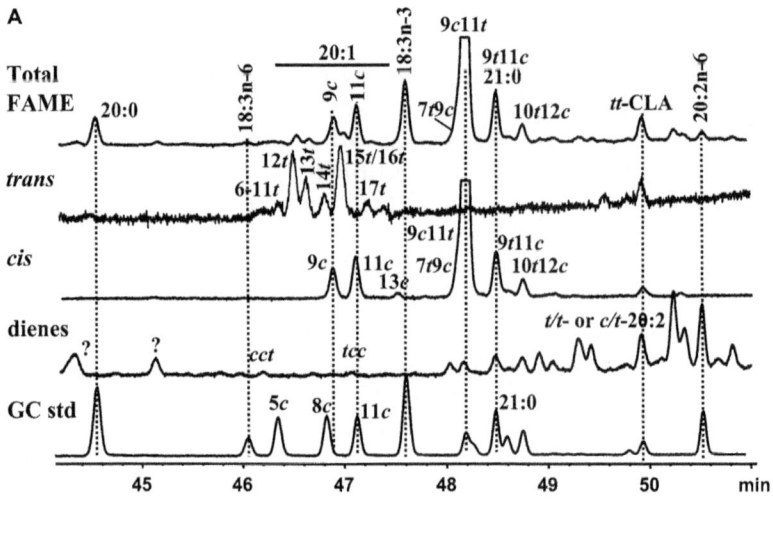

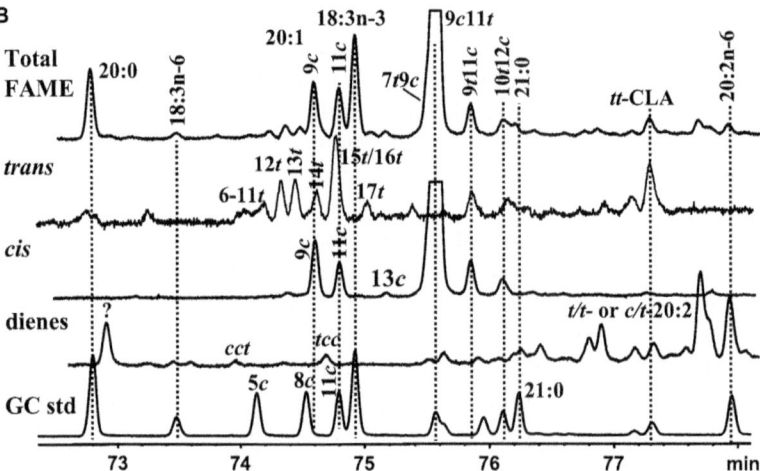

Figure 7.18   Partial GC chromatograms of the 20 : 1/18 : 3 region of the same milk fat FAMEs using a 100 m SP-2560 column and two temperature programs with a plateau at 175 °C (A) and 150 °C (B). In each case, the top graphs are that of total milk fat FAMEs followed by the *trans*, *cis* and diene fractions obtained from a Ag⁺-SPE separation; the GLC-463 standard is shown on the bottom.[59]
Reproduced with permission of *Lipids* and the authors.

few representative samples by Ag⁺-HPLC or MS, and then applied this ratio to the remaining samples of that study.[96] Such analyses ignore the variations between biological samples, which may be the key to the understanding the metabolic processes. In a survey we conducted of beef steaks in North America, large variations were observed between the quantities of these two isomers.[97,98]

7.6.4 Separation of FAME Using Very Polar (Ionic Liquid) GC Columns

7.6.4.1 Column Properties

Long fused silica capillary columns coated with ionic liquid stationary phases are available from Supelco Inc. Among those, the SLB-IL100 and SLB-IL111 have a higher polarity than 100% CPS columns and a greater stability (up to 270 °C)[99] The novel SLB-IL111 capillary column in particular has provided new opportunities to test the possibility of better resolutions of individual geometric and positional isomers of unsaturated FAMEs and increased separation between different groups of FAMEs to prevent overlaps. The improved resolution of the positional/geometric isomers of the C18 CDAs and *c*- and *t*-MUFAs was tested using synthetic reference standards.[5,6,10] The ionic liquid phase of SLB-IL111 was shown to be sensitive to changes in elution temperature, which made it possible to select conditions that optimized the desired separations.

In our first study, a 100 m SLB-IL111 column was operated isothermally at 168 °C with hydrogen as carrier gas at a flow rate of 1.0 mL/min. The elution profile of authentic synthetic standards of C18 CDAs and MUFAs were examined.[52] In the following two studies, two 100 m SLB-IL111 columns were combined into a 200 m column and operated isothermally for the first 50 min at 170 °C, followed by a temperature program to 185 °C and an increased column flow rate to allow for the more rapid elution of the long-chain PUFAs. The 200 m GC column provided enhanced resolution of the FAMEs prepared from milk fat[53] and a marine oil (menhaden oil).[54] In each case, the FAMEs prepared from the test samples were compared to the reference FAME mixture GLC-463 (Nu-Chek Prep Inc.), and the synthetic mixtures of MUFAs, geometric isomers of linoleic and linolenic acids, and C18 CDAs. Supporting evidence for the FAME identifications was also obtained from a prior Ag^+-HPLC separation of the total FAMEs.[53,54] The chromatographic conditions described by Delmonte *et al.*,[53] using hydrogen as carrier gas, require inlet pressures lower than 100 psi and are compatible with most modern GCs.

7.6.4.2 FAME

The higher polarity of the SLB-IL111 ionic column compared to 100 m 100% CPS columns was evident in the elution of 18:3*n*-3 well after the 20:1 isomers and just before 22:0. The 200 m SLB-IL111 column provided a number of improved separations of the individual *t*- and *c*-18:1 and 16:1 isomers[53] compared to a 100 m SP-2560 column,[10] The separation of the 18:0 to 18:2*n*-6 region of milk fat showed the capability of the ionic column to almost separate the clusters of *t*- and *c*-18:1 isomers, except for the overlap of *c*6/*c*7- with *t*12-18:1 and *t*15- with *c*8-18:1 (Figure 7.19).[53,54] The same occurred for the 16:1 isomers where only *c*6/*c*7-coeluted with *t*13-16:1 (see Figure 3 in reference 53). In addition, a single separations with the 200 m

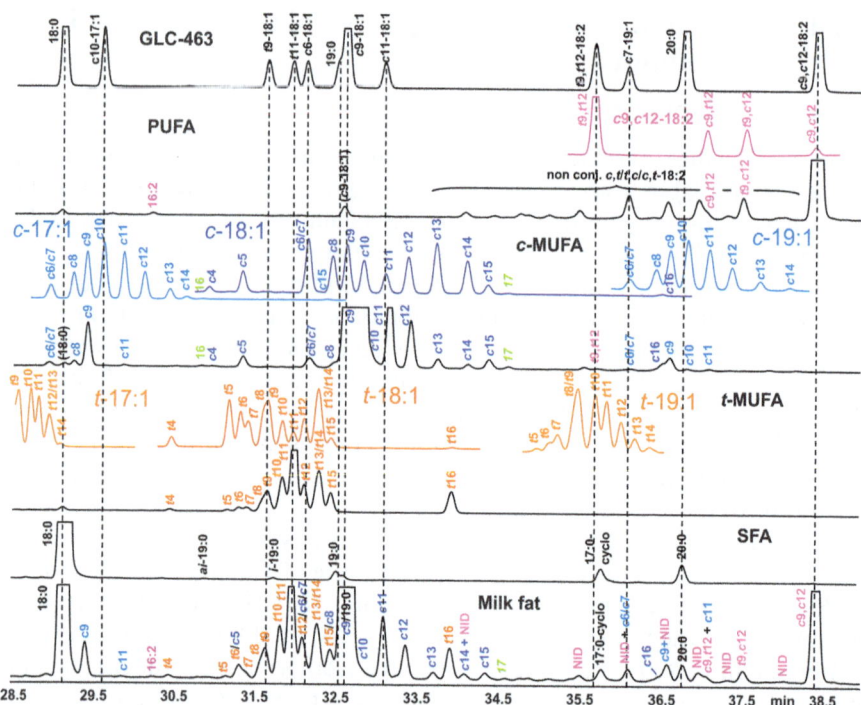

Figure 7.19 Partial GC chromatogram of the 18:0 to 18:2n-6 region of milk fat FAMEs together with the SFA, *trans*, *cis* and PUFA fractions isolated using Ag⁺-HPLC. The reference FAME mixture GLC-463 plus several synthetic *trans* (red) and *cis* (blue) isomer mixtures of different chain length were inserted into the figure as reference. FAMEs of the n-1 series are labeled in green. A 200 m ionic liquid column (SLB-IL111, Supelco Inc.) was operated at 170 °C for 50 min, increased by 6 °C/min to 185 °C, and then maintained for 35 min.[53]
Reproduced with permission of *J. Chromatogr. A* and the authors.

SLB-IL111 column included for the first time the separation of *t*6- from *t*7-18:1, *t*15- from *c*9-18:1, *t*16- from *c*14-18:1, and *c*8- from *c*6/*c*7-18:1, and provided a good separation between the two main *trans*-18:1 isomers in ruminant fats, *i.e.*, *t*10- and *t*11-18:1 (Figure 7.19).[53] Two sets of unresolved 18:1 FAMEs remained, *t*13- eluting with *t*14-18:1 and *c*9- eluting with *c*10-18:1. These coelutions could only be resolved by operating the 100 m 100% CPS columns at 120 °C.[58,68] It has not been determined what elution temperature would be required for the SLB-IL111 column to resolve these pairs of 18:1 isomers. In the 16:1 region, the described conditions provided the separation of *t*6- from *t*7-16:1, and *c*8- from *c*6/*c*7-16:1. There are a few C16 PUFAs eluting in the 18:1/18:2 region, most of which were well resolved from the other FAMEs in the analysis of menhaden oil (Figure 7.20).[54] The separation of 16:4n-3 and 18:2n-6 can be improved with a minor adjustment to the temperature program. The geometric and positional isomers of

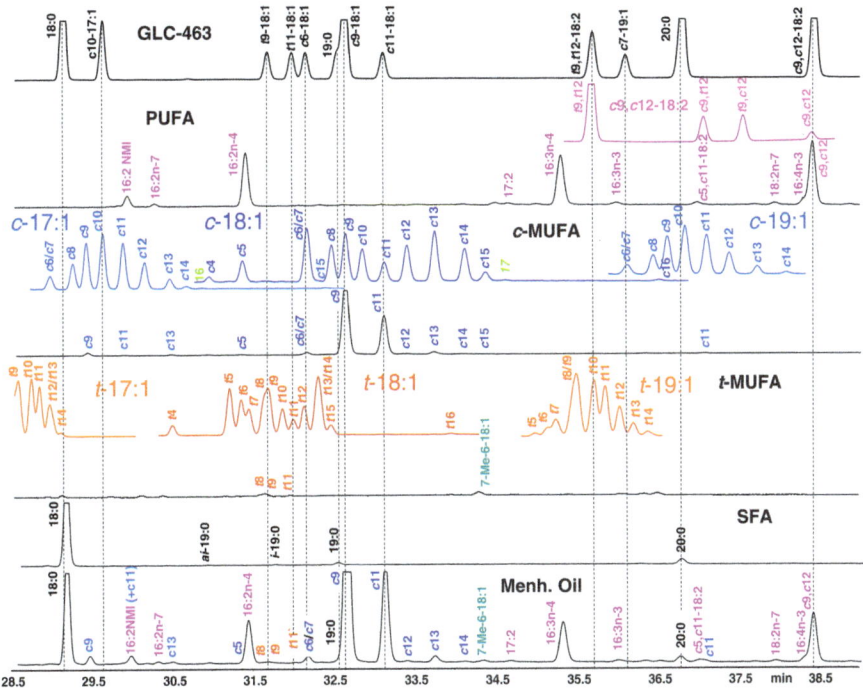

Figure 7.20 Partial GC chromatogram of the 18:0 to 18:2n-6 region of menhaden oil FAMEs together with the SFA, *trans*, *cis* and PUFA fractions isolated using Ag⁺-HPLC. The reference FAME mixture GLC-463 plus several synthetic *trans* (red) and *cis* (blue) isomer mixtures of different chain length were inserted into the figure as reference. FAMEs of the n-1 series are labeled in green. A 200 m ionic liquid column (SLB-IL111, Supelco Inc.) was operated at 170 °C for 50 min, then 6 °C/min to 185 °C and then maintained for 35 min.[54]
Reproduced with permission of *Lipids* and the authors.

20:1 and 18:3 were almost separated except for the *t,t,t* and *c/t/t* isomers of 18:3n-3; however, all the geometric isomers of 18:3n-6 extensively overlapped with the 20:1 FAMEs (Figure 5 in reference[53]).

An increase in column length of the SLB-IL111 column from 100 to 200 and 300 m resulted in a marked improvement of the separation of the 18:1 and 18:2 isomers in total milk fat FAMEs (Figure 7.21). Near baseline resolution between most 18:1 isomers was obtained with the 300 m column, except for *t*6- from *c*5-18:1, *t*13- from *t*14-18:1, *t*15- from *c*8-18:1, and 19:0 coeluting with *c*9-18:1. Many of the 18:2 isomers were not identified since currently there are no standards available and definitive characterizations have not been conducted.

The higher polarity of the SLB-IL111 column was instrumental to demonstrate for the first time the presence of an entire series of even carbon number *n*-1 MUFAs from 9-10:1 to 17-18:1 FA in milk fat.[53,54] The *n*-1

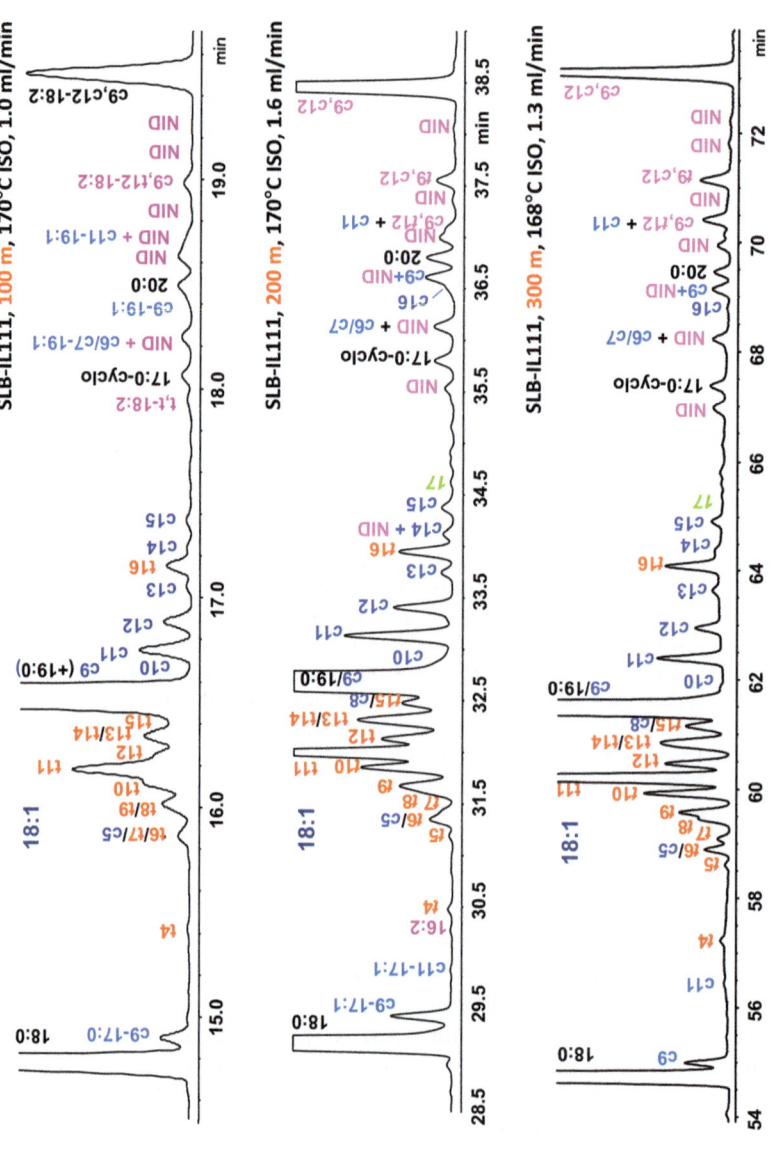

Figure 7.21 Partial GC chromatogram of the 18:0 to 18:2n-6 region of total milk fat FAMEs using a 100 m (top), 200 m (center) and 300 m SLB-IL111 column (bottom). The column length, isothermal operating condition, and flow rates are shown (Delmonte and Fardin-Kia, unpublished data).

MUFAs were not resolved using the 100 m 100% CPS columns.[59,61] The 100 m SLB-IL111 column was also used to confirm the identity of *t*6-16 : 1 in menhaden oil, which was separated from the *t*7- and *t*8-16 : 1 isomers.[54] This separation of *t*7- and *t*8-16 : 1 was not possible with a 100 m SP-2560 column.[76] The ability to separate *c*6- from *c*8-18 : 1 on the SLB-IL111 ionic liquid columns allowed others to identify the Δ6-MUFAs series in human hair,[100] and resolve *t*11,*c*15-18 : 2 from *t*10,*c*15-18 : 2.[101] The 200 m SLB-IL111 column was also used to complement the GC results of fast food samples using a 100 SP-2560 column.[102,103] The combination of the ionic column and GC-time of flight-MS operated in the CI$^+$ mode with isobutane as ionization reagent, made it possible to identify a new minor branched-chain FAME, 7-methyl-6-18 : 1 in menhaden oil, and a whole series of furanoid FAMEs in menhaden oil.[54] The GC-time of flight-MS method also provided an abundant molecular ion for all FAMEs even when present in trace amounts (Figure 7.22).

7.6.4.3 Conjugated Dienoic Acid (CDA)

The C18 CDAs have been notoriously difficult to resolve by GC including the 100 m 100% CPS columns except for a few isomers, which did not include some of the major CDAs of interest. A complementary Ag$^+$-HPLC separation was always necessary for the analysis of the C18 CDAs. Thus, it came as a complete surprise that both the 100 and 200 m SLB-IL111 columns were able to provide a separation of several major C18 CDA isomers. It has now made it possible to directly analyse the *t*7,*c*9- and *c*9,*t*11-18 : 2 isomers using GC without relying on complementary Ag$^+$-HPLC separations (Figure 7.23).[52-54] Several of the other C18 CDAs of interest in food analysis separated on the 200 m SLB-IL111 column, such as *t*9,*c*11-, *t*10,*c*12-, *t*11,*c*13-, *t*12,*c*14-, *t*12,*t*14-, and *t*11,*t*13-18 : 2. The elution of the C18 CDAs occurred in a very different GC region using the SLB-IL111 column compared with the 100 m 100% CPS columns. The FAMEs that interfere with the separation of C18 CDAs using the SLB-IL111 column include 18 : 4*n*-3 eluting just before *c*9,*t*11-18 : 2 and 20 : 2*n*-6 shortly thereafter, while the *cis*-22 : 1 isomers elute in the same region as the *t*,*t*-C18 CDAs.

Figure 7.24 shows the separation of the complete series of methyl esters of C18 CDAs with double bond positions from 6,8- to 13,15-18 : 2, GLC-463, and the isomerized mixture of UC-59M using the 100 m SLB-IL111 column operated isothermally at 168 °C.[52] The elution order of the geometric isomers of C18 CDAs with the same double bond position was *c*,*t* < *t*,*c* < *c*,*c* < *t*,*t*. The retention times of the *c*/*t*-C18 CDA isomers increased with an increase in the Δ values. The difference was sufficient to separate several of the major C18 CDA isomers commonly found in ruminant fat, including the separation of *t*7,*c*9- from *c*9,*t*11-18 : 2. The elution order of several of the *t*,*t*-C18 CDA isomers was opposite to that observed with *c*/*t*-C18 CDA isomers, except for the *t*,*t*-C18 CDA isomers of 6,8-, 7,9- and 12,14-18 : 2 that did not follow this

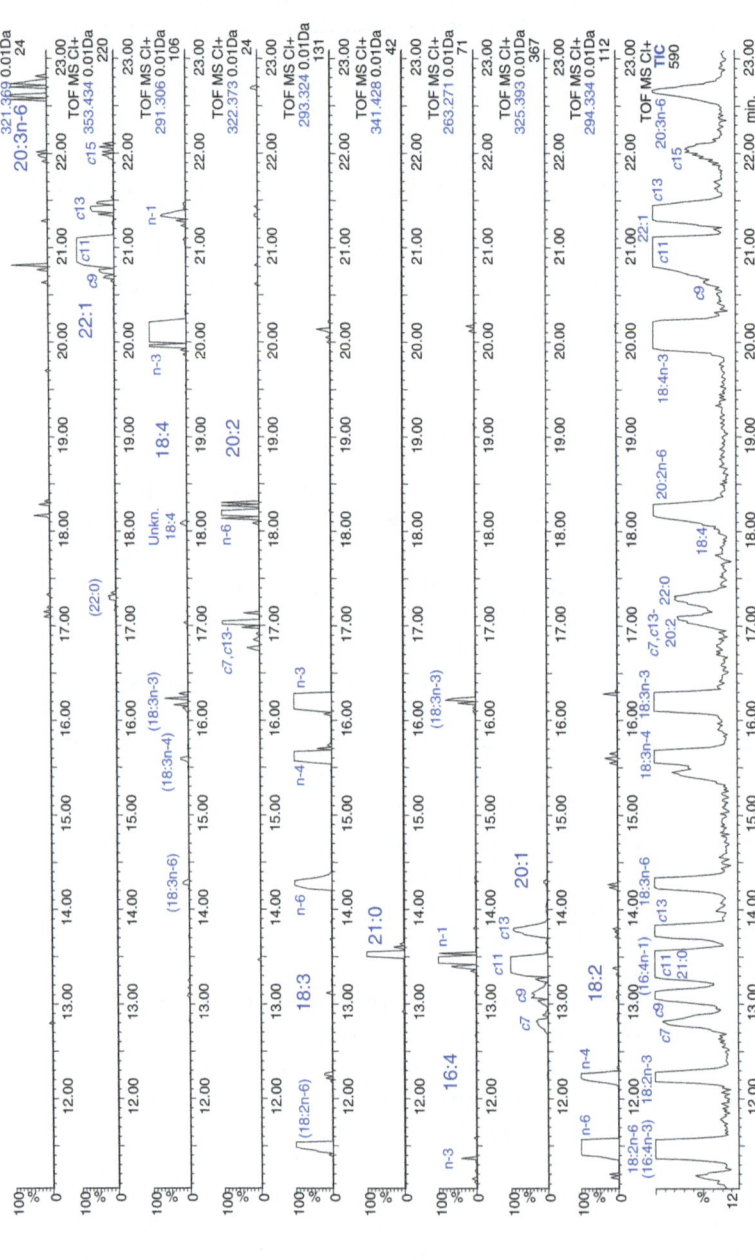

Figure 7.22 Partial total ion current chromatogram of a GC-time of flight-MS separation of menhaden oil FAMEs from 18:2n-6 to 20:3n-6 (bottom graph). Molecular ions (M⁺) were selected to demonstrate the presence of 18:2 (*m/z* 294.334), 20:1 (*m/z* 325.393), 16:4 (*m/z* 263.271), 21:0 (*m/z* 341.428), 18:3 (*m/z* 293.320), 20:2 (*m/z* 322.373), 18:4 (*m/z* 291.306), 22:1 (*m/z* 353.434), and 20:3 (*m/z* 321.369) in this chromatographic region. A 100 m SLB-IL111 column was used at a constant elution temperature of 170 °C, and the MS was operated in the CI⁺ mode with isobutane as the chemical ionization reagent gas.[54] Reproduced with permission of *Lipids* and the authors.

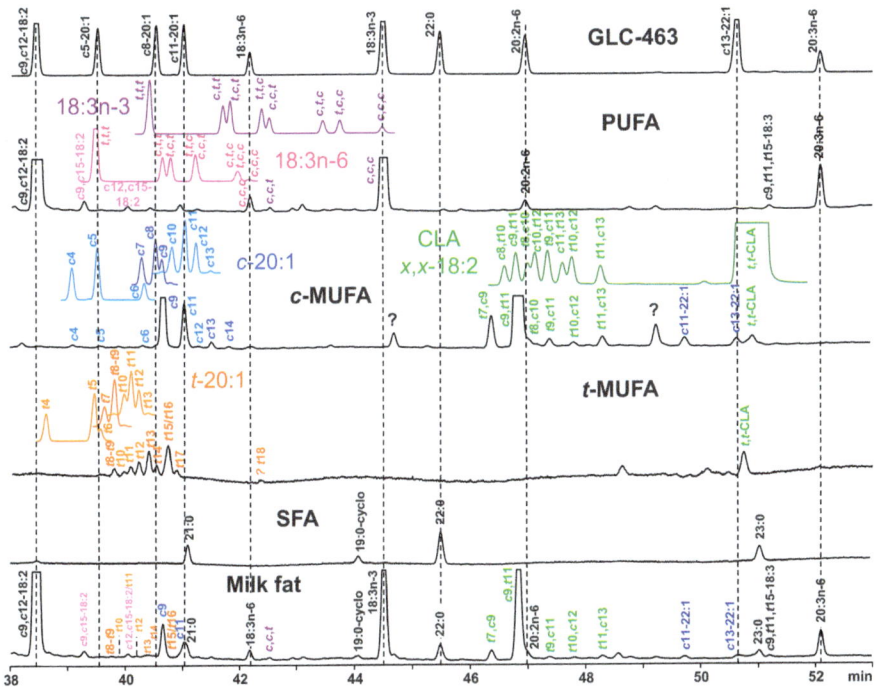

Figure 7.23 Partial GC chromatogram of the 18:2n-6 to 20:3n-6 region of milk fat FAMEs together with the SFA, *trans*, *cis* and PUFA fractions isolated using Ag⁺-HPLC. The reference FAME mixture GLC-463 plus several synthetic *trans* (red) and *cis* (blue) MUFAs, a positional and geometric isomer mixtures of CDA (green), 18:3n-6 (pink) and 18:3n-3 (violet) were inserted onto the figure as reference. The column and operating condition were as in Figure 7.19.[53]

Reproduced with permission of *J. Chromatogr. A* and the authors.

trend. A partial separation of several *t,t*-C18 CDA isomers was achieved when the separation was conducted at 130 °C.

The separation of the *t*7,*c*9- and *c*9,*t*11-18:2 isomers in ruminant fats was also evaluated using a 30 m SLB-IL111 column.[104] Although the two major C18 CDA isomers separated on the short column, the results showed extensive overlap of the CDAs with other FAMEs eluting in this region.

7.6.4.4 Mono- and Diconjugated Trienoic Acid (MCTA and DCTA)

On the 200 m SLB-IL111 column, the two MCTA isomers *c*9,*t*11,*t*15-18:3 and *c*9,*t*11,*c*15-18:3 eluted between *c*13-22:1 and 20:3n-6 (Figure 7.23) and between 20:3n-6 and 20:3n-3 (see Figure 6 in reference 53), respectively. The DCTA were not analysed using this column. The higher polarity of this column might help separate these isomers that were not resolved on columns of medium polarity.

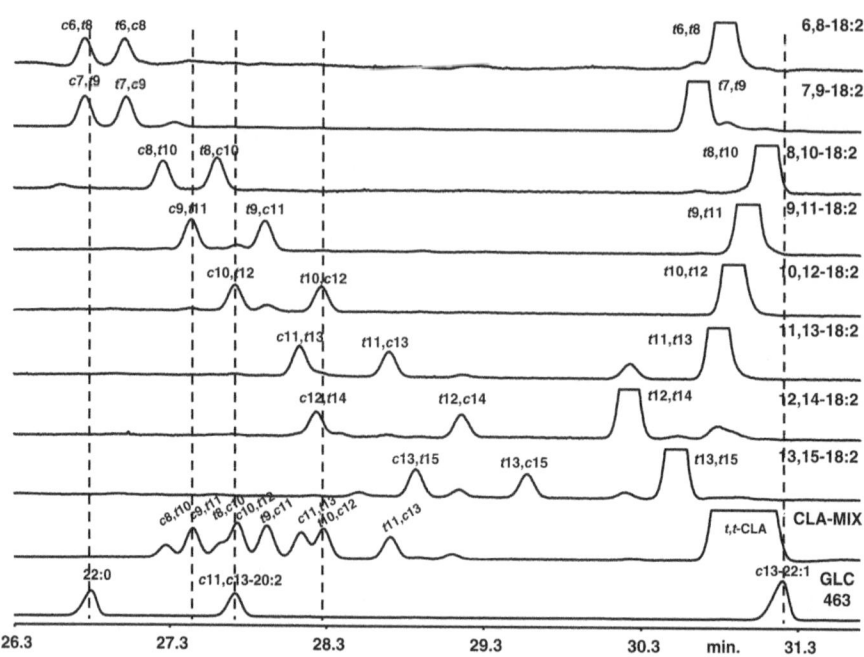

Figure 7.24 GC separation of the individual synthetic geometric and positional isomers of the C18 CDAs from 6,8- to 13,15-18 : 2 using a 100 m SLB-IL111 column and an isothermal temperature condition at 168 °C. An isomerized mixture of UC-59M and GC reference standard GLC-463 were included on the bottom.[52]
Reproduced with permission of *J. Chromatogr. A.* and the authors.

7.6.4.5 Limitations

- The high polarity of SLB-IL111 resulted in the elution of the SFAs between the *c*- and *t*-MUFA isomers with one less carbon (Figures 7.19 and 7.20). In the case of even-chain SFAs the coelution was with odd-chain MUFAs that are generally less abundant, while the odd-chain SFAs coelute with the more abundant even-chain MUFAs in natural products. As seen in Figures 7.19 and 7.20, 19:0 coeluted just ahead of *c*9-18:1, which was not resolved in samples containing a predominant *c*9-18:1 isomer, as in ruminant fats. Figure 7.23 shows 21:0 coeluting with *c*11-20 : 1.
- The relative elution of FAMEs with different number of double bonds was observed to be markedly temperature dependent, and selected coeluting pairs of FAMEs can be resolved by modifying the elution temperature at the expense of others.
- The coelution of *t*11/*t*12-16 : 1, *t*13/*t*14-18 : 1, *t*15/*t*16-20 : 1 could not be resolved on the 100 m CPS column, or the 100 m and 200 m SLB-IL111 columns at 168 °C or 170 °C, respectively.[53,54] A separation of these isomers will require a lower temperature, which has not been determined.

- There was an overlap of the t,t,t and t,t,c geometric isomers of 18 : 3n-3 with the 20 : 1 isomers (Figure 7.23). This overlap was not as extensive as with the 100 m 100% CPS columns. Only GC columns with moderate polarity clearly separate the 18 : 3 from the 20 : 1 isomers, but not the isomers within each group.[61] This requires more polar GC columns.
- The SLB-IL111 column has both advantages and disadvantages with regard to the separation of CDAs. Although several C18 CDA positional/ geometric isomers are well separated including the $t7,c9$- from $c9,t11$- 18 : 2, 20 : 2n-6 can interfere with $c9,t11$-18 : 2 and the 22 : 1 isomers can interfere with t,t-C18 CDAs (Figure 7.23). By lowering the column temperature to 160 °C, $c9,t11$-18 : 2 was resolved from the other FAMEs but it led to the coelution of 22 : 0 and $t7,c9$-18 : 2 (Delmonte unpublished data). When adjusting the elution temperature one needs to consider the comprehensive separation of all FAMEs in the GC chromatogram.
- To date, Ag$^+$-HPLC remains the method of choice to obtain the best and most complete separation of all the C18 CDAs. However, the results of using a SLB-IL111 column are compelling since it resolves most of the major C18 CDA isomers and provides a good separation for most other FAMEs in any sample when optimum experimental conditions are applied.
- The 100 m 100% CPS and the 200 m ionic liquid columns truly complement each other and combining their results provides an excellent GC method for a complete analysis of most lipid matrices.

7.7 Separation of Conjugated FA Using Silver Ion HPLC Columns

7.7.1 As FAME

The separation of the methyl esters of C18 CDAs on Ag$^+$-HPLC using a DAD detector with a UV setting at 234 nm was the first breakthrough to effectively resolve most of their geometric and positional isomers.[105] The DAD detector was used to detect possible interfering unsaturated FAMEs in the CDA region by examining the UV absorption at 205 nm. It was subsequently shown that three analytical ChromSpher 5 Lipid columns (250 × 4.6 mm; 5 μm particle size; Agilent Technologies) in series provided the most effective means of separating up to 20 individual C18 CDA isomers in cheese and milk lipids (Figure 7.25).[93] The separation was conducted with an isocratic elution of 0.1% acetonitrile in hexane. As a result of the poor solubility of acetonitrile in hexane, the retention times (volumes) were not reproducible. In addition, solvents were added to keep the acetonitrile in solution, such as 0.5% diethyl ether,[58,106] tertiary butyl ether,[29,107] or 0.05% and 0.1% 2-propanol.[108] Several different alkanes and nitrile modifiers were also

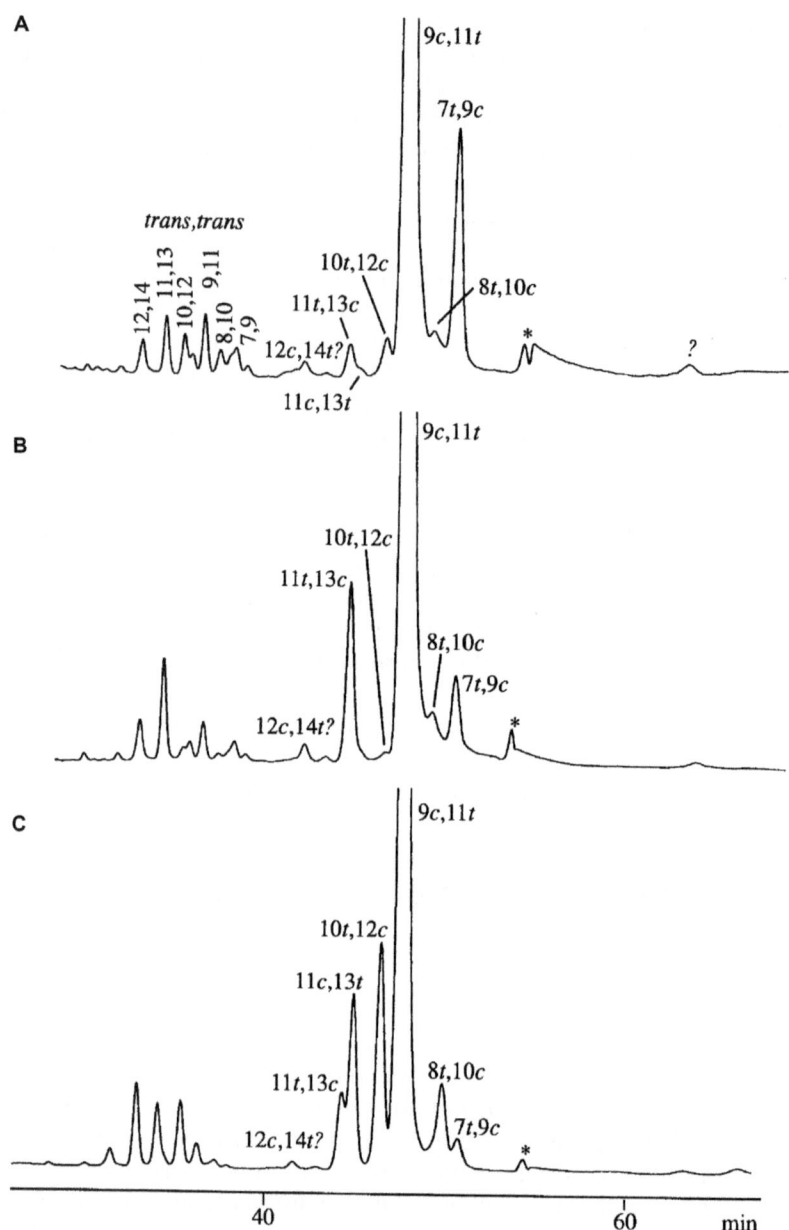

Figure 7.25 Ag⁺-HPLC separation of the FAMEs from cheese (A), bovine milk (B), and a spiked sample of bovine milk with the C18 CDA mixture UC-59M (C). Three silver-ion HPLC columns were used in series and a mobile phase consisting of 0.1% acetonitrile and 99.9% hexane.[11] Reproduced with permission of *Lipids* and the authors.

investigated at 0.1 and 0.2% concentrations, and the mobile phase of hexane and 0.2% propionitrile was reported to give the highest stability because of the greater solubility of propionitrile in hexane.[109]

The above method of using 0.1% acetonitrile in hexane as the mobile phase proved to be successful to analyse the distribution of C18 CDAs in the tissue lipids of pigs fed a commercial mixture of C18 CDAs that contained four positional isomers (Figure 7.26A).[32,107] Tissue, lipid class and isomer specific distributions were identified in the pig. Of great concern was the high incorporation of the *c*11,*t*13-18:2 isomer into the diphosphatidylglycerol (cardiolipin) fraction of both the liver and heart lipids of pigs (Figure 7.26A, center two panels). Cardiolipin is found primarily in the inner mitochondrial membrane, and is intrinsically involved in many of the enzymes of bioenergetics of mitochondria. Therefore, the selective incorporation of the *c*11,*t*13-18:2 isomer was considered a possible concern since changes in FA composition are known to adversely affect the activity of key enzymes.[32] This information lead to the removal of the *c*11,*t*13-18:2 isomer and accompanying *t*8,*c*10-18:2 isomer from the commercial production of CLA mixtures.[110]

An interesting application of Ag⁺-HPLC separation of C18 CDAs was the evaluation of the products of enzymatic methylation using the *Geotrichum candidum* lipase to convert the commercial mixture of free C18 CDAs (UC-59A; Nu-Chek Prep Inc.) into their methyl esters (Figure 7.26B), or the reverse not shown here.[111] Of the four C18 CDA positional isomers in the UC-59A mixture, only the *c*9,*t*11-18:2 isomer was methylated demonstrating the specificity of this enzyme (Figure 7.26B).[111]

Figure 7.27A shows the separation of the complete series of methyl esters of C18 CDAs from 6,8- to 13,15-18:2 using three Ag⁺-HPLC columns in series, a mobile phase of 0.1% acetonitrile and 0.5% diethyl ether in hexane, and UV detection at 233 nm.[51] To more accurately compare the C18 CDAs, a small quantity of *c*9,*t*11-18:2 was added to each positional isomer mixture to be able to convert all the values to relative retention volume (RRV). Under these conditions, the RRV for each positional C18 CDA isomer was: *t*/*t* < *c*/*t* < *c*,*c*. Within the *t*/*t*, *c*/*t* and *c*,*c* isomers, the RRV values differ with the position of the conjugated system along the FA acyl chain, and the relative elution order of the *c*/*t* geometric isomers changes after the 10,12-18:2 isomer; at the crossover the two 10,12-18:2 *c*/*t* isomers coelute (Figure 7.27A). The RRV values of the C18 CDAs are graphically presented in Figure 7.27B.[51] By expressing the results as RRV relative to *c*9,*t*11-18:2, the authors were able to establish that the drift in RV was due to the loss of acetonitrile in the mobile phase.

The authors also evaluated the separation of the synthetic C18 CDA methyl esters using 2% acetic acid in hexane as the mobile phase (Figure 7.28A).[51] The separations were similar to Figure 7.27A, except for the relative elution of some of the *c*/*t* isomers. The advantage was that the two *c*/*t* 10,12 isomers were resolved and there was a partial resolution of the *c*6,*t*8 and *t*7,*c*9 isomers. A graphic representation of the RRV values shows the changes in the relative elution of the *c*/*t* isomers with increasing Δ value (Figure 7.28B). However, the new mobile phase resulted in significantly longer RTs, and monitoring of unsaturated FAMEs at 205 nm was no longer possible because of the presence of acetic acid. By simultaneously using this mobile phase

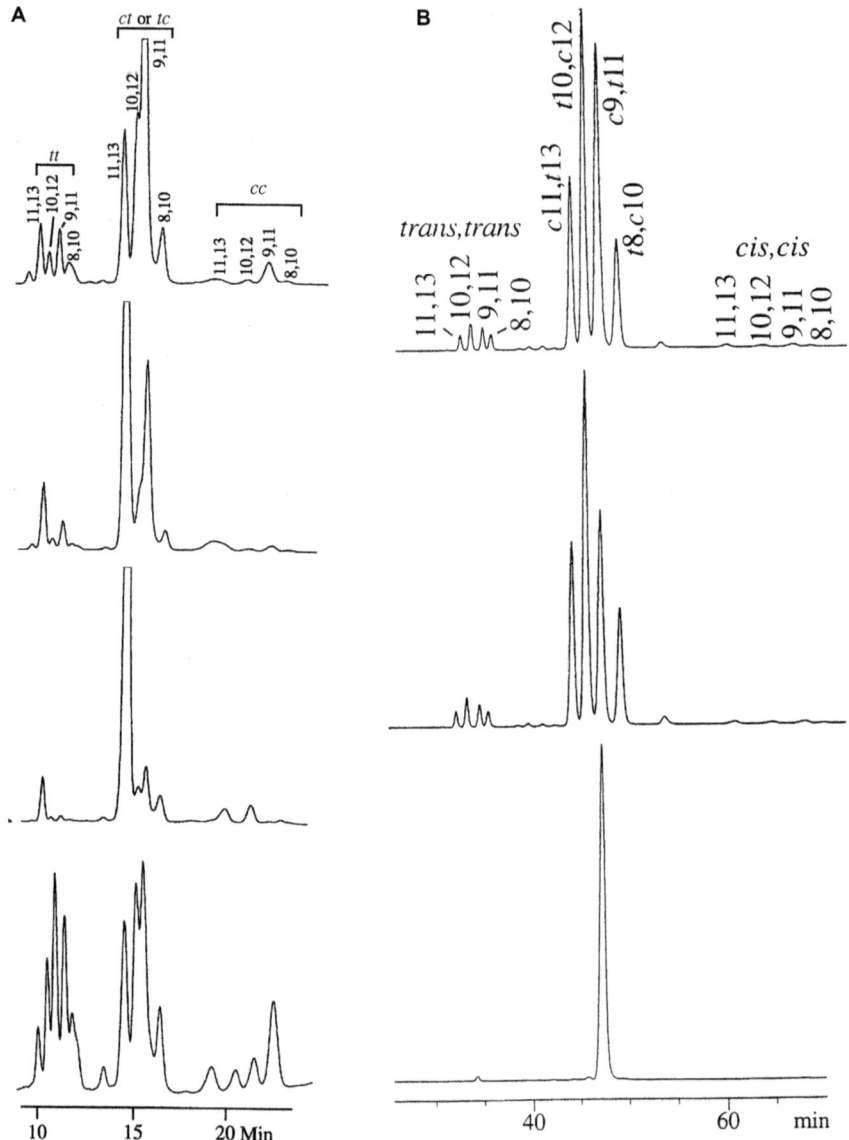

Figure 7.26   Ag$^+$-HPLC separation of the C18 CDA isomers present in different lipid fractions of pigs fed a CDA mixture containing all four positional C18 CDA isomers (right side). (A, top) liver phosphatidylcholine, (A, second) heart phosphatidylcholine, (A, third) heart diphosphadidyl-glycerol (cardiolipin), (A, bottom) liver FFAs.[32] On the left side, a lipase-mediated methylation of the free FA mixture of the four C18 CDAs (UC-59A; Nu-Chek Prep Inc.) using *G. candidum lipase B*. (B, top) Original UC-59A, (B, middle) FFA residue after 22.5 h incubation, (B, bottom) FAME fraction generated during 22.5 h of reaction that was extracted with hexane.[111] Three Ag$^+$-HPLC columns were used in each analysis with a mobile phase consisting of 0.1% acetonitrile and 99.9% hexane, and UV detection at 233 nm.

Reproduced with permission of *Lipids* and the authors.

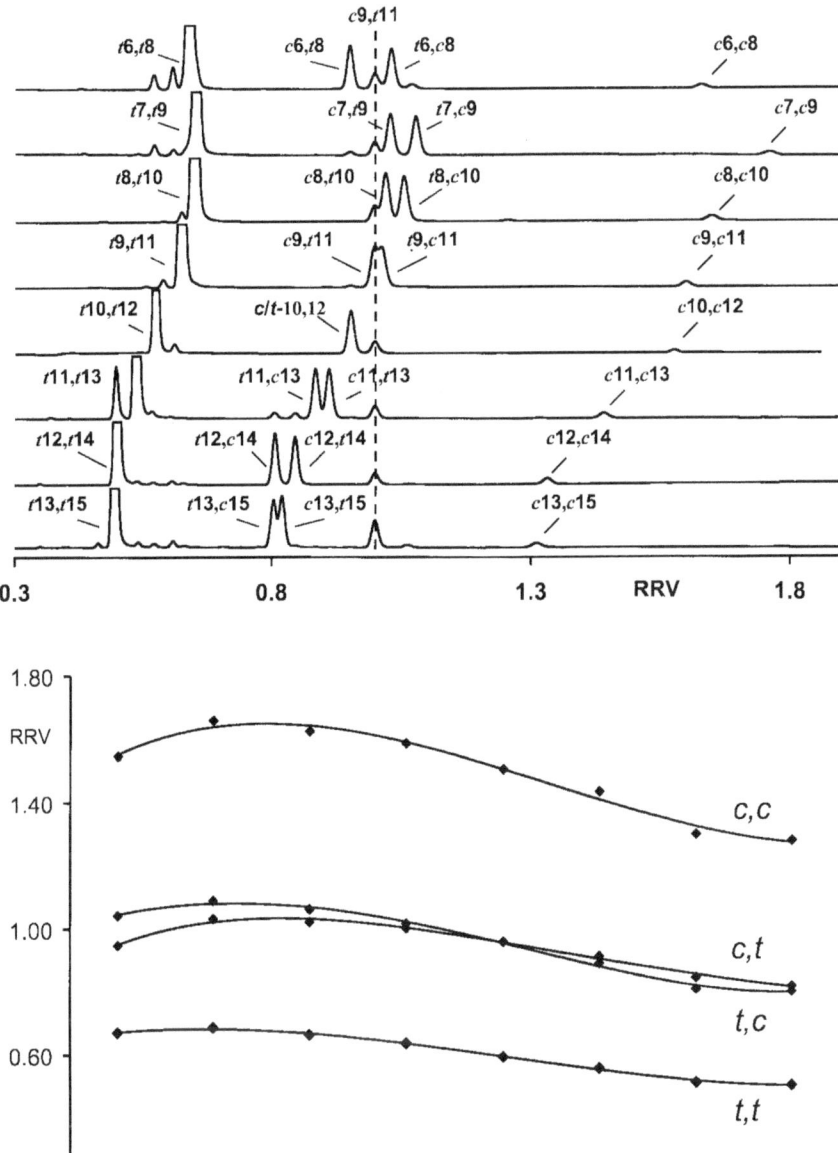

Figure 7.27 Ag⁺-HPLC separation of all geometric C18 CDA isomers from the 6,8 to the 13,15 using three ChromSpher 5 Lipids columns in series, a mobile phase of 0.1% acetonitrile/0.5% diethyl ether/99.4% hexane, and UV detection at 233 nm (top). A small amount of c9,t11-18:2 was added to each CDA mixture to make it possible to express the elution of all the isomers as relative retention volume (RRV) to c9,t11-18:2. The plot of RRV *vs* carbon–carbon double bond position for all geometric CLA isomers is shown on the bottom.[51] Reproduced with permission of *Lipids* and the authors.

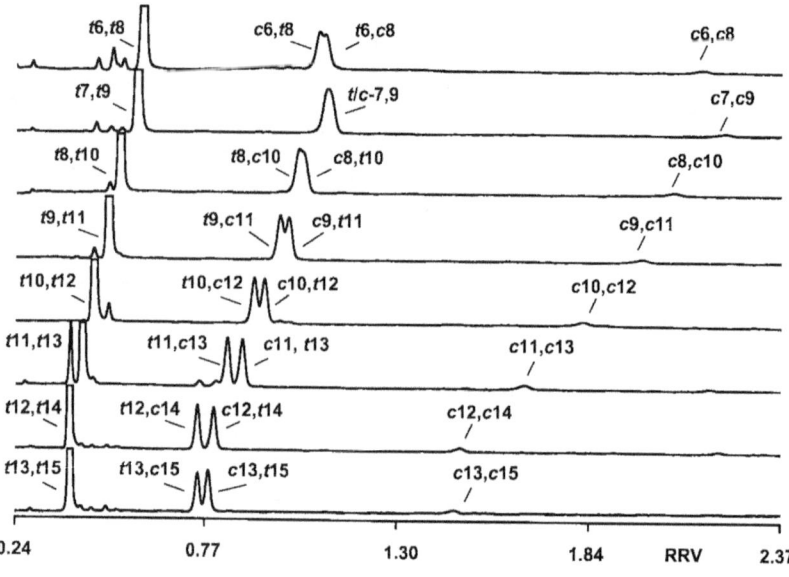

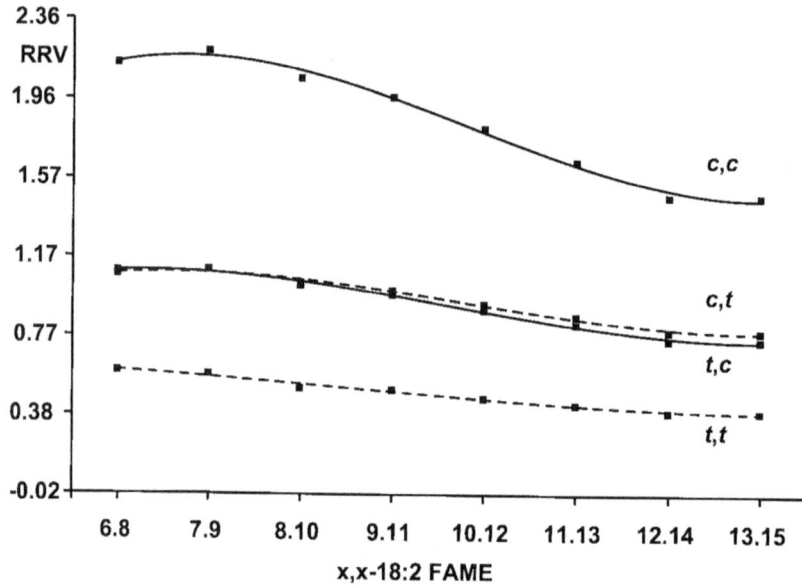

Figure 7.28   Ag⁺-HPLC separation of all geometric C18 CDA as FAME isomers from the 6,8 to the 13,15 using three ChromSpher 5 Lipids columns in series, operated at 25 °C, a mobile phase of 2% acetic acid in hexane, and UV detection at 233 nm (top). A small amount of *c*9,*t*11-18 : 2 was added to each CDA mixture to make it possible to express the elution of all the isomers as relative retention volume (RRV) to *c*9,*t*11-18 : 2. The plot of RRV *vs* carbon–carbon double bond position for all geometric CLA isomers is shown on the bottom.[51]
Reproduced with permission of *Lipids* and the authors.

and decreasing the elution temperature, the resolution among isomers was increased, resulting in a separation of *t*9,*c*11- and *c*9,*t*11-18 : 2 when the column was operated at 0°C instead of at room temperature.[8]

7.7.2 As Free FA

The separation of CDAs as their FFAs was also possible by using the same Ag⁺-HPLC column (ChromSpher 5 Lipid columns; 250×4.6 mm; 5 μm particle size), except in this case a similar separation could be achieved using a single column.[33] Figure 7.29A shows the separation of a synthetic mixture of four positional C18 CDAs as FFAs (Sigma-Aldrich Inc.), while Figure 7.29B shows a typical separation of pig muscle lipids. Analysing CDAs as FFAs instead of FAMEs has advantages and disadvantages. The analysis of FFAs does not require the methylation step,[33] which can lead to isomerization if

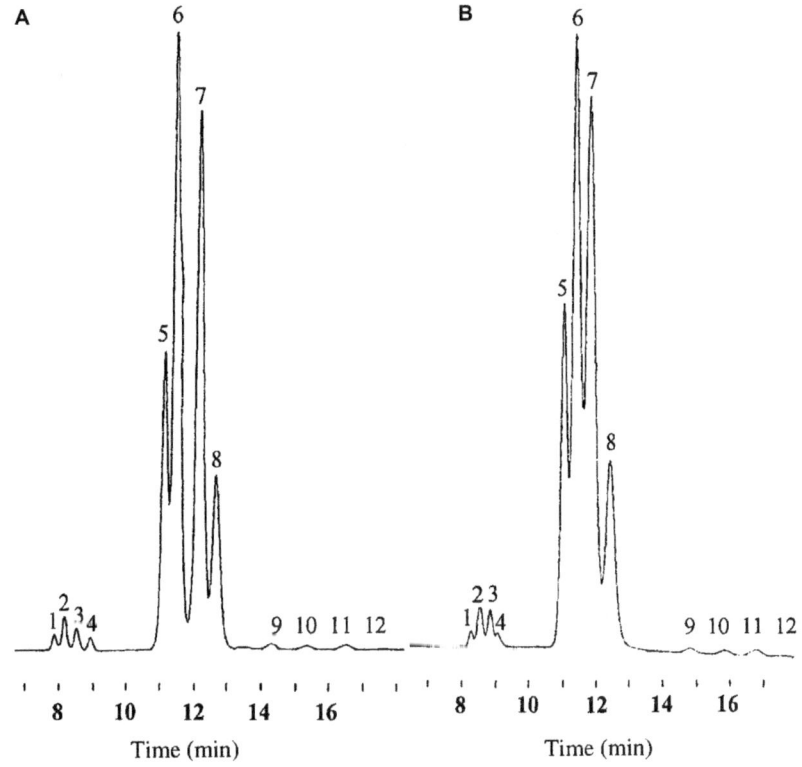

Figure 7.29 Ag⁺-HPLC separation of C18 CDAs as their FFAs using one ChromSpher 5 Lipids column, a mobile phase of 2.5% acetic acid and 0.025 acetonitrile in hexane, and UV detection at 234 nm. In (A) the mixture consisting of four positional C18 CDA isomers is resolved, while (B) shows a typical separation of pig muscle from pigs fed the commercial CDA mixture.[33]

Reproduced with permission of *Lipids* and the authors.

an acid catalyst is used.[40] However, lipids need to be hydrolysed to FFAs. If that hydrolysis step involves an acid catalyst it can equally lead to iso-merization, while a base hydrolysis/saponification step is not quantitative and will lead to the underestimation of the total lipid content and a change in FA composition, particularly when in samples contain sphingomyelin or plasmalogen.[12,29,30,44]

7.7.3   Limitations

- It does not resolve CDAs with different chain lengths.[11]
- The lack of reproducibility in the retention times (volumes) can be discouraging. It is based on the poor solubility of acetonitrile in hexane. Adding a small amount of some ethers helps to retain the acetonitrile in solution. Alternatively, the identification of C18 CDAs can also be improved by expressing the data in terms of RRT or RRV relative to the most common and abundant $c9,t11$-18:2 isomer.
- There is a reversal of the elution order of c/t-CDA isomers with increasing Δ value. The C18 CDA isomers at the position where the crossover occurs are not resolved. For elution with 0.1% acetonitrile in hexane the two 10,12-18:2 isomers do not separate (Figure 7.27), and when 2% acetic acid in hexane was used as the mobile phase, the two 7,9-18:2 isomers coeluted (Figure 7.28).[8]

7.8   Separation of FA Using Reverse Phase HPLC Columns

While Ag^+-HPLC is generally accepted as the most suitable technique for the analysis of the positional and geometric isomers of CDAs, since it lacks selectivity toward the chain length of FAs. Therefore, this technique is not suitable for the analysis of the elongation or β-oxidation metabolites of CDAs. These compounds have instead been separated and measured by RP-HPLC coupled with DAD detection, taking advantage of the selectivity provided by UV spectroscopy at ~ 234 nm for CDAs, at ~ 270 nm for DCTAs, and at ~ 200 nm for unsaturated FAs. Furthermore, the sensitivity of UV detection for CFAs is greater than that of GC-FID, which has made the RP-HPLC method with DAD the preferred technique for the detection, quantification and isolation of CFA metabolites.[20–22,30,31] Reference materials of CFA metabolites are not available, and therefore, the development of chromatographic methods for their measurement must include their identification by spectroscopic (HPLC-DAD, second derivatives) and/or spectrometric techniques (HPLC-MS).[31] The CFAs can then be separated by RP-HPLC as FFAs or as FAMEs.

7.8.1   As Free FA

The benefit of analysing CFAs as FFAs is to eliminate the methylation step, which has been shown to be a source of isomerization of CFAs under acidic conditions.[40] FFAs can be prepared by mild saponification of FA esters followed by neutralization and extraction with acidified hexane.[112] Phospholipids can also be hydrolysed enzymatically to FFAs using phospholipase A2.[113] However, it should be noted that any base-catalysed hydrolysis or saponification step does not quantitate all lipids. In many cases the preparation of FAMEs simplifies the combination of analyses by GC and HPLC, and it also enhances the stability of FA for long term storage. The analysis of FFA by RP-HPLC requires the addition of an acidic modifier (as 0.1–0.2% acetic acid) to the mobile phase to maintain FFAs in their neutral form.

Several RP columns might be suitable for the separation of CFA metabolites, but the Inertsil 5 ODS-2 column (150 mm×4.6 mm i.d. stainless steel column, 5 μm particle size, Phenomenex Inc., Torrance, CA) is preferred since it takes advantage of previous identifications.[21,22,30,112,114,115] The separation was achieved using an acetonitrile/water/acetic acid (70/30/0.12, by volume) mobile phase at the flow rate of 1.5 mL min^{-1}. These chromatographic conditions provided partial resolution of the *c/t*- and *c,c*-18:2 CDAs with *t,t*-18:2 eluting last, but could not separate individual isomers.[112] Figure 7.30 shows a typical separation of the FFAs obtained from sheep muscle lipids.[30] Using this approach several desaturated and elongated metabolites of C18 CDA were identified in rat and lamb tissues such as the MCDAs of 18:3, 20:3, 20:4 and 22:4, as well as products of peroxisomal β-oxidation (16:2 and 16:3 MCDAs).[21,22,30,112,114] UV spectra acquired by DAD allow the identification of the conjugated double bonds by their unique absorption. In the case of CDA metabolites, UV spectra and their second derivatives were reported to provide unique confirmatory information.[21]

7.8.2 As FAME

While CFA metabolites have been preferably analysed by RP-HPLC as FFAs, positional and geometric C18 CDA isomers are instead analysed as FAMEs. The rationale for this choice was to complement data acquired by GC, since the latter was not able to separate *c*9,*t*11- from *t*7,*c*9-18:2. The progressive increase in retention and selectivity observed by lowering the elution temperature below the water freezing point is amplified when using a specific brand of polymeric C18 column. This was used to develop unique RP-HPLC methodologies with comparable separation capabilities to Ag^{+}-HPLC.[8]

The separation of geometric and positional C18 CDA isomers required optimizing the RP-HPLC selectivity toward the location and geometry of double bonds, rather than for the FA chain length. The Vydac 201TP54 column (4.6 mm i.d. × 250 mm; 5 μm particle size; Vydac, Hesperia, CA) was previously used for the separation of carotenoids and showed a unique selectivity for the separation of C18 CDA isomers.[8] The selectivity and retention provided by the Vydac 201TP54 column appeared to be markedly affected by the elution temperature, and what was at first considered a factor affecting the reproducibility of these separations instead became the most relevant characteristic for improving the separation of C18 CDA isomers. At 10 °C, three of

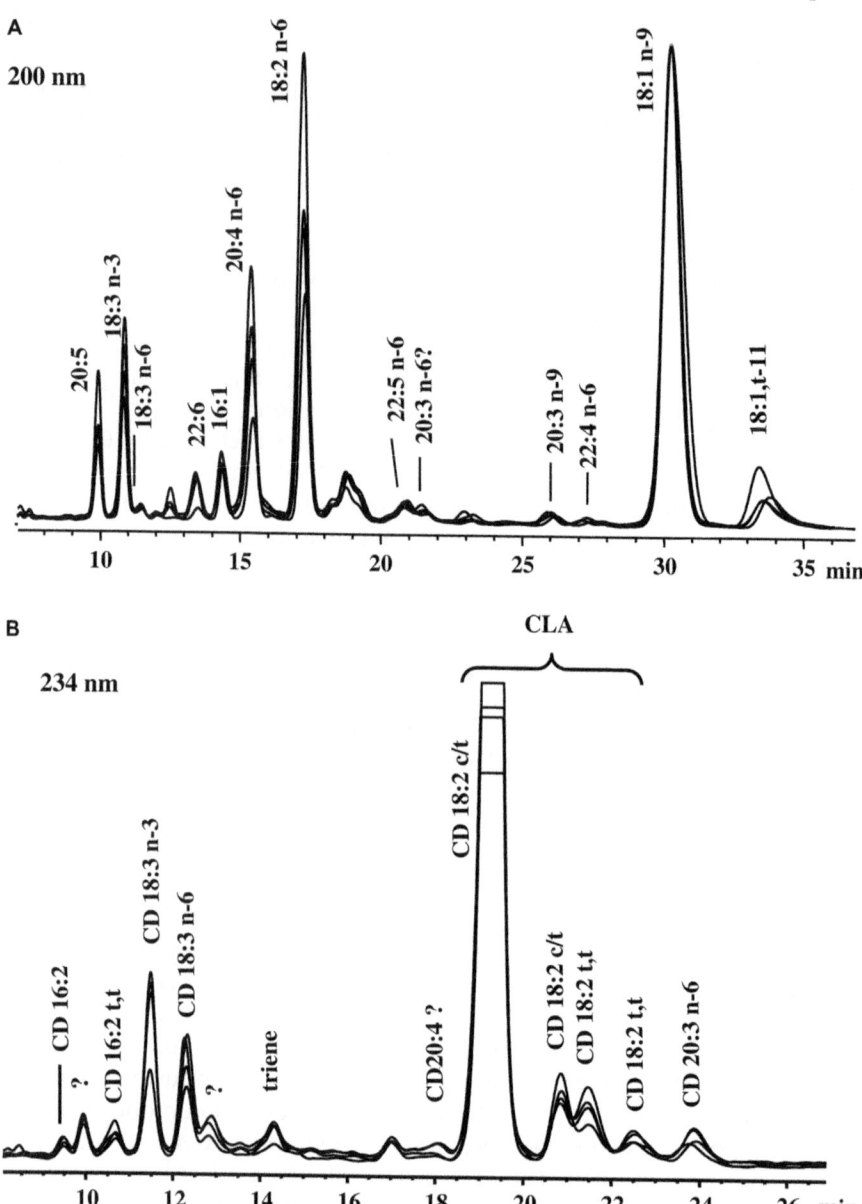

Figure 7.30   Partial RP-HPLC chromatograms of four representative sheep muscle samples recorded at 200 nm for unsaturated FAs (A) and at 234 nm for mono-conjugated FAs (B).[30] A C18 reversed phase column (Inertsil ODS-2, 5 µm particle size, 150 × 4.6 mm; Phenomenex Inc., Torrance, CA) was used with a mobile phase of acetonitrile/water/acetic acid (70/30/0.12).
Reproduced with permission of *Lipids* and the authors.

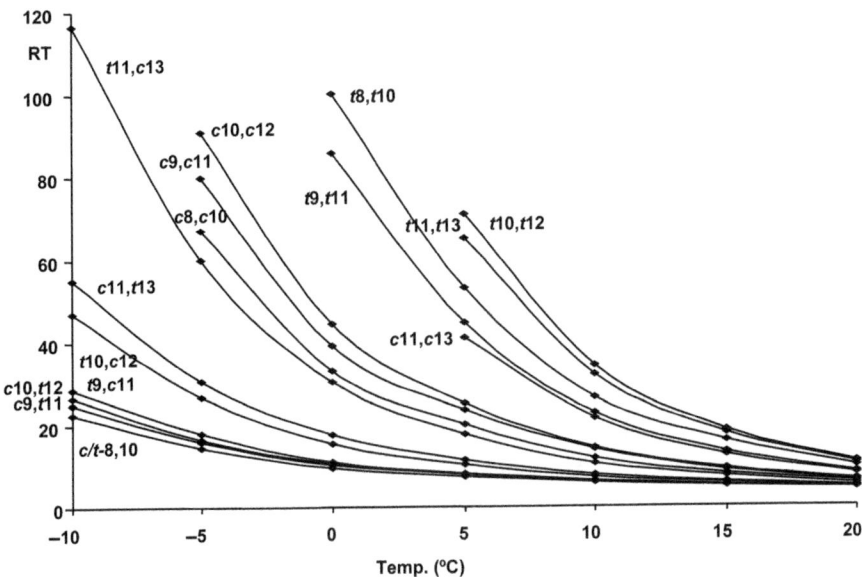

Figure 7.31 Plot of retention times (RT) *vs.* column temperature of all the C18 CDAs present in the iodine isomerization mixture of UC-59M.[8] One Vydac 201TP54 column was used with 100% acetonitrile as mobile phase and UV detection at 233 nm.
Reproduced with permission from authors and AOCS Press.

the four *c/t* C18 CDA isomers present in the mixture UC-59M were separated in 12 min using 100% acetonitrile as the mobile phase and UV detection at 233 nm, while the *c,c*- and *t,t*-C18 CDA FAMEs eluted in 30 to 50 min. While this separation of C18 CDA isomers was a marked improvement, further optimization was needed to make RP-HPLC a viable alternative to Ag$^+$-HPLC. Figure 7.31 shows the dependence of C18 CDA geometric/positional isomer retention times (from 8,10- to 11,13-18:2) on the elution temperature using a Vydac 201TP54 column and a mobile phase of 100% acetonitrile. None of the *t,t*-C18 CDA isomers eluted in less than 120 min at an elution temperature equal or lower than −5 °C, while the *c,c*-C18 CDAs were not eluted in the same time frame when the elution temperature was −10 °C or lower.

To avoid peak tailing caused by further lowering the elution temperatures, two columns were used in series with an elution gradient of acetonitrile, hexane and tetrahydrofuran (Figure 7.32).[8] Positional C18 CDA isomers were eluted in reverse order compared with Ag$^+$-HPLC. All of the *c/t*-C18 CDA isomers between the 8,10- and 11,13-18:2 were at least partially separated from each other, with the exception of *c8,t10*- coeluting with *t8,c10*-18:2 (Figure 7.33).[8] The resolving capabilities of RP-HPLC for C18 CDA isomers appears inferior compared to that of Ag$^+$-HPLC, but if the elution temperature is adequately controlled, RP-HPLC will provide more reproducible retention times than Ag$^+$-HPLC.

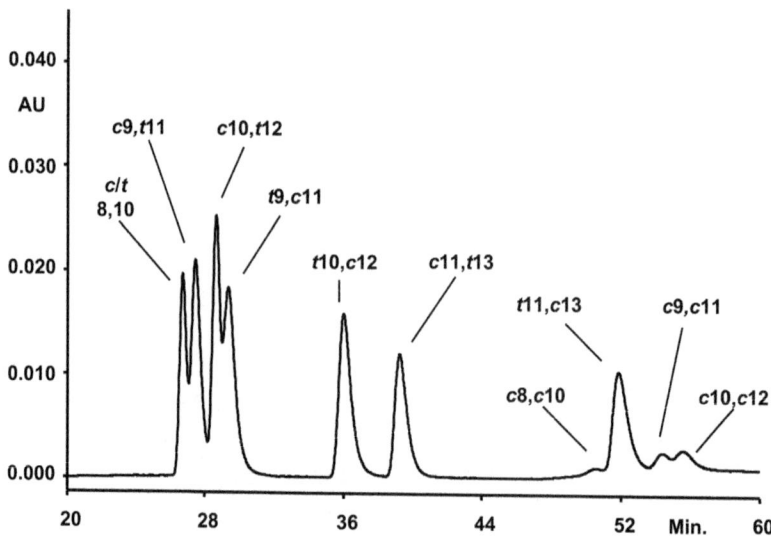

Figure 7.32 Partial RP-HPLC chromatogram of the iodine isomerized mixture UC-59M (Nu-Chek Prep, Inc.).[8] Two Vydac 201TP54 columns were combined in series, maintained at −5 °C, elution with an acetonitrile/tetrahydrofuran/hexane gradient, and UV detection 233 nm. Details of the gradient were: 100% acetonitrile to 3:97 hexane/acetonitrile in 40 min, a second gradient to 5:15:80 tetrahydrofurane/hexane/acetonitrile at 50 min, maintained until 80 min, then a gradient to 25:15:60 tetrahydrofurane/hexane/acetonitrile at 85 min.
Reproduced with permission from authors and AOCS Press.

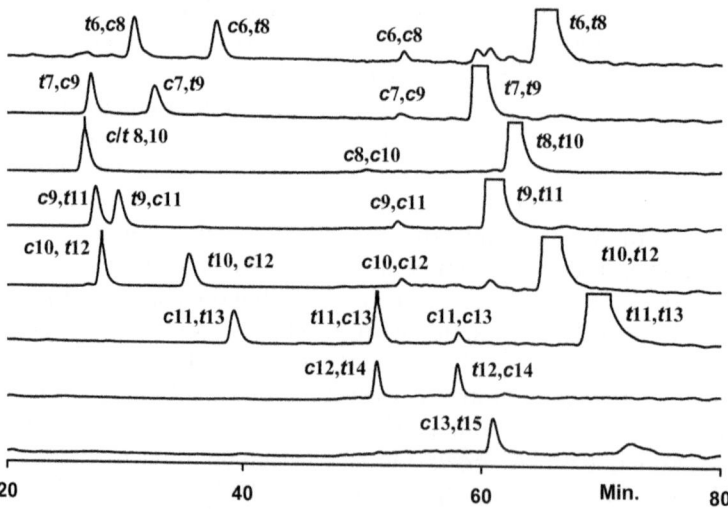

Figure 7.33 Partial RP-HLC chromatogram of iodine isomerized positional CDA isomers from 6,8- to 13,15-18:2 as their FAMEs.[8] Chromatographic conditions were the same as in Figure 31.
Reproduced with permission from authors and AOCS Press.

7.8.3 Limitations

- Very limited separations of CDA geometric/positional isomers are possible when they are analysed as FFAs. Analysis as FAMEs, and with reduced elution temperatures, some separation can be achieved.[8]
- The separation of CDAs by RP-HPLC is column specific, which means a re-evaluation of the relative elution pattern of the CDAs as well as the other FAs is necessary when changing the separation column.
- CFAs differ not only in their elution order but also for their UV spectrum and absorption maxima. This has been used to identify CFA geometrical isomers and their metabolites in complex biological samples. However, quantitation using this technique is only possible when the structure is correctly identified.
- RP-HPLC is not a standalone method even for measuring CFA metabolites, since most of the non-conjugated FAME metabolites are not adequately analysed.
- There is a lack of CFA and other FA standards.

7.9 Combining Methods

The challenge remains of how best to analyze the whole range of FAMEs from $4:0$ to $22:6n$-$3/26:0$ (and possibly longer chain FAMEs) including all geometric and positional isomers of MUFAs, PUFAs and CFAs. In addition, the methods need to be suited to the samples being analysed. For instance, dairy fats contain short-chain, branch-chain, *trans*, unsaturated, conjugated and long chain PUFAs, while meats and animal/human tissues contain additional plasmalogenic lipids. On the other hand, marine lipids contain mainly PUFAs and branch-chain FAs, while partially hydrogenated fats mainly geometric and positional isomers of mono- and di-unsaturated FAs. CFAs present in synthetic mixtures or from plant/algae sources will require more than one GC separation for their analyses. Currently, there is no single GC separation for the analysis of FAs contained in natural and synthetic products. For this reason, results from different GC columns and conditions as well as other analytical techniques need to be combined to resolve specific complex mixtures of FAMEs. Obviously, there are more combinations than those listed below, and certainly this contribution will not be the last chapter on this subject. However, the hope is to stimulate further development of methods to obtain a more complete analysis of complex lipid mixtures that include CFAs.

7.9.1 Single Column without and with a Temperature Program

A single analysis would obviously be preferred because it requires less labor. In fact, most official methods use this approach with a 100 m 100% CPS column to determine the total content of *trans* FA, SFA, PUFA and CDA.[36]

However, as pointed out above there are a number of GC regions in which the FAMEs are not well resolved with these GC columns specifically the geometric/positional isomers of mono-, di-, and conjugated FAs. Without the benefit of a Ag[+]-HPLC separation, the major C18 CDAs cannot be resolved.[11,29] Furthermore, without the benefit of a prior Ag[+]-TLC (Figure 7.8)[56] or Ag[+]-SPE[59] separation analysts can lead to make claims that may not be well supported, such as *t*9-16 : 1 being associated with an increased risk in coronary heart disease[56,94] and lower insulin resistance.[95] The highly polar 200 m SLB-IL111 ionic liquid column provides the separation of many more FAMEs from milk fats and marine oils compared to the 100 m 100% CPS columns.[52-54] Furthermore, even this column has limitations if used by itself to resolve short-chain SFAs from some *cis*- and *trans*-MUFAs.

The separation of *trans* and *cis* isomers using low temperature isothermal GC condition is an interesting adaptation based on the principle that lowering the column temperature of polar columns increases the separation between members of a homologous FAME series. This principle was used to separate the positional isomer pairs of *t*11/*t*12-16 : 1, *t*13/*t*14-18 : 1, *t*15/*t*16-20 : 1 and *t*17/*t*18-22 : 1 at 120 °C that could not be resolved at higher elution temperatures using a 100 m 100% CPS column (Figure 7.11)[56,57,59,60,68] or on the 200 m SLB-IL111 ionic liquid column.[52,53] However, the 100 m 100% CPS column cannot resolve many of the low Δ MUFAs such as *t*6/*t*7-16 : 1 or *t*6/*t*7/*t*8-18 : 1 even when using an elution temperature of 120 °C (Figure 7.8), nor does it separate all four positional C18 CDA isomers in the UC-59M mixture at 120 °C (Figure 7.13C).[61,91]

7.9.2 Two GC Columns of Different Polarities

There are several combinations that have been used, and each requires that the same sample be analysed twice. One popular approach is to combine the benefits of a rapid analysis on a GC column of moderate polarity to ease the identification of PUFAs, with the benefits of an isothermal separation at ∼175 °C on a more polar column that maximizes the separation of the *cis* and *trans* 16 : 1 and 18 : 1 isomers.[57,82,88,116] This procedure involves conducting two separate GC analyses, but it does not addresses the lack of separation of C18 CDAs, and the overlapping 16 : 1, 18 : 1, 18 : 2, 20 : 1/18 : 3, and PUFA isomers observed with the 100 m 100% CPS columns.[57,88,116] Similarly, the limited reports using the 200 m CP Select for FAME column[82] suggest the same conclusion. Combining the results of a 70 m BPX-70 with a 100 m CPS column[71] is also subject to the same limitations, since the BPX-70 column provides only a limited separation of the geometric FAME isomers. Instead of only an isothermal separation with the polar column, some have conducted a complete separation with the polar column and combined the results with those of the PEG column.[11,71] This approach provides more valuable separations, but the characteristic limitations of each column remains.

An alternate approach would be to use two polar GC columns with different polarities. This situation was recently made possible with the availability of the new highly polar SLB-IL111 ionic liquid column to complement the results obtained with 100 m 100% CPS columns.[102,103] The 100 and 200 m SLB-IL111 ionic liquid columns have provided remarkably improved separations of the geometric and positional isomers of mono- and diunsaturated FAs, PUFAs and C18 CDAs in PHVOs,[52] milk fats (Figures 7.19, 7.21 and 7.23)[53] and menhaden oil (Figure 7.20).[54] The separation of the major C18 CDAs in ruminant fats has made it possible for the first time to measure separately t7,c9- and c9,t11-18 : 2 by GC (Figures 7.23 and 7.24).[52,53] The identity of the FAME isomers was established by using authentic MUFA and C18 CDA standards, conducting prior Ag⁺-HPLC separations, and confirming the FAMEs using GC-time of flight-MS (Figure 7.22). The 200 m SLB-IL111 capillary column, operated under a combined temperature and effluent flow gradient, successfully resolved up to 160 FAMEs in milk fat and marine oil in a single chromatographic separation in 80 and 90 min, respectively. It appears very promising to more fully analyse complex lipid mixtures by combining the results of a 100 m CPS and a 200 m SLB-IL111 ionic liquid column to resolve most of the geometric and positional isomers of mono-, di-, and conjugated FAMEs in any matrix by GC.

Although not really a comparison of two GC columns, the results using the same GC instrument and single polar 100 m 100% CPS column and operated using two different temperature programs has proven to be a very effective method to separate FAMEs.[36,59] This method requires no extra procedures or work-up since the same instrument is used, except the FAMEs need to be matched up after the analysis. The success of this approach depends on selecting two temperature settings that are sufficiently different to result in an altered elution profile of the different types of FAMEs, so that peaks that coelute or are poorly separated at one temperature are resolved at the second temperature (Figure 7.10). This approach was also used to resolve the 18 : 3 and 20 : 1 isomers in PHVOs (Figure 7.9),[69] and 11-cyclohexylundecanoic acid from the 18 : 2 FAMEs prepared from milk fat.[70] The two MCTAs of c9,t11,c15-18 : 3 and c9,t13,c15-18 : 3 could also be confirmed[117] and resolved.[118] In addition, analysing milk fats using two temperature programs enabled the identification of most 16 : 1, 18 : 1 and 20 : 1 FAMEs.[59] Selected temperature settings also provided improved resolution of t13/t14- from c9-18 : 1, and t10- from t11-18 : 1.[119]

7.9.3 Combining GC with Polar Columns and Ag⁺ Separations

This approach combines the GC separation with a 100 m 100% CPS column recommended for the analysis of *trans* FAs with two types of Ag⁺ methods that have proven to be most helpful in combination with GC analyses of these FAMEs. First is a prior Ag⁺ fractionation using Ag⁺-TLC (Figure 7.8), Ag⁺-SPE (Figures 7.11 and 7.12) or Ag⁺-HPLC followed by GC (Figures 7.19, 7.20 and 7.23) This is needed to sort out and identify overlapping SFAs, the geometric isomers of MUFAs, and different PUFA fractions. The second approach uses Ag⁺-HPLC to separate most of the C18 CDA isomers including t7,c9- and c9,t11-18 : 2 (Figures 7.1, 7.14, 7.16 and 7.25–7.29). This separation was the breakthrough to resolving the complex mixture of C18 CDA isomers,[11,105] and still remains the gold standard for separating and quantifying them. However, at room temperature Ag^+-HPLC does not resolve c9,t11-18:2 from t9,c11-18:2, which are well separated using 100 m 100% CPS columns (Figures 7.14 and 7.16).[29,58] For this reason, the GC separations on 100% CPS columns and Ag^+-HPLC results need to be combined to provide a more complete analysis of FAs and C18 CDAs, particularly for ruminant products.[11,29,39,58,120]

7.9.4 Combining RP-HPLC Separations and Polar GC Columns

The RP-HPLC method was successfully used to identify CFA metabolites in rat and sheep tissue lipids by monitoring the eluate at ∼234 nm (Figure 7.30).[20,22,31,34,76,115] After isolating the fractions from HPLC, the CFA metabolites were further identified either directly by HPLC-MS[22,34] or by conversion to FAMEs and analysed by GC and GC-MS.[20] The total *trans* 18:1 FAME fraction can also be isolated by RP-HPLC and subsequently analysed by GC or GC-MS.[121,122]

7.9.5 Combining GC Separation, Cryogenic Trapping, and Infrared Analysis

This procedure combined the efficiency of separation using a 100 m 100% CPS column with cryogenic trapping (matrix isolation or direct deposition) and FTIR analysis of the isolated FAMEs. In matrix isolation each FAME or DMOX component eluting from the GC column was individually frozen as a solid peak on a slowly rotating gold plated disk held at a cryogenic temperature of ∼12 °K under vacuum (10^{-5} torr).[35,38] The carrier gas was a mixture of helium (98.5%) and argon (1.5%). While the helium was pumped off under vacuum, the IR-transparent argon gas condensed with the separated FAME components on the cold disk forming a solid matrix that could subsequently be examined by FTIR. Post GC run FTIR spectral data acquisition of each frozen FAME or DMOX peak was possible, which gave rise to detailed IR spectra, and led to improved signal to noise ratios. This method was used to confirm the identity of geometric isomers of MUFAs, mono *trans* dienes, CFAs in a number of matrices.[35-40] This combination allowed the discrimination between FAMEs with the same and different chain lengths, and with different double-bond configurations and degrees of unsaturation. Figure 7.2 shows the characteristic FTIR absorptions of the c/t-, c,c- and t,t-CDAs obtained using this method.

Abbreviations

Ag^+	silver ion
AME	alk-1-enyl methyl ether

CDA	conjugated dienoic acid
CFA	conjugated fatty acid
CLA	conjugated linoleic acid
CPS	100% cyanopropylsiloxane
DAD	diode array detector
DCTA	di-conjugated trienoic acid
DMA	dimethylacetal
DMOX	4,4-dimethyloxazoline
FA	fatty acid
FAME	fatty acid methyl ester
FFA	free fatty acid
FID	flame ionization detection
FTIR	Fourier transform infrared
GC	gas chromatography
GLA	γ-linolenic acid
HPLC	high performance liquid chromatography
IR	infrared
MCTA	mono-conjugated trienoic acid
MI	methylene interrupted
MS	mass spectrometry
MUFA	monounsaturated fatty acid
NMR	nuclear magnetic resonance
NMI	non-methylene interrupted
PEG	polyethylene glycol
PHVO	partially hydrogenated vegetable oil
PUFA	polyunsaturated fatty acid
RP	reverse phase
RT	retention time
RRT	relative retention time
RRV	relative retention volume
SFA	saturated fatty acid
SPE	solid phase extraction
TLC	thin layer chromatography
UV	ultraviolet

References

1. M. W. Pariza, Y. Park and M. E. Cook, *Progr. Lipid Res.*, 2001, **40**, 283–298.
2. K. Eulitz, M. P. Yurawecz and Y. Ku, *Advances in Conjugated Linoleic Acid Research*, ed. M. P. Yurawecz, M. M. Mossoba, J. K. G. Kramer, M. W. Pariza, and G. J. Nelson, AOCS Press, Champaign, IL, 1999, vol. 1, ch. 5, 55–63.
3. R. Suzuki, R. Noguchi, T. Ota, M. Abe, K. Miyashita and T. Kawada, *Lipids*, 2001, **36**, 477–482.

4. T. Tsuzuki, Y. Tokuyama, M. Igarashi and T. Miyazawa, *Carcinogenesis*, 2004, **25**, 1417–1425.

5. P. Delmonte, J. A. G. Roach, M. M. Mossoba, K. M. Morehouse, L. Lehmann and M. P. Yurawecz, *Lipids*, 2003, **38**, 579–583.

6. P. Delmonte, J. A. G. Roach, M. M. Mossoba, G. Losi and M. P. Yurawecz, *Lipids*, 2004, **39**, 185–191.

7. P. Delmonte, M. P. Yurawecz, M. M. Mossoba, C. Cruz-Hernandez and J. K. G. Kramer, *J. AOAC Int.*, 2004, **87**, 563–568.

8. P. Delmonte, J. K. G. Kramer, S. Banni and M. P. Yurawecz, *Advances in Conjugated Linoleic Acid Research*, ed. M. P. Yurawecz, J. K. G. Kramer, O. Gudmundsen, M. P. Pariza and S. Banni, AOCS Press, Champaign, IL, 2006, vol. 3, ch. 5, 95–118.

9. P. Delmonte, H. Qing, A.-R. Fardin Kia and J. I. Rader, *J. Chromatogr. A*, 2008, **1214**, 30–36.

10. P. Delmonte, A.-R. Fardin Kia, Q. Hu and J. I. Rader, *J. AOAC Int.*, 2009, **92**, 1310–1326.

11. N. Sehat, R. Rickert, M. M. Mossoba, J. K. G. Kramer, M. P. Yurawecz, J. A. G. Roach, R. O. Adlof, K. M. Morehouse, J. Fritsche, K. D. Eulitz, H. Steinhart and Y. Ku, *Lipids*, 1999, **34**, 407–413.

12. N. Aldai, M. de Renobales, L. J. R. Barren and J. K. G. Kramer, *Eur. J. Lipid Sci. Technol.*, 2013, **115**, 1378–1401.

13. F. Destaillats, J. P. Trottier, J. M. G. Galvez and P. Angers, *J. Dairy Sci.*, 2005, **88**, 3231–3239.

14. P. Gómez-Cortés, C. Tyburczy, J. T. Brenna, M. Juárez and M. A. de la Fuente, *Lipid Res.*, 2009, **50**, 2412–2420.

15. C. Y. Hopkins and M. J. Chisholm, *J. Am. Oil Chem. Soc.*, 1964, **41**, 42–44.

16. M. P. Yurawecz, A. A. Molina, M. M. Mossoba and Y. Ku, *J. Am. Oil Chem. Soc.*, 1993, **70**, 1093–1099.

17. F. D. Gunstone, J. L. Harwood and F. B. Padley, *The Lipid Handbook*, Chapman & Hall, London, UK, 2nd edn, 1994, ch. 1.4, 7–8.

18. G. Sassano, P. Sanderson, J. Franx, P. Groot, J. van Straalen and J. Bassaganya-Riera, *J. Sci. Food Agric.*, 2009, **89**, 1046–1052.

19. M. Kýralan, M. Gölükcü and H. Tokgöz, *J. Am. Oil Chem. Soc.*, 2009, **86**, 985–990.

20. J.-L. Sébédio, P. Juanéda, G. Dobson, I. Ramilison, J. C. Martin, J. M. Chardigny and W. W. Christie, *Biochim. Biophys Acta*, 1997, **1345**, 5–10.

21. S. Banni, G. Carta, E. Angioni, E. Murru, P. Scanu, M. P. Melis, D. E. Bauman, S. M. Fischer and C. Ip, *J. Lipid Res.*, 2001, **42**, 1056–1061.

22. S. Banni, A. Petroni, M. Blasevich, G. Carta, E. Angioni, E. Murru, B. W. Day, M. P. Melis, S. Spada and C. Ip, *Biochim. Biophys. Acta*, 2004, **1682**, 120–127.

23. B. Narayan, M. Hosokawa and K. Miyashita, *Nutraceutical and Specialty Lipids and their Co-Products*, ed. F. Shahidi, CRC Press, Boca Raton, FL, 2006, ch. 13, 201–218.

24. J. K. G. Kramer, P. W. Parodi, R. G. Jensen, M. M. Mossoba, M. P. Yurawecz and R. O. Adlof, *Lipids*, 1998, **33**, 835.
25. I. Wasowska, M. R. G. Maia, K. M. Niedźwiedzka, M. Czauderna, J. M. C. Ramalho Ribeiro, E. Devillard, K. J. Shingfield and R. J. Wallace, *Br. J. Nutr.*, 2006, **95**, 1199–1211.
26. F. Warnaar, *Lipids*, 1977, **12**, 707–710.
27. S. Ö. Yücel, *J. Am. Oil Chem., Soc.*, 2005, **82**, 893–897.
28. Y. Cao, L. Yang, H.-L. Gao, J.-N. Chen, Z.-Y. Chen and Q.-S. Ren, *Chem. Phys. Lipids*, 2007, **145**, 128–133.
29. C. Cruz-Hernandez, J. K. G. Kramer, J. Kraft, V. Santercole, M. Or-Rashid, Z. Deng, M. E. R. Dugan, P. Delmonte and M. P. Yurawecz, *Advances in Conjugated Linoleic Acid Research*, ed. M. P. Yurawecz, J. K. G. Kramer, O. Gudmundsen, M. P. Pariza, S. Banni, AOCS Press, Champaign, IL, 2006, vol. 3, ch. 4, 45–93.
30. V. Santercole, R. Mazzette, E. P. L. De Santis, S. Banni, L. Goonewardene and J. K. G. Kramer, *Lipids*, 2007, **42**(361–382).
31. S. Banni, G. Carta, M. S. Contini, E. Angioni, M. Deiana, M. A. Dessí, M. P. Melis and F. P. Corongiu, *J. Nutr. Biochem.*, 1996, 7, 150–155.
32. J. K. G. Kramer, N. Sehat, M. E. R. Dugan, M. M. Mossoba, M. P. Yurawecz, J. A. G. Roach, K. Eulitz, J. L. Aalhus, A. L. Schaefer and Y. Ku, *Lipids*, 1998, **33**, 549–558.
33. E. Ostrowska, F. R. Dunshea, M. Muralitharan and R. F. Cross, *Lipids*, 2000, **35**, 1147–1153.
34. S. Banni, B. W. Day, R. W. Evans, F. P. Corongiu and B. Lombardi, *J. Nutr. Biochem.*, 1995, **6**, 281–289.
35. M. M. Mossoba, R. E. McDonald, D. L. Armstrong and S. W. Page, *J. Chromatogr. Sci.*, 1991, **29**, 324–330.
36. M. M. Mossoba and J. K. G. Kramer, *Official Methods for the Determination of Trans Fatty Acids*, AOCS Press, Champaign, IL, 2009.
37. M. M. Mossoba, R. E. McDonald, J. A. G. Roach, D. D. Fingerhut, M. P. Yurawecz and N. Sehat, *J. Am. Oil Chem. Soc.*, 1997, **74**, 125–130.
38. M. M. Mossoba, M. P. Yurawecz, J. K. G. Kramer, K. D. Eulitz, J. Fritsche, N. Sehat, J. A. G. Roach and Y. Ku, *Advances in Conjugated Linoleic Acid Research*, ed. M. P. Yurawecz, M. M. Mossoba, J. K. G. Kramer, M. P. Pariza and G. J. Nelson, AOCS Press, Champaign, IL, 1999, vol. 1, ch. 10, 141–151.
39. M. P. Yurawecz, J. A. G. Roach, N. Sehat, M. M. Mossoba, J. K. G. Kramer, J. Fritsche, H. Steinhart and Y. Ku, *Lipids*, 1998, **33**, 803–809.
40. J. K. G. Kramer, V. Fellner, M. E. R. Dugan, F. D. Sauer, M. M. Mossoba and M. P. Yurawecz, *Lipids*, 1997, **32**, 1219–1228.
41. M. P. Yurawecz, J. K. Hood, J. A. G. Roach, M. M. Mossoba, D. H. Daniels, Y. Ku, M. W. Pariza and S. F. Chin, *J. Am. Oil Chem. Soc.*, 1994, **71**, 1149–1155.
42. J. Kraft, J. K. G. Kramer, F. Schoene, J. R. Chambers and G. Jahreis, *J. Agric. Food Chem.*, 2008, **56**, 4775–4782.

43. N. Aldai, M. E. R. Dugan, J. K. G. Kramer, A. Martínez, O. López-Campos, A. R. Mantecón and K. Osoro, *Animal*, 2011, **5**, 1643–1652.
44. N. Aldai, J. K. G. Kramer, C. Cruz-Hernandez, V. Santercole, P. Delmonte, M. M. Mossoba and M. E. R. Dugan, *Fats and Fatty Acids in Poultry Nutrition and Health*, ed. G. Cherian and R. Poureslami, Context Products Ltd, Leicestershire, UK, 2012, ch. 12, 249–290.
45. AOCS Official Method Ce 1j-07, AOCS, Peoria, IL, Amended 2012.
46. L. A. Horrocks, *Ether Lipids, Chemistry and Biology*, ed. F. Snyder, Academic Press, New York, 1972, ch. 9, 177–272.
47. M. Winterfeld and H. Debuch, *Hoppe-Seyler Z. physiol. Chem.*, 1966, **345**, 11–21.
48. K. Eulitz, M. P. Yurawecz, N. Sehat, J. Fritsche, J. A. G. Roach, M. M. Mossoba, J. K. G. Kramer, R. O. Adlof and Y. Ku, *Lipids*, 1999, **34**, 873–877.
49. J. K. G. Kramer, N. Sehat, J. Fritsche, M. M. Mossoba, K. Eulitz, M. P. Yurawecz and Y. Ku, *Advances in Conjugated Linoleic Acid Research*, ed. M. P. Yurawecz, M. M. Mossoba, J. K. G. Kramer, M. W. Pariza and G. J. Nelson, AOCS Press, Champaign, IL, 1999, vol. 1, ch. 7, 83–109.
50. C. Kellersmann, L. Lehmann, W. Krancke and H. Steinhart, *Advances in Conjugated Linoleic Acid Research*, ed. M. P. Yurawecz, J. K. G. Kramer, O. Gudmundsen, M. W. Pariza and S. Banni, AOCS Press, Champaign, IL, 2006, vol. 3, ch. 3, 27–43.
51. P. Delmonte, A. Kataoka, B. A. Corl, D. E. Bauman and M. P. Yurawecz, *Lipids*, 2005, **40**, 509B514.
52. P. Delmonte, A.-R. Fardin Kia, J. K. G. Kramer, M. M. Mossoba, L. Sidisky and J. I. Rader, *J. Chromatogr. A*, 2011, **1218**, 545–554.
53. P. Delmonte, A.-R. Fardin Kia, J. K. G. Kramer, M. M. Mossoba, L. Sidisky, C. Tyburczy and J. I. Rader, *J. Chromatogr. A*, 2012, **1233**, 137–146.
54. A.-R. Fardin-Kia, P. Delmonte, J. K. G. Kramer, G. Jahreis, K. Kuhnt, V. Santercole and J. I. Rader, *Lipids*, 2013, **48**, 1279–1295.
55. (a) J. Harynuk, P. M. Wynne and P. J. Marriott, *Chromatographia Suppl.*, 2006, **63**, S61; (b) www.sge.com/products/columns/gc-columns/bpx90.
56. D. Precht and J. Molketin, *Eur. J. Lipid Sci. Technol.*, 2000, **102**, 102–113.
57. D. Precht and J. Molkentin, *Nahrung*, 2000, **44**, 222–228.
58. C. Cruz-Hernandez, Z. Deng, J. Zhou, A, R. Hill, M. P. Yurawecz, P. Delmonte, M. M. Mossoba, M. E. R. Dugan and J. K. G. Kramer, *J. AOAC Int.*, 2004, **87**, 545–562.
59. J. K. G. Kramer, M. Hernandez, C. Cruz-Hernandez, J. Kraft and M. E. R. Dugan, *Lipids*, 2008, **43**, 259–273.
60. J. Kraft, J. K. G. Kramer, J. M. Hernandez, J. Letarte, N. Aldai, V. Santercole, R. Mohammed, F. Mayer, M. M. Mossoba and P. Delmonte, *Lipid Technol.*, 2014, **26**, 39–42.
61. J. K. G. Kramer, C. B. Blackadar and J. Zhou, *Lipids*, 2002, **37**, 823–835.
62. S. P. Alves and R. J. B. Bessa, *Eur. J. Lipid Sci. Technol.*, 2007, **109**, 879–883.

63. C. R. Kepler, K. P. Hirons, J. J. McNeill and S. B. Tove, *J. Biol. Chem.*, 1966, **241**, 1350–1354.
64. Y. L. Ha, N. K. Grimm and M. W. Pariza, *J. Agric. Food Chem.*, 1989, **37**, 75–81.
65. T. Tsuzuki, Y. Tokuyama, M. Igarashi, T. Miyazawa, K. Nakagava, Y. Ohsaki, M. Komai and T. Miyazawa, *J. Nutr.*, 2004, **134**, 2634–2639.
66. J. A. G. Roach, M. P. Yurawecz, J. K. G. Kramer, M. M. Mossoba, K. Eulitz and Y. Ku, *Lipids*, 2000, **35**, 797–802.
67. J. A. G. Roach, M. M. Mossoba, M. P. Yurawecz and J. K. G. Kramer, *Anal. Chim. Acta*, 2002, **465**, 207–226.
68. D. Precht and J. Molkentin, *Kieler Milchwirtschaftliche Forschungsberichte*, 1997, **49**, 17–34.
69. R. L. Wolff, *J. Am. Oil Chem. Soc.*, 1994, **71**, 907–909.
70. D. Precht and J. Molkentin, *Milchwissenschaft*, 2003, **58**, 30–34.
71. S. A. Mjøs and B. O. Haugsgjerd, *J. Agric. Food Chem.*, 2011, **59**, 3520–3531.
72. J. K. G. Kramer, R. C. Fouchard and K. J. Jenkins, *J. Chromatogr. Sci.*, 1985, **23**, 54–56.
73. N. Vingering and M. Ledoux, *Eur. J. Lipid Sci. Technol.*, 2009, **111**, 669–677.
74. R. G. Ackman, *Analysis of Oils and Fats*, ed. R. J. Hamilton and J. B. Rossell, Elsevier Applied Science Publishers, New York, 1986, ch. 4, 137–206.
75. R. H. Thompson, *J. Chromatogr. Sci.*, 1996, **34**, 495–504.
76. V. Santercole, P. Delmonte and J. K. G. Kramer, *Lipids*, 2012, **47**, 329–344.
77. K. Stenerson, M. R. Halpenny, L. M. Sidisky and M. D. Buchanan, *Supelco Reporter*, 2013, **31.1**, 10–11.
78. AOCS Official Method Ce 1i-07, AOCS, Urbana, IL, Revised 2007.
79. V. Spitzer, F. Marx and K. Pfeilsticker, *J. Am. Oil Chem. Soc*, 1994, **71**, 873–876.
80. J. J. Peene, J. de Zeeuw, F. Biermans and L. Joziasse, #P-147, 2004 http://www.gulfcoastconference.com/pdfs/2003_gcc_program3.pdf (accessed 01/24/2014).
81. J. Kuipers, http://www.chem.agilent.com/Library/applications/SI-02197.pdf (accessed 01/ 24/2014).
82. A. Jaudszus, R. Kramer, M. Pfeuffer, A. Roth, G. Jahreis and K. Kuhnt, *Am. J. Clin. Nutr.*, 2014, **99**, DOI: 10. 3945/ajcn.113.076117.
83. K. Kuhnt, M. Baehr, C. Rohrer and G Jahreis, *Eur, J. Lipid Sci. Technol.*, 2011, **113**, 1281–1292.
84. M. M. Mossoba, J. Moss and J. K. G. Kramer, *J. AOAC Int.*, 2009, **92**, 1284 1300.
85. A. K. Vickers, Agilent Technologies 2009, www.chem-agilent.com/cimg/5989-1817EN.pdf, (accessed 01, 14, 2014).
86. D. Precht and J. Molkentin, *Int. Dairy J.*, 1996, **6**, 791–809.
87. B. M. Weber and J. J. Harynuk, *J. Chromatogr. A*, 2013, **1271**, 170–175.
88. J. Molkentin and D. Precht, *Chromatographia*, 1995, **41**, 267–272.

89. G. Van Poppel, M.-A. Van Erp-Baart, T. Leth, E. Gevers, J. Van Amelsvoort, D. Lanzmann-Petithory, A. Kafatos and A. Aro, *J. Food Compos. Anal.*, 1998, **11**, 112–136.
90. R. L. Wolff and D. Precht, *Lipids*, 2002, **37**, 627–629.
91. J. K. G. Kramer, C. Cruz-Hernandez and J. Zhou, *Eur. J. Lipid Sci. Technol.*, 2001, **103**, 600–609.
92. N. Sehat, J. K. G. Kramer, M. M. Mossoba, M. P. Yurawecz, J. A. G. Roach, K. D. Eulitz, K. M. Morehouse and Y. Ku, *Lipids*, 1998, **33**, 963–971.
93. M. E. R. Dugan, N. Aldai, J. L. Aalhus, D. C. Rolland and J. K. G. Kramer, *Can. J. Anim. Sci.*, 2011, **91**, 545–556.
94. L. H. Thomas, *J. Epidemiol. Community Health*, 1992, **46**, 78–82.
95. D. Mozaffarian, H. Cao, I. B. King, R. N. Lemaitre, X. Song, D. S. Siscovick and G. S. Hotamisligil, *Ann. Intern. Med.*, 2010, **153**, 790–799.
96. K. J. Shingfield, C. K. Reynolds, G. Hervas, J. M. Griinari, A. S. Grandison and D. E. Beever, *J. Dairy Sci.*, 2006, **89**, 714–732.
97. N. Aldai, M. E. R. Dugan, D. C. Rolland and J. K. G. Kramer, *Can. J. Anim. Sci.*, 2009, **89**, 315–329.
98. N. Aldai, J. K. G. Kramer and M. E. R. Dugan, Proceeding of the 55th International Congress of Meat Science and Technology (ICoMST), Copenhagen, Denmark 2009, pp. 1598–1603, http://www.icomst2009.dk/index.php? id$\frac{1}{4}$4048.
99. Supelco Ionic Liquid GC Columns http://www.sigmaaldrich.com/content/dam/sigma-aldrich/docs/Supelco/Posters/1/ionic_liquid_gc_columns.pdf, (accessed 01/14/2014).
100. F. Destaillats, M. Guitard and C. Cruz-Hernandez, *J. Chromatogr. A*, 2011, **1218**, 9384–9389.
101. S. P. Alves and R. J. B. Bessa, *Lipids*, 2014, **49**, DOI: 10.1007/s11745-014-3897-4.
102. C. Tyburczy, P. Delmonte, A.-R. Fardin Kia, M. M. Mossoba, J. K. G. Kramer and J. I. Rader, *J. Agric. Food Chem.*, 2012, **60**, 4567–4577.
103. C. Tyburczy, M. M. Mossoba, A.-R. Fardin Kia and J. I. Rader, *Anal. Bioanal. Chem.*, 2012, **404**, 809–819.
104. T. Turner, D. C. Rolland, N. Aldai and M. E. R. Dugan, *Can. J. Anim. Sci.*, 2011, **91**, 711–713.
105. N. Sehat, M. P. Yurawecz, J. A. G. Roach, M. M. Mossoba, J. K. G. Kramer and Y. Ku, *Lipids*, 1998, **33**, 217–221.
106. B. A. Corl, L. H. Baumgard, J. M. Griinari, P. Delmonte, K. M. Morehouse, M. P. Yurawecz and D. E. Bauman, *Lipids*, 2002, **37**, 681–688.
107. J. K. G. Kramer, C. Cruz-Hernandez, Z. Deng, J. Zhou, G. Jahreis and M. E. R. Dugan, *Am. J. Clin. Nutr.*, 2004, **79**(Suppl.), 1137S–1145S.
108. K. Kuhnt, C. Degen and G. Jahreis, *J. Chromatogr. B*, 2010, **878**, 88–91.
109. A. Müller, K. Düsterloh, R. Ringseis, K. Eder and H. Steinhart, *J. Sep. Sci.*, 2006, **29**, 358–65.
110. A. Saebo, *Lipid Technol. Newsletter*, February, **001**, 9–13.

111. M. J. Haas, J. K. G. Kramer, G. McNeill, K. Scott, T. A. Foglia, N. Sehat, J. Fritsche, M. M. Mossoba and M. P. Yurawecz, *Lipids*, 1999, **34**, 979–987.

112. M. P. Melis, E. Angioni, G. Carta, E. Murru, P. Scanu, S. Spada and S. Banni, *Eur. J. Lipid Sci. Technol.*, 2001, **103**, 594–632.

113. S. A. Iversen, P. Cawood and T. L. Dormandy, *Ann. Clin. Biochem.*, 1985, **22**, 137–140.

114. E. Murru, E. Angioni, G. Carta, M. P. Melis, S. Spada and S. Banni, *Advances in Conjugated Linoleic Acid Research*, ed. J.-L. Sebedio, W, W. Christie and R. O. Adlof, AOCS Press, Champaign, IL, 2003, vol. 2, Ch. 7, 94–100.

115. M. C. Mele, G. Cannelli, G. Carta, L. Cordeddu, M. P. Melis, E. Murru, C. Stanton and S. Banni, *Prostag. Leukotr. Essent. Fatty Acids*, 2013, **89**, 115–119.

116. S. P. Alves and R. J. B. Bessa, *J. Chromatogr. A*, 2009, **1216**, 5130–5139.

117. M. Plourde, F. Destaillats, P. Y. Chouinard and P. Angers, *J. Dairy Sci.*, 2007, **90**, 5269–5275.

118. R. Gervais and P. Y. Chouinard, *J. Dairy Sci.*, 2008, **91**, 3568–3578.

119. S. P. Alves, J. Santos-Silva, A. R. J. Cabrita, A. J. M. Fonseca and R. J. B. Bessa, *PLOS ONE*, 2013, **8**, e58386.

120. K. J. Shingfield, C. K. Reynolds, B. Lupoli, V. Toivonen, M. P. Yurawecz, P. Delmonte, J. M. Griinari, A. S. Grandison and D. E. Beever, *Anim. Sci.*, 2005, **80**, 225–238.

121. P. Juanéda, *J. Chromatog. A*, 2002, **954**, 285–289.

122. D. Bauchart, E. B. Villar, A. Thomas, B. Lyan, M. Hăbeanu, D. Gruffat and D. Durand, *Archiva Zootechnica*, 2010, **13**, 5–11.

Subject Index

References to tables and charts are in **bold** type.